by
Paul and Shirley Smith

Published by:
Travel Publishing Ltd
7a Apollo House, Calleva Park
Aldermaston, Berks, RG7 8TN

ISBN 1-902-00747-6
© Travel Publishing Ltd

First Published: 2000

Printing by: Ashford Press, Gosport

Maps by: © MAPS IN MINUTES ™ (2000)

Authors: © Paul and Shirley Smith

Cover Design: Lines & Words, Aldermaston

Cover Photograph: The M6 motorway in Cumbria © Britain on View/Stockwave.

This book is dedicated to Elaine and Louise

Preface

A system of high-speed roads to link all the main areas of population was first considered as long ago as in the 1920s and some detailed planning was undertaken in the following decade. World War II curtailed any advances along these lines but, in the late 1940s, surveys of routes were undertaken and the section of motorway between Junctions 4 and 8 on the M5 was the first in the country to have its line established by an order under the Special Roads Act of 1949. The first section of motorway was not built for another nine years until the Preston By-Pass (between Junctions 29 and 32 on the M6) was opened on December 5th, 1958.

Since those early days, motorway and trunk road construction has continued apace but, with the meteoric growth in personalized transport, large stretches of motorway are regularly utilized well beyond the designed capacity. Nonetheless, despite the jams and hold-ups, many travellers still consider this form of transport to be the most cost-effective and motorway travel is more popular than ever.

With this in mind it was felt that a travel guide concentrating specifically on motorways would prove to be of significant benefit to the millions of travellers who use these routes every year. **Off the Motorway** is a comprehensive guide to the facilities to be found at each junction as well as between junctions, coupled with a detailed description of landmarks and places of interest along the route. The guide is designed to not only direct motorway drivers to the many different types of facilities along the way but also to provide interesting information on places close to the motorway, many of which can be visited without significant diversion should the traveller wish to relieve a tedious journey.

Coming off at a junction can sometimes prove to be something of a lottery, especially if running short of fuel, and the primary purpose of this guide is to inform travellers of all food, fuel and accommodation available within a radius of about one mile from each junction, with precise details of opening times and the facilities available at each site. From a practical viewpoint, junctions on urban motorway stretches have been omitted as there are many facilities freely available and within easy reach. At junctions within urban areas only the main roads have been included.

Minor injuries not necessitating a 999 call, can happen all too often and the nearest **Accident and Emergency Hospital** to each junction has been identified, along with the route. In our experience very few hospitals are signed from junctions and trying to locate one with an active A&E department can prove to be time-consuming and frustrating. We are indebted to Ambulance Control Centres up and down the country for their help in compiling this information.

The **Places of Interest** sections - located with the information for each junction - describe some of the attractions within easy reach of the motorway and which could conveniently be used for journey breaks. The **Views from the Motorway** section comprises information on the major landmarks visible (listing them in sequence as if travelling from Junction 1) and, in many cases, identifying the marker post at which they may be viewed (see below).

To keep this book within a reasonable, and affordable, size a great deal of consideration was given on which motorways to include or omit. The result was a guide which embraces the principal north-south and east-west routes within England and Wales; hence our inclusion of the M1, M4, M5, M6 and M62 primary routes along with the inter-connecting M40, M42, M50 and M69 motorways which would, hopefully, serve the needs of the majority of motorway users.

We would like to thank staff at Motorway Maintenance Departments and Highways Agency offices across the country for their generous assistance in supplying us with information.

Note:

Marker Posts are found at 100m intervals along motorways and major trunk routes. With the widespread use of mobile telephones, when calling for assistance at the side of the motorway it is essential that the operator is told which motorway you are on, the direction being travelled and the nearest marker post to the vehicle.

Contents

How to Use

Off the Motorway has been specifically designed as an easy to use guide so there is no need for complicated instructions.

LOCATING A MOTORWAY

Each of the motorways covered in the guide can be found in it's own self-contained section and the information presented in exactly the same way. Simply refer to the contents page for the page number.

LOCATING A MOTORWAY JUNCTION

All junctions are listed in numerical order from Junction 1 upwards within each motorway section so locating a junction couldn't be simpler.

LOCATING FACILITIES, PLACES OF INTEREST AND HOSPITALS

All facilities, places of interest and A & E Hospitals are presented after the map of each junction and the facilities are pinpointed on the junction map. The location of the places of interest can be found on the motorway maps at the beginning of the sections covering each motorway.

IDENTIFYING LANDMARKS AND POINTS OF INTEREST VISIBLE FROM THE MOTORWAY

Passengers in the vehicle may wish to identify the many interesting landmarks and points of interest whilst travelling up and down the motorway. These are located and explained in easy-to-use maps contained in the **View from the Motorway** section starting on page 268.

EXPLANATORY NOTES ON THE SYMBOLS USED IN THE GUIDE

Short descriptions of the symbols used to describe the facilities at each junction may be found at the bottom of each page. You may however find the following more detailed explanations overleaf of use when deciding where to stop.

Explanation of Symbols

Ⱦ Licensed premises serving alcoholic beverages

¶ Food and non-alcoholic drinks available

⛽ Petrol and/or diesel fuel

24 Open 24 hours. (Some petrol stations may close for a short period to effect staff changeovers)

£ Cash dispensing machine available on site, or adjacent (as at supermarkets)

WC Toilets available for customers use at petrol stations. Please note that at some of those premises indicated as open for 24 hours toilets may not be available for the total duration if access is only gained through a shop area which is closed overnight for security reasons

♿ At petrol station sites, this indicates that the toilets, although they may not have been specifically adapted, are considered suitable for the use of wheelchair customers. At pubs, restaurants and hotels, this either indicates that the buildings have been fully adapted for disabled access or that the buildings are considered suitable to accommodate wheelchair customers

⛹ Children are welcome in public houses for meals if accompanied by parents and entirely at the discretion of the management. Local licensing laws may apply regarding the hours during which children may be accommodated

🛏B Guest house or Bed and Breakfast accommodation

🛏H Medium to large sized hotel

🛏h Small hotel

🛏M Motel or budget accommodation

🐎 Pets allowed at the discretion of the management

Notes:

Not all Public Houses within a mile of the junction are listed. Although our policy is to be as comprehensive as possible, only those establishments offering food or accommodation at the time of the survey are incorporated within this guide.

OFF THE MOTORWAY

M1

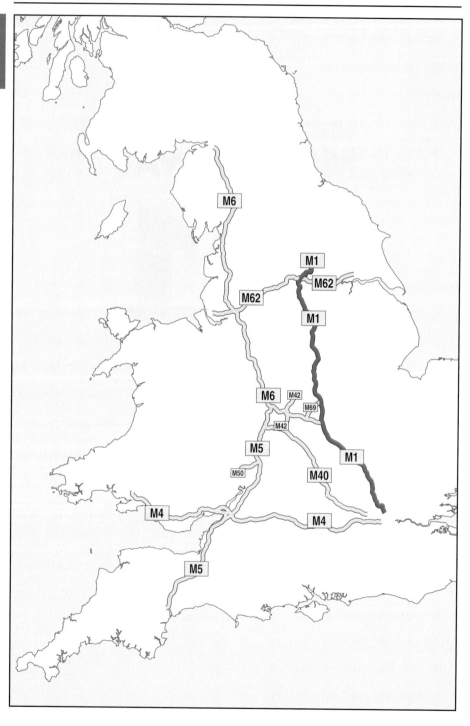

The M1 Motorway

Although not the first stretch of motorway to be constructed in this country, it was undoubtedly the opening of the first 72 mile section of the M1 by Mr Ernest Marples, the Transport Minister, on November 2nd, 1959 that heralded the long-awaited Motorway Age.

Commencing in Mill Hill, North London, the motorway heads north out of the capital, skirting past Watford and the hat-making town of Luton and passes the grounds of Woburn Abbey on the west side before reaching Milton Keynes. Just north of the footwear town of Northampton, three modes of transport run parallel together, the Grand Union Canal, the M1 and the former London & Birmingham Railway line which nowadays is a high speed electrified railway and forms part of the West Coast Main Line.

Beyond this point the radio masts at Rugby have been a feature of the skyline for many years but most of those visible are now redundant and may well be removed in the future. Leicester is passed on the east side before the motorway cuts through Charnwood Forest, now part of the National Forest, and continues north towards Kegworth where aircraft can be seen crossing the carriageway as they land at East Midlands Airport to the west.

The cooling towers of Radcliffe on Soar Power Station dominate the skyline to the east as the motorway crosses a series of lakes and weirs in the Trent Valley and then slips between Nottingham and Derby before bearing north through what was an extensive coalfield. Whilst redundant installations, including the former Markham Colliery, are in view at various points along the length, relics of England's older history can also be seen; Hardwick Hall, of Tudor origin, overlooks the carriageways whilst, further north, the remains of the 17thC castle at Bolsover are visible to the east.

As the motorway continues northwards through Yorkshire it takes a sharp turn to the west to pass between Rotherham and Sheffield and it is here where one of the major construction projects on the M1 was undertaken - to cross the River Don at Tinsley. Meadowhall Shopping Centre, typical of the redevelopments on disused steelworks sites which used to proliferate throughout this area, can be seen to the south and the cooling towers for Blackburn Meadows Power Station are alongside the viaduct to the north.

Continuing north westwards, Wentworth Castle, Stainsborough Castle and Queen Anne's Obelisk are visible to the west before Barnsley is passed to the east. The carriageway continues northwards and passes Wakefield to the east as it approaches Leeds. When originally constructed, the M1 used to terminate in Leeds, but the city is now by-passed with the motorway skirting around the south east side and passing the grounds of Temple Newsam before forming an end-on junction with the A1 just east of Garforth.

Location of Places of Interest

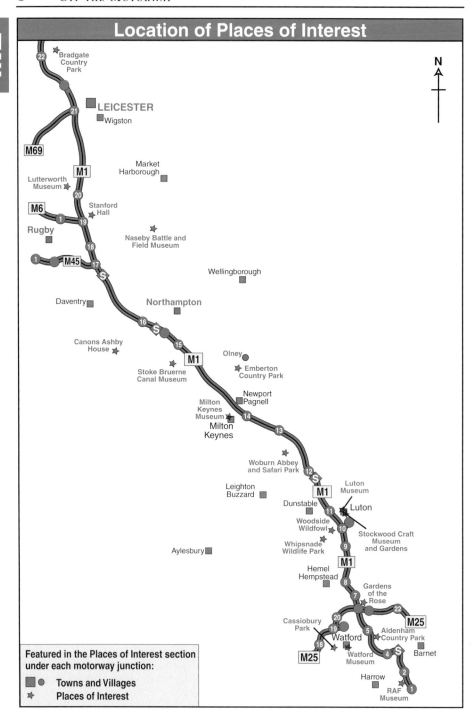

N

M1

22 Bradgate Country Park

LEICESTER
Wigston

21

M69

Market Harborough

M1
Lutterworth Museum

20
Stanford Hall

M6
19
1

Rugby

18

1 M45 17 S

Daventry

Naseby Battle and Field Museum

Wellingborough

Northampton

16 S

Canons Ashby House

15

M1

Olney

Stoke Bruerne Canal Museum

Emberton Country Park

Newport Pagnell

Milton Keynes Museum

14

Milton Keynes

13

Woburn Abbey and Safari Park

12 S

Leighton Buzzard

M1

Luton Museum

Dunstable

Luton

11

Woodside Wildfowl

10

Stockwood Craft Museum and Gardens

Whipsnade Wildlife Park

9

Aylesbury

M1

Hemel Hempstead

8

7

Gardens of the Rose

Cassiobury Park

20

22

M25

19

Watford

5

Aldenham Country Park

M25

18

4 S

Barnet

Watford Museum

2

Harrow

1

RAF Museum

Featured in the Places of Interest section under each motorway junction:

▪ ● Towns and Villages
✦ Places of Interest

M1

Location of Places of Interest

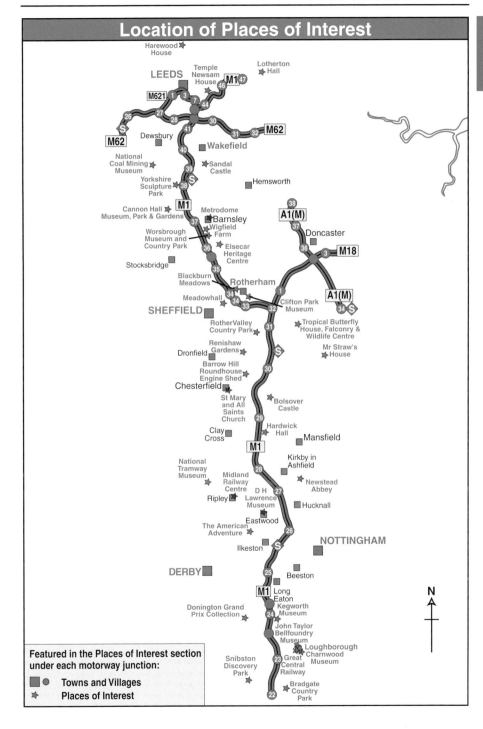

Harewood House
Temple Newsam House
Lotherton Hall
LEEDS
M1 47
46
M621
1
3
7
44
27
26
28
41
30
31 32 M62
40
M62
Dewsbury
Wakefield
S
National Coal Mining Museum
39
Sandal Castle
Yorkshire Sculpture Park
38 S
Hemsworth
Cannon Hall Museum, Park & Gardens
M1
Metrodome
38
Barnsley
A1(M)
37
Wigfield Farm
37
Doncaster
Worsbrough Museum and Country Park
36
Elsecar Heritage Centre
36
3 M18
Stocksbridge
35
Blackburn Meadows
Rotherham
1
A1(M)
Meadowhall
34 34
Clifton Park Museum
33
32
34 S
SHEFFIELD
RotherValley Country Park
31
Tropical Butterfly House, Falconry & Wildlife Centre
Renishaw Gardens
Dronfield
S
Mr Straw's House
Barrow Hill Roundhouse Engine Shed
30
Chesterfield
St Mary and All Saints Church
Bolsover Castle
Clay Cross
29
Hardwick Hall
Mansfield
National Tramway Museum
M1
Kirkby in Ashfield
28
Midland Railway Centre
Ripley
D H Lawrence Museum
27
Newstead Abbey
Eastwood
Hucknall
The American Adventure
26
NOTTINGHAM
Ilkeston
S
DERBY
25
Beeston
M1
Long Eaton
Donington Grand Prix Collection
Kegworth Museum
24
John Taylor Bellfoundry Museum
Loughborough
Charnwood Museum
Snibston Discovery Park
23
Great Central Railway
22
Bradgate Country Park

N

Featured in the Places of Interest section under each motorway junction:

■ ● Towns and Villages

✳ Places of Interest

M1 JUNCTION 2

THIS IS A RESTRICTED ACCESS JUNCTION

Vehicles can only exit from the southbound lanes and travel south along the A1

Vehicles can only enter the motorway along the northbound lanes from the A1 north carriageway

NEAREST NORTHBOUND A&E HOSPITAL

Watford General Hospital, Vicarage Road, Watford WD1 8HB Tel: (01923) 244366

Proceed to Junction 5 and take the A41 south and turn right along the A4008 towards Bushey. After about 1 mile turn right at the roundabout to Watford and continue along the A4145 to West Watford. The hospital is on the south side of Vicarage Road (A4145). (Distance Approx 11.2 miles)

NEAREST SOUTHBOUND A&E HOSPITAL

Barnet General Hospital, Wellhouse Lane, Barnet EN5 3DJ Tel: (020) 8216 4000

Take the A1 south and turn left along the B552 (Holders Hill Road). Follow the B552 to its end and turn right along the A411. After about 1.5 miles turn right into Wellhouse Lane and the hospital is on the left. (Distance Approx 5.4 miles)

M1 JUNCTION 3

THERE IS NO JUNCTION 3

SERVICE AREA

BETWEEN JUNCTIONS 2 AND 4

Scratchwood Services (Welcome Break)

Tel: (020) 8906 0611 The Granary Restaurant, KFC, Burger King, La Brioche Doree, Welcome Lodge & Shell Fuel.

M1 JUNCTION 4

THIS IS A RESTRICTED ACCESS JUNCTION

Vehicles can only exit from the southbound lanes and travel south along the A41

Vehicles can only enter the motorway along the northbound lanes from the A41 north carriageway.

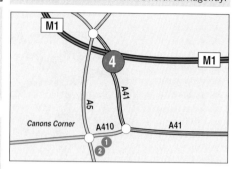

NEAREST NORTHBOUND A&E HOSPITAL

Watford General Hospital, Vicarage Road, Watford WD1 8HB Tel: (01923) 244366

Proceed to Junction 5 and take the A41 south and turn right along the A4008 towards Bushey. After about 1 mile turn right at the roundabout to Watford and continue along the A4145 to West Watford. The hospital is on the south side of Vicarage Road (A4145). (Distance Approx 7.2 miles)

NEAREST SOUTHBOUND A&E HOSPITAL

Barnet General Hospital, Wellhouse Lane, Barnet EN5 3DJ Tel: (020) 8216 4000

Follow the A41 south and turn left at the second roundabout north along the A1. Turn right at the first roundabout along the A411 and after about 1.8 miles turn right into Wellhouse Lane and the hospital is on the left. (Distance Approx 4.6 miles)

FACILITIES

1 Canons Corner Service Station (Esso)

🍴 ♿ WC 24   Tel: (020) 8958 8166

1 mile south along the A410, on the left. Access, Visa, Overdrive, All Star, Switch, Dial Card, Mastercard, Amex, Diners Club, Delta, AA Paytrak, UK Fuelcard, Shell Gold, BP Supercharge, Esso Cards.

2 McDonald's

🍴 ♿   Tel: (020) 8958 3482

1 mile south along the A410, on the left. Open; 07.00-23.00hrs daily. "Drive-Thru" closes at 0.00hrs daily.

| H Large Hotel | h Small Hotel | M Motel | B Guest House/ Bed & Breakfast | Disabled Facilities | Childrens Facilities |

PLACES OF INTEREST

FACILITIES

RAF Museum

Grahame Park Way, London NW9 5LL
Tel: (020) 8205 2266. 24 hour Information Line
Tel: 0891 6005 633.
website; http://www.rafmuseum.org.uk
Follow the A41 south and the route is signposted.
(4 miles)

Occupying 10 acres of what was once the historic Hendon Aerodrome over 70 aircraft are contained within 260,000ft^2 of exhibition halls. Specially constructed walkways and platforms enable close inspection of the planes and other attractions include a Red Arrows Flight Simulator, "Touch and Try" Jet Provost cockpit, interactive gallery and a walk-through Sunderland Flying Boat. The Battle of Britain Hall exhibit enables visitors to experience the sights and sounds of Britain's finest hour. Shop. Licensed Restaurant. Disabled access. Large grounds with indoor and outdoor picnic area

1 The Old Red Lion

Tel: (01923) 225826
0.5 miles south along the A41, on the left. (Bass) Open all day. Meals served; 12.00-22.00hrs daily

2 Hilton National Watford Hotel

Tel: (01923) 235881
0.6 miles south along the A41, on the left. Restaurant Open; Breakfast; Mon-Fri; 07.00-10.00hrs, Sat & Sun; 07.30-10.30hrs, Lunch; Sun-Fri; 12.30-14.00hrs, Sat; Closed, Dinner; Mon-Fri; 19.00-22.00hrs, Sat & Sun; 17.30-22.00hrs.

3 Elton Way Filling Station (Elf)

Tel: (01923) 248919
0.7 miles south along the A41, on the right. Access, Visa, Overdrive, All Star, Switch, Dial Card, Mastercard, Amex, Diners Club, Delta, Keyfuels, Elf Cards, Total Fina Cards.

4 Quinceys Bar & Restaurant

Tel: (01923) 229137
0.7 miles south along the A41, on the right. (Scottish & Newcastle) Open all day; Meals served; Mon-Sat; 12.00-23.00hrs, Sun; 12.00-22.30hrs

5 Sainsbury's Petrol Station

Tel: (01923) 681984
1 mile north along the A41, on the right. Access, Visa, Overdrive, All Star, Switch, Dial Card, Mastercard, Amex, Delta, UK Fuelcard, JS Fuelcard. Toilets in adjacent store.

6 BP Dome Filling Station

Tel: (01923) 672129
1 mile north along the A41, on the right. Access, Visa, Overdrive, All Star, Switch, Dial Card, Mastercard, Amex, Diners Club, Delta, Routex, AA Paytrak, Shell Agency, BP Cards.

M1   JUNCTION 5

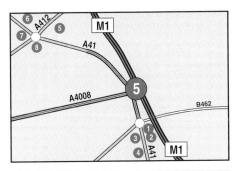

NEAREST A&E HOSPITAL

Watford General Hospital, Vicarage Road,
Watford WD1 8HB Tel: (01923) 244366

Take the A41 south and turn right along the A4008 towards Bushey. After about 1 mile turn right at the roundabout to Watford and continue along the A4145 to West Watford. The hospital is on the south side of Vicarage Road (A4145). (Distance Approx 2.7 miles)

 Pets Welcome   Cash Dispenser   24 Hour Facilities   Toilets   Alcoholic Drinks   Food and Drink   Fuel

M1

7   Shell Sceptre Service Station

[icons] WC [icons] 24   Tel: (01923) 800920

1 mile north along the A41, on the left. Access, Visa, Overdrive, All Star, Switch, Dial Card, Mastercard, Amex, Diners Club, Delta, UK Fuelcard, Shell Cards, Smartcard, BP Supercharge

8   Asda Petrol Station

[icon]   Tel: (01923) 250380

1 mile north along the A41, on the left. Access, Visa, Overdrive, All Star, Switch, Dial Card, Mastercard, Amex, Diners Club, Delta, BP Supercharge, Asda Business Card. Open; Continuously between 06.00hrs Mon and 23.00hrs Sat, Sun; 09.00-18.00hrs. Toilets in adjacent store

PLACES OF INTEREST

Watford Museum

194 High Street, Watford WD1 2HG
Tel: (01923) 232297 Follow the A4008 into Watford
(Signposted 2.2 Miles)

Displaying the town's history, the exhibits include a superb art collection, displays relating to printing and brewing and a tribute to the local football club.

Cassiobury Park

Cassiobury Avenue, Watford WD1 7SL
Tel: (01923) 235946 Follow the A41 north and
A412 southwest through Watford (4.8 Miles)

With over 190 acres of open space and woodland it provides an area of natural beauty and historic interest. Visitors have access to the canal and lock, children's paddling pools, miniature railway and tennis courts. Cafe. Some disabled access.

Aldenham Country Park

Elstree WD6 3AT Tel: (020) 8953 9602
Follow the A41 south and turn left along the A411
towards Elstree. (4.1 Miles)

Consisting of a large reservoir surrounded by 175 acres of woodland and parkland with an extensive network of footpaths. There is a rare breeds farm, Winnie-the-Pooh's 100 Acre

Wood, children's adventure play area and picnic site. Fishing available. Disabled access.

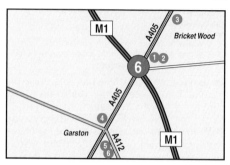

M1   JUNCTION 6

NEAREST NORTHBOUND A&E HOSPITAL

Hemel Hempstead General Hospital,
Hillfield Road HP2 4AD Tel: (01442) 213141

Proceed to Junction 8 and follow the A4147 to the second roundabout and the hospital is signposted to the right along the A4147 continuation. (Distance Approx 7.2 miles)

NEAREST SOUTHBOUND A&E HOSPITAL

Watford General Hospital, Vicarage Road,
Watford WD1 8HB Tel: (01923) 244366

Take the A412 south into Watford town centre and follow the A4145 south to West Watford. The hospital is on the south side of Vicarage Road (A4145). (Distance Approx 4.2 miles)

FACILITIES

1   Chequers Service Station (Esso)

[icon] WC [icon]   Tel: (01923) 672051

0.1 miles north along the A405, on the right. Access, Visa, Overdrive, All Star, Switch, Dial Card, Mastercard, Amex, Diners Club, Delta, Shell Gold, BP Supercharge, Esso Cards. Open; 06.00-0.00hrs daily.

2   Little Chef

[icon] [icon]   Tel: (01923) 661842

0.1 miles north along the A405, on the right. Open; 07.00-22.00hrs daily.

| Large Hotel | Small Hotel | Motel | Guest House/ Bed & Breakfast | Disabled Facilities | Childrens Facilities |

3 Classic Service Station (Elf)

🅿 WC   Tel: (01923) 680024
0.7 miles north along the A405, on the right.
Access, Visa, Overdrive, All Star, Switch, Dial
Card, Mastercard, Amex, AA Paytrak, Diners
Club, Delta, Elf Cards, Total Fina Card. Open;
06.00-23.00hrs daily

4 The Three Horseshoes

🍷 🍴 ♿ 🐾   Tel: (01923) 672061
0.8 miles south along the A405, in Garston,
on the right. (Bass) Open all day. Meals served;
12.00-22.00hrs daily

5 McDonald's

🍴 🐾 ♿   Tel: (01923) 671550
0.9 miles south along the A412, in Garston,
on the right. Open; 07.00-23.00hrs Daily.

6 Calendars Cafe Bar & Restaurant

🍷 🍴 ♿ 🐾   Tel: (01923) 672310
1 mile south along the A412, in Garston, on
the right. (Greene King) Open all day. Meals
served; 12.00-23.00hrs daily.

PLACES OF INTEREST

The Gardens of the Rose

The Royal National Rose Society, Chiswell Green,
St Albans AL2 3NR Tel: (01727) 850461
website; http://www.roses.co.uk
e-mail; mail@mrs.org.uk
Follow the A405 north (2.2 Miles)

Over 30,000 roses create a stunning spectacle
from early Spring to late Summer. These are
complemented by a rich variety of companion
plants including Spring bulbs, herbaceous
borders, Lavenders and over 100 varieties of
Clematis. Disabled access.

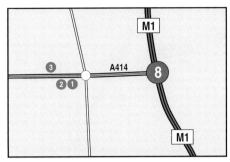

M1   JUNCTION 8

NEAREST A&E HOSPITAL
**Hemel Hempstead General Hospital,
Hillfield Road HP2 4AD Tel: (01442) 213141**
Follow the A4147 to the second roundabout and
the hospital is signposted to the right along the
A4147 continuation. (Distance Approx 3.3 miles)

FACILITIES

1 Breakspear Way Service Station (BP)

🅿 ♿ WC 24 🍴 £   Tel: (01442) 269003
0.4 miles west along the A414, on the left.
Access, Visa, Overdrive, All Star, Switch, Dial
Card, Mastercard, Amex, Diners Club, Delta,
Routex, AA Paytrak, Shell Agency, BP Cards,
Mobil Cards. Coffee Shop.

2 Posthouse Hemel Hempstead

🛏H 🍷 🍴 ♿ 🐾 🛏   Tel: 0870 400 9041
0.4 miles west along the A414, on the left.
Traders Bar & Restaurant; Open all day.
Breakfast 06.30-10.00hrs, Lunch 12.30-
14.30hrs, Evening Meals 17.30-22.30hrs.

3 Shell Hemel Hempstead

🅿 ♿ WC 24   Tel: (01442) 275010
0.5 miles west along the A414, on the right.
Access, Visa, Overdrive, All Star, Switch, Dial
Card, Mastercard, Amex, Diners Club, Delta,
AA Paytrak, Shell Cards, Smartcard, BP Agency,
Eurocard.

M1   JUNCTION 6A & JUNCTION 7

**JUNCTIONS 6A AND 7 ARE MOTORWAY
INTERCHANGES ONLY AND THERE IS NO
ACCESS TO ANY FACILITIES**

| Pets Welcome | £ Cash Dispenser | 24 Hour Facilities | WC Toilets | Alcoholic Drinks | Food and Drink | Fuel |

M1 JUNCTION 9

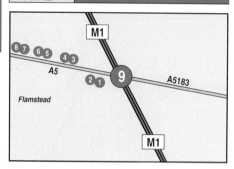

NEAREST NORTHBOUND A&E HOSPITAL

**Luton & Dunstable Hospital, Lewsey Road,
Luton LU4 0D Tel: (01582) 491122**

Proceed north to Junction 11 and take the A505 west. The hospital is on the north side of this road, adjacent to the motorway junction.(Distance Approx 6.1 miles)

NEAREST SOUTHBOUND A&E HOSPITAL

**Hemel Hempstead General Hospital,
Hillfield Road HP2 4AD Tel: (01442) 213141**

Proceed to Junction 8, follow the A4147 to the second roundabout and the hospital is signposted to the right along the A4147 continuation.(Distance Approx 7.7 miles)

FACILITIES

1 The Harvester Flamstead

Tel: (01582) 842800

0.1 miles west along the A5, on the left. (Bass Taverns) Open all day. Meals served; 12.00-22.00hrs daily

2 Holiday Inn Express, Luton-Hemel

Tel: (01582) 841332

0.2 miles west along the A5, on the left.

3 Watling Street Filling Station (Shell Diesel)

Tel: (01582) 840215

0.4 miles west along the A5, on the right. Access, Visa, Overdrive, Switch, Mastercard,

Amex, Diners Club, UK Fuelcard, Shell Cards, Texaco Fastfuel, Securicor Fuelserv, IDS, Keyfuels, AS24. Open; 05.45-22.00hrs daily (Card operated pumps 24 hours)

4 Watling Street Cafe

Tel: (01582) 840270

0.4 miles west along the A5, on the right. Open; Mon-Fri; 06.00-22.00hrs, Sat; 07.00-13.00hrs, Sun; 08.00-13.00hrs

5 Flamstead Service Station (Shell)

Tel: (01582) 842051

0.5 miles west along the A5, on the right. Mastercard, Visa, Switch, Amex, Diners Club, Dial Card, Overdrive, All Star, Shell Cards. Open; Mon-Thurs; 24hrs, Fri & Sat; 08.00-22.00hrs, Sun; 08.00-0.00hrs.

6 Waggon & Horses

Tel: (01582) 841932

0.7 miles west along the A5, on the right. (Whitbread) Open all day. Meals served; Mon-Sat; 12.00-21.00hrs, Sun; 12.00-17.00hrs

7 Little Chef

Tel: (01582) 840302

0.9 miles west along the A5, on the right. Open; 07.00-22.00hrs daily.

8 Hertfordshire Moat House

Tel: (01582) 449988

1 mile west along the A5, on the right. Borders Restaurant & Bar; Meals served 07.00-23.00hrs daily

PLACES OF INTEREST

Whipsnade Wild Animal Park

Whipsnade, Nr Dunstable, Bedfordshire LU6 2LF
Information Hotline Tel; 0990 200 123
Follow the A5 north (Signposted 6 Miles)

One of Europe's largest Conservation Parks with 2500 animals on view within a 600 acre parkland. Visitors may drive around or take one

| Large Hotel | Small Hotel | Motel | Guest House/ Bed & Breakfast | Disabled Facilities | Childrens Facilities |

of the free Safari Tour Buses. Miniature Steam Railway. Children's Play Area. Cafe. Disabled Access

M1 JUNCTION 10 & JUNCTION 10A

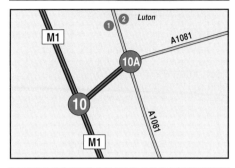

NEAREST A&E HOSPITAL

Luton & Dunstable Hospital, Lewsey Road, Luton LU4 0D Tel: (01582) 491122

Proceed north to Junction 11 and take the A505 west. The hospital is on the north side of this road, adjacent to the motorway junction.(Distance Approx 3.7 miles)

FACILITIES

1   Stockwood Hotel

`B` Tel: (01582) 721000
0.9 miles north along the A6 (Castle Street), in Luton, on the left.

2   Shell Motorway Station

`WC` `24` Tel: (01582) 747430
1 mile north along the A6 (Castle Street), in Luton, on the right. Access, Visa, Overdrive, All Star, Switch, Dial Card, Mastercard, Amex, Diners Club, Delta, UK Fuelcard, Shell Cards, Smartcard.

PLACES OF INTEREST

Stockwood Craft Museum & Gardens

Farley Hill, Luton LU1 4BH Tel: (01582) 738714
e-mail; burgessl@luton.gov.uk
Follow the A6 north towards Luton (Signposted 1.8 Miles)

A craft museum featuring rural trades, and set within beautifully restored Period Gardens. The gardens represent over 1000 years of English gardening and feature Knot, Mediaeval, Victorian and Italian styles. Other attractions include the Mossman Collection of Horse Drawn Vehicles and a Bee Gallery. Childrens Play Area. Gift Shop. Disabled Access.

Woodside Wildfowl Park

Woodside Road, Slip End Village, Luton LU1 4DG
Tel: (01582) 841044. Follow the A1081 south, turn first right east towards Slip End and then left along the B4540 (Signposted 2.9 Miles)

One of England's largest poultry centres incorporating very rare breeds. Visitors can stroke, handle and feed the animals and poultry. Children's Farm. Leisure Complex. Farm and Gift Shops. Coffee Shop. Disabled Access.

Luton Museum & Art Gallery

Warndown Park, Luton LU2 7HA
Tel: (01582) 546721 or Tel: (01582) 546739
e-mail; burgessl@luton.gov.uk
Follow the A6 north through Luton (3.3 Miles)

The history of Luton and the surrounding district is featured in this Victorian mansion set in Warndown Park. Saxon jewellery, mediaeval guild books for Luton and Dunstable and the development of the hat industry are just some of the exhibits on display. Gift Shop. Tea Room. Disabled facilities.

| Pets Welcome | Cash Dispenser | 24 Hour Facilities | Toilets | Alcoholic Drinks | Food and Drink | Fuel |

M1 JUNCTION 11

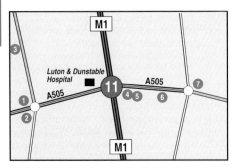

NEAREST A&E HOSPITAL

Luton & Dunstable Hospital, Lewsey Road, Luton LU4 0D Tel: (01582) 491122

Take the A505 west and the hospital is on the north side of this road, adjacent to the motorway junction. (Distance Approx 0.4 miles)

FACILITIES

1 The Halfway House

Tel: (01582) 609938

0.5 miles west along the A505, on the right. (Beefeater) Open all day. Breakfast; Mon-Fri; 07.00-09.00hrs, Sat & Sun; 08.00-10.00hrs. Meals served all day; Sun-Thurs; 12.00-22.30hrs, Fri & Sat; 12.00-23.00hrs.

2 Tesco Petrol Station

Tel: (01582) 687500

0.5 miles west along the A505, on the left. Access, Visa, Overdrive, All Star, Switch, Dial Card, Mastercard, Amex, Delta, AA Paytrak, Tesco Fuelcard. Open; Continuously between 06.00hrs Mon and 0.00hrs Sat, Sun; 06.00-0.00hrs. Toilets in adjacent store

3 Shell Wilbury Service Station

Tel: (01582) 479950

1 mile along Poynters Road, on the left. Access, Visa, Overdrive, All Star, Switch, Dial Card, Mastercard, Amex, Diners Club, Delta, Shell Cards, BP Supercharge, Smartcard.

4 Travelodge Luton

Tel: (01582) 575955

0.1 miles east along the A505, on the right. Bar Cafe; Open; Breakfast; 07.00-10.00hrs daily, Lunch; 12.00-14.00hrs daily, Dinner; 18.00-22.00hrs daily.

5 Dunstable Road Filling Station (Jet)

Tel: (01582) 574366

0.1 miles east along the A505, on the right. Access, Visa, Overdrive, All Star, Switch, Dial Card, Mastercard, Amex, Diners Club, Delta, Routex, UK Fuelcard, BP Supercharge, Jet Cards. Open; Mon-Fri; 06.30-22.00hrs, Sat & Sun; 07.00-22.00hrs.

6 The Leicester Arms Harvester

Tel: (01582) 572718

0.5 miles east along the A505, on the right. (Bass Taverns) Meals served; Mon-Fri; 12.00-14.30hrs & 17.00-21.30hrs, Sat & Sun; 12.00-22.00hrs

7 Empire Petrol Station (Esso)

Tel: (01582) 593886

0.8 miles east along the A505, on the left. Esso Cards, Visa, Access, Diners Club, Overdrive, Switch, All Star, Dial Card, Amex. Open; 07.00-23.00hrs daily.

PLACES OF INTEREST

Luton Museum & Art Gallery

Warndown Park, Luton LU2 7HA
Follow the A505 east and turn left along the A5228 north (Signposted 2.5 Miles)

For details please see Junction 10 information

| Large Hotel | Small Hotel | Motel | Guest House/ Bed & Breakfast | Disabled Facilities | Childrens Facilities |

M1

SERVICE AREA

BETWEEN JUNCTIONS 11 AND 12

Toddington Services (Northbound) (Granada)
Tel: (01525) 878400 Fresh Express Self Service Restaurant, Burger King, Little Chef, Harry Ramsden's & BP Fuel

Toddington Services (Southbound) (Granada)
Tel: (01525) 878424 Fresh Express Self Service Restaurant, Burger King, Little Chef, Travelodge & BP Fuel

Footbridge connection between sites

M1   JUNCTION 12

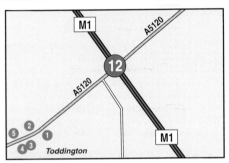

NEAREST NORTHBOUND A&E HOSPITAL

Bedford General Hospital, South Wing, Kempston Road MK42 9DJ Tel: (01234) 355122

Proceed north to Junction 13 and take the A421 east to Bedford. Continue along the A6 and then the A5141 and the hospital is along this road. (Distance Approx 16.3 miles)

NEAREST SOUTHBOUND A&E HOSPITAL

Luton & Dunstable Hospital, Lewsey Road, Luton LU4 0D Tel: (01582) 491122

Proceed south to Junction 11 and take the A505 west. The hospital is on the north side of this road, adjacent to the motorway junction. (Distance Approx 5.5 miles)

FACILITIES

1   The Griffin Hotel

Tel: (01525) 872030
0.9 miles south along the A5120, on the left.

(Greene King) Open all day. Lunch; 12.00-14.00hrs daily, Evening meals; Mon-Sat; 19.00-21.00hrs

2   The Bell at Toddington

Tel: (01525) 872564
0.9 miles south along the A5120, on the right. (Freehouse) Open all day. Meals served; Mon-Sat; 12.00-22.00hrs, Sun; 12.00-21.30hrs

3   Oddfellows Arms

Tel: (01525) 872021
1 mile south along the A5120, in Market Place, Toddington. (Free House) Lunch; Thurs-Tues; 12.00-14.00hrs, Evening Meals; Mon-Sat; 18.00-20.30hrs.

4   The Bombay Tandoori Restaurant

Tel: (01525) 872928
1 mile south along the A5120, in Market Place, Toddington. Open; 18.30-23.30hrs daily.

5   The Red Lion

Tel: (01525) 872524
1 mile south along the A5120, in Market Place, Toddington. (Free House) Open all day on Sat & Sun. Closed until 18.00hrs on Mon. Meals served; Tues-Sun; 12.00-14.30hrs & 19.00-21.30hrs.

PLACES OF INTEREST

Whipsnade Wild Animal Park

Whipsnade, Nr Dunstable, Bedfordshire LU6 2LF
Follow the A5120 south (Signposted 9 Miles)

For details please see Junction 9 Information

| Pets Welcome | Cash Dispenser | 24 24 Hour Facilities | WC Toilets | Alcoholic Drinks | Food and Drink | Fuel |

M1 · JUNCTION 13

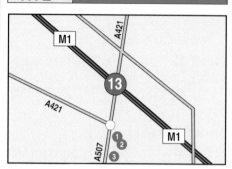

NEAREST A&E HOSPITAL

**Bedford General Hospital, South Wing,
Kempston Road MK42 9DJ Tel: (01234) 355122**

Take the A421 east to Bedford. Continue along the A6 and then the A5141 and the hospital is along this road. (Distance Approx 9.5 miles)

OR ALTERNATIVELY

**Milton Keynes General, Standing Way,
Eaglestone MK6 5LD Tel: (01908) 660033**

Take the A421 west to Milton Keynes and continue into the A4146. The hospital is along this road. (Distance Approx 5.8 miles)

FACILITIES

1   Crawley Crossing (Keyfuels)

🍴 WC 24   Tel: (01908) 281084

0.3 miles east along the A507, on the left. Diesel Direct, IDS, Keyfuels

2   Trunkers Cafe

🍴

0.3 miles east along the A507, on the left. Open; Mon-Fri; 06.00-22.00hrs, Sat; 07.00-12.00hrs.

3   Guise Motors (Q8)

🍴 WC   Tel: (01908) 281333

0.3 miles east along the A507, on the left. Access, Visa, Overdrive, All Star, Switch, Dial Card, Mastercard, Amex, Diners Club, Delta, AA Paytrak, BP Supercharge, Q8 Cards. Open; 07.00-21.00hrs daily.

PLACES OF INTEREST

Woburn Safari Park

Woburn, Bedfordshire MK17 9QN
Tel: (01525) 290407
e-mail; WobSafari@aol.com
Follow the A507 south. (Signposted 2 Miles)

Featuring a drive-through Safari Park, with extensive reserves, and a Worldwide Leisure area with adventure playgrounds, and animal encounters and demonstration features. Boating. Miniature Railway. Gift Shop. Restaurant. Disabled access.

Woburn Abbey

Woburn, Bedfordshire MK43 0TP
Tel: (01525) 290666
Adjacent to Woburn Safari Park.

The home of the Marquess and Marchioness of Tavistock, the house contains one of the most important private collections of works of art in the country. Surrounded by a 3000 acre park containing nine species of deer, there are two gift shops, an antiques centre, pottery and the Flying Duchess Pavilion serves light lunches and teas.

M1 · JUNCTION 14

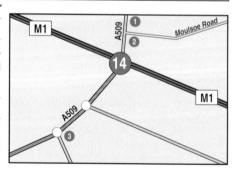

NEAREST A&E HOSPITAL

**Milton Keynes General, Standing Way,
Eaglestone MK6 5LD Tel: (01908) 660033**

Take the A509 towards Milton Keynes and at the crossroads bear right along the A4146. At the roundabout junction with the A421 turn right along the A421 and this leads to the hospital. (Distance Approx 4.1 miles)

FACILITIES

1 Courtyard by Marriott Hotel

Tel: (01908) 613688

0.3 miles north along the A509, on the right. Restaurant Open; Breakfast; Mon-Fri; 07.00-09.30hrs, Sat & Sun; 07.00-11.00hrs, Lunch; 12.00-14.00hrs daily, Dinner; 19.00-22.00hrs daily.

2 The Old Stables Executive B&B at Heritage Farm

Tel: (01908) 217766

0.1 miles east along Moulsoe Road, on the right.

3 Total Filling Station

Tel: (01908) 354240

0.7 miles south along the A509, on the left. Visa, Access, Amex, Diners Club, All Star, Overdrive, Dial Card, Switch, Total Fina Cards

PLACES OF INTEREST

Milton Keynes Museum

Southern Way, Wolverton MK12 5EJ
Tel: (01908) 316222
website; http://www.artizan.co.uk/mkm
Follow the A505 east through Central Milton Keynes, turn right along the A5 and right along the A422 (6.4 Miles)

Housed in a beautiful Victorian farmstead and grounds and sited in Wolverton, England's first railway town. The exhibits and displays reflect the agricultural and industrial heritage of the area. Shop. Tea Rooms. Disabled access.

Emberton Country Park

Nr Olney, Buckinghamshire MK46 5DB
Tel: (01234) 711575
Take the A509 north towards Olney (5.8 Miles)

Originally a gravel works, in 1965 it was transformed into an attractive parkland setting covering 205 acres. It caters for a wide variety of uses and interests including caravanning and camping, fishing, sailing, walking, a nine hole pitch and putt golf course and picnic areas. Shop. Tea Rooms. Disabled access.

Olney

Olney Chamber of Trade Tel: (01234) 241140.
website;www.olney.org.uk e-mail; info@olney.org.uk
Follow the A509 north (Signposted 7.1 Miles)

A beautiful market town, with an attractive wide high street and market, it is home to a fascinating collection of architecturally varied buildings dating back many centuries. Although the earliest documentary evidence as to the existence of Olney is contained in a Saxon Charter of 979 a number of archaeological finds indicate that the area was occupied as early as 1600BC. The Church of St Peter & St Paul dates from the 14thC and the bridge from the 19thC. The Cowper & Newton Museum (Tel: 01234-711516) in the Market Place was once the home of William Cowper from 1768 to 1786 and is dedicated to preserving the memory and artifacts of Cowper and his friend the Rev.John Newton, reformed slave trader and curate of Olney between 1764 and 1780. They collaborated in writing the Olney Hymns, of which "Amazing Grace" is the most well known.

SERVICE AREA

BETWEEN JUNCTIONS 14 AND 15

Newport Pagnell (Northbound) (Welcome Break)
Tel: (01908) 217722 Red Hen, The Granary Restaurant, KFC, Burger King, Welcome Lodge & Texaco Fuel

Newport Pagnell (Southbound) (Welcome Break)
Tel: (01908) 217722 The Granary Restaurant & Texaco Fuel

Footbridge connection between sites

	Pets Welcome	£	Cash Dispenser	24	24 Hour Facilities	WC	Toilets		Alcoholic Drinks		Food and Drink		Fuel

M1   JUNCTION 15

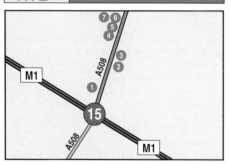

NEAREST A&E HOSPITAL

Northampton General Hospital, Cottarville,
Off Billing Road NN1 5BD Tel: (01604) 634700

Take the A508 north into Northampton and turn right along the A5123. The hospital is along this route. (Distance Approx 3.8 miles)

FACILITIES

1   Stakis Country Court Hotel

Tel: (01604) 700666

0.2 miles north along the A508, on the left. Restaurant Open; Breakfast; Mon-Sat; 07.00-10.00hrs, Sun; 08.00-10.30hrs, Lunch; Mon-Fri; 12.30-14.00hrs, Sun; 13.00-14.30hrs, Dinner; 19.00-22.00hrs daily. Bar snacks available throughout the day.

2   Grange Farm Services (Save)

Tel: (01604) 700498

0.5 miles north along the A508, on the right. Keyfuels, Access, Visa, Overdrive, All Star, Switch, Dial Card, Mastercard, AA Paytrak, Delta, BP Supercharge, Save Card

3   Little Chef

Tel: (01604) 705722

0.5 miles north along the A508, on the right. Open; 07.00-22.00hrs daily.

4   Midway Hotel

Tel: (01604) 769676

0.9 miles north along the A508, on the left.

(Toby Restaurant) Open all day. Meals served; Mon-Sat; 12.00-14.00hrs & 17.30-22.00hrs, Sun; 12.00-22.00hrs.

5   Burger King

Tel: (01604) 701078

0.9 miles north along the A508, on the left. Open; Mon-Fri; 11.00-21.00hrs, Sat & Sun; 10.00-22.00hrs

6   Little Chef

Tel: (01604) 701078

0.9 miles north along the A508, on the left. Open; 07.00-22.00hrs daily.

7   Shell Northampton

Tel: (01604) 664940

0.9 miles north along the A508, on the left. Access, Visa, Overdrive, All Star, Switch, Dial Card, Mastercard, Amex, Diners Club, Delta, AA Paytrak, Shell Cards, BP Agency, Smartcard.

PLACES OF INTEREST

Northampton

Follow the A508 north (Signposted 4.3 Miles)

For details please see Junction 16 information.

Within the town can be found ...

Central Museum and Art Gallery

Guildhall Road, Northampton NN1 1DP

For details please see Junction 16 information.

Stoke Bruerne Canal Museum

Bridge Road, Stoke Bruerne, Towcester NN12 7SE
Tel: (01604) 862229
Follow the A508 south (3.8 Miles)

The Canal Museum portrays over 200 years of colourful waterways history, whilst the picturesque village of Stoke Bruerne contains canalside pubs. Waterside walks, boat trips to Blisworth Tunnel, restaurants and tearooms. Disabled access.

M1 JUNCTION 15A

APART FROM ROTHERSTHORPE SERVICES THERE ARE NO OTHER FACILITIES WITHIN ONE MILE OF THIS JUNCTION

SERVICE AREA

ACCESSED FROM THIS JUNCTION

Rothersthorpe Services (Northbound) (RoadChef)
Tel: (01604) 831888 Wimpy Bar & BP Fuel

Rothersthorpe Services (Southbound) (RoadChef)
Tel: (01604) 831888 Food Fayre Self-Service Restaurant, & BP Fuel

Footbridge connection between sites

NEAREST A&E HOSPITAL

Northampton General Hospital, Cottarville, Off Billing Road NN1 5BD Tel: (01604) 634700

Take the A43 north and continue north along the A45. Turn right along the A5123 and the hospital is signposted along this route. (Distance Approx 4.5 miles)

PLACES OF INTEREST

Northampton

Follow the A43 north (Signposted 4.2 Miles)

For details please see Junction 16 information.

Within the town can be found ...

Central Museum and Art Gallery

Guildhall Road, Northampton NN1 1DP

For details please see Junction 16 information.

M1 JUNCTION 16

NEAREST A&E HOSPITAL

Northampton General Hospital, Cottarville, Off Billing Road NN1 5BD Tel: (01604) 634700

Proceed to Junction 15A, take the A43 north and continue north along the A45. Turn right along the A5123 and the hospital is signposted along this route. (Distance Approx 8 miles)

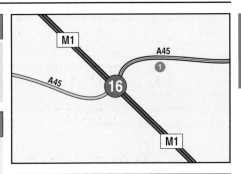

FACILITIES

1   The Red Lion Pub & Cafe

Tel: (01604) 831914
0.5 miles east along the A45, on the right. (Freehouse) Open all day. Meals served; Mon-Fri; 06.00-21.30hrs, Sat & Sun; 12.00-21.00hrs

PLACES OF INTEREST

Northampton

Northampton Tourist Information Office, St Giles Square, Northampton NN1 1DA
Tel: (01604) 622677
e-mail; tic@Northampton.gov.uk
Follow the A45 east (Signposted 5 Miles)

Synonymous with the manufacture of footwear, the history of Northampton goes back to pre-Roman times with Iron Age settlements having been found in and around the town. Ever since King John bought a pair of shoes here for nine pence in the early 13thC the boot and shoe trade has flourished greatly and a world wide reputation established. Cromwell's Parliamentarians, with whom the town sided during the Civil War, were sent 1500 pairs and many other armies, over the years, have been supplied with footwear made here. Lace-making, the other industry here, was probably established during the 17thC in the wake of the influx of Protestant refugees from the Continent. The town was all but destroyed by the Great Fire of 1675 but was rebuilt in such a spacious and well-planned way that Daniel Defoe called it "the handsomest and best built town in all this part of England".

Today there is a fascinating mix of historic buildings; Norman churches, Victorian architecture, museums and an art gallery, and modern amenities. Reflecting on the major trade in the town, there are numerous factory shoe shops including Barker Shoes in Station Road, Earls Barton (Tel: 01604-980387) and Barratts in Barrack Road (Tel: 01604-718632).

Within the town can be found ...

Central Museum and Art Gallery

Guildhall Road, Northampton NN1 1DP
Tel: (01604) 639415

With the largest collection of boots and shoes in the world, visitors can see shoe fashions across the centuries and the machines that made them. There are exhibits showing the history of Northampton from the Stone Age to the present day and the Art Gallery has a fine collection of Italian 15thC to 18thC paintings and British art. Disabled access.

Canons Ashby House (NT)

Canons Ashby, Nr Daventry NN11 3SD
Tel: (01327) 860044
Follow the A45 west, turn left through Nether Heyford and left along the A5. Canons Ashby is signposted along this route. (10.4 Miles)

One of the oldest and most romantic of Northamptonshire's great houses and home of the Dryden family since the 1550's. The house contains fascinating Elizabethan wall paintings and sumptuous Jacobean plasterwork, formal gardens with terraces, mediaeval church and parkland. Shop. Tea Room. Some disabled access.

M1 JUNCTION 17

JUNCTION 17 IS A MOTORWAY INTERCHANGE ONLY AND THERE IS NO ACCESS TO ANY FACILITIES

SERVICE AREA

SERVICE AREA

Watford Gap Services (Northbound) (RoadChef)
Tel: (01327) 879001 Food Fayre Self-Service Restaurant, Wimpy Bar & BP Fuel

Watford Gap Services (Southbound) (RoadChef)
Tel: (01327) 879001 Food Fayre Self-Service Restaurant, Wimpy Bar, RoadChef Lodge & BP Fuel

Footbridge connection between sites

M1 JUNCTION 18

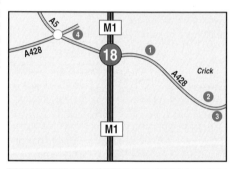

NEAREST A&E HOSPITAL

Hospital of St.Cross, Barby Road, Rugby CV22 5PX Tel: (01788) 572831
Take the A5 west and the A428 into Rugby. Turn left into Barby Road and the hospital is on the left. (Distance Approx 5.4 miles)

FACILITIES

1 Posthouse Rugby/Northampton

Tel: 0870 400 9059
0.3 miles east along the A428, on the left. Junction Restaurant; Open; Breakfast; Mon-Fri; 07.00-10.00hrs, Sat & Sun; 07.30-10.30hrs, Lunch; Mon-Fri; 12.30-14.30hrs, Sun; 12.30-15.00hrs, Dinner; Mon-Fri; 18.30-22.30hrs, Sun; 19.00-22.00hrs.

2 The Wheatsheaf

Tel: (01788) 822284
1 mile east along the A428, in Crick, on the

left. (Free House) Open all day. Meals served; Mon-Sat; 12.00-14.30hrs & 18.00-21.30hrs, Sun; 12.00-21.30hrs

---

3   The Red Lion

▮ ▮ ▮   Tel: (01788) 822342

1 mile east along the A428, in Crick, on the right. (Free House) Meals served; Lunch; 12.00-14.00hrs daily, Evening Meals; Mon-Fri; 19.00-21.00hrs, Sat; 19.00-21.30hrs.

---

4   Holiday Inn Express, Crick

▮H ▮   Tel: (01788) 550333

0.3 miles west along the A5, on the right.

PLACES OF INTEREST

James Gilbert Rugby Football Museum

5 St.Matthews Street, Rugby CV21 3BY
Tel: (01788) 542426.
Follow the A4328 into Rugby (5.2 Miles)

The museum is on the site where Gilbert footballs have been made since 1842. The birthplace of rugby union, visitors can find out how Rugby gave its name to what became an international game. Contains many artefacts and there are practical demonstrations of the art of manufacturing the oval shaped ball. There is also "The Total Rugby Experience", an audio visual show, and a Factory Outlet shop.

Also, within the town can be found ...

Rugby School Museum

10 Little Church Street, Rugby CV21 3AW
Tel: (01788) 556109 (Museum)
or Tel: (01788) 556227 (Tours).

Famously, the scene for "Tom Brown's Schooldays" and the birthplace of Rugby football, the museum tells the story of Rugby School, its pupils and the ethos that led it to the forefront of Public Schools and its outstanding influence on the development of the game. Music and sound bring to life the displays which feature memorabilia of the game and of the many famous people who were pupils here. Tours of the school itself are arranged regularly.

Stanford Hall

Lutterworth, Leicestershire LE17 6DH
Follow the A428 east (Signposted 5.2 Miles)

For details please see Junction 19 information.

M1   JUNCTION 19

THIS IS A MOTORWAY INTERCHANGE WITH THE M6 AND A RESTRICTED ACCESS JUNCTION

Vehicles can only exit from the southbound lanes.

Vehicles can only enter the motorway along the northbound lanes

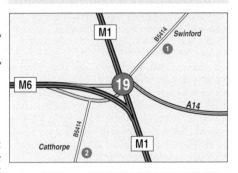

NEAREST A&E HOSPITAL

Hospital of St.Cross, Barby Road, Rugby CV22 5PX Tel: (01788) 572831

Take the B5414 into Rugby and turn right along the A428. Turn left into Barby Road and the hospital is on the left. (Distance Approx 5.7 miles)

FACILITIES

1   The Chequers

▮ ▮ ▮ ▮   Tel: (01788) 860318

0.5 miles east along the B5414, in Swinford, on the right. (Pubmaster) Meals served; Tues-Sat; 12.00-14.00hrs & 18.30-21.00hrs, Sun; 12.00-14.00hrs

---

2   The Cherry Tree

▮ ▮ ▮ ▮   Tel: (01788) 860430

0.8 miles south along the Catthorpe Road, on the left. (Free House) Open all day on Sat & Sun. Meals served; Thurs-Sun; 12.00-15.00hrs

---

| Pets Welcome | £ Cash Dispenser | 24 24 Hour Facilities | WC Toilets | Alcoholic Drinks | Food and Drink | Fuel |

PLACES OF INTEREST

Stanford Hall

Lutterworth, Leicestershire LE17 6DH
Tel: (01788) 860250
Signposted from the junction (2.7 Miles)

Built by the Smiths of Warwick in the 1690's and still occupied by the Cave family, it contains antique furniture, fine pictures and family costumes. The grounds include a walled rose garden and nature trail and there is also a Motorcycle Museum. Gift Shop.

Naseby Battle & Farm Museum.

Naseby Post Office, Tel: (01604) 740213
Follow the A14 east, at the first junction turn right along the A50 and then first left to Naseby.
(Signposted off the A14, 9.7 Miles)

The decisive battle of the Civil War, Naseby has played a significant part in English history. It was here on June 14th, 1645 that Cromwell defeated Charles I's Royalist army and brought about his surrender some 11 months later in Newark. Today, Naseby is one of the least spoilt of English battlefields and the adjacent museum contains a miniature layout of the battlefield with commentary as well as relics from the battle.

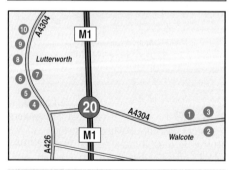

M1   **JUNCTION 20**

NEAREST NORTHBOUND A&E HOSPITAL

Leicester Royal Infirmary, Infirmary Square, Leicester LE1 5WW Tel: (0116) 254 1414

Proceed to Junction 21 and take the exit to Leicester. Follow the A5460 into the city and turn right along Upperton Road. Turn left at the end and the hospital is on the left. (Distance Approx 15 miles)

NEAREST SOUTHBOUND A&E HOSPITAL

Hospital of St.Cross, Barby Road, Rugby CV22 5PX Tel: (01788) 572831

Take the A426 into Rugby and turn left along the A428. Turn right into Barby Road and the hospital is on the left. (Distance Approx 7.6 miles)

FACILITIES

---

1   The Tavern Inn

Tel: (01455) 553338

0.9 miles east along the A4304, in Walcote, on the left. (Free House) Open all day. Meals served; 11.30-21.30hrs daily

---

2   Walcote Service Station (Total Fina)

Tel: (01455) 553911

1 mile east along the A4304, in Walcote, on the right. Access, Visa, Overdrive, All Star, Switch, Dial Card, Mastercard, Amex, Diners Club, Delta, Elf Cards, Total Fina Cards. Open; 07.00-22.00hrs.

---

3   The Black Horse

Tel: (01455) 552684

1 mile east along the A4304, in Walcote, on the left. (Free House) Meals served; Lunch; Fri;12.00-14.00hrs & Sun;12.00-14.30hrs, Evening Meals; 19.00-21.30hrs daily

---

4   The Fox Inn

Tel: (01455) 552677

0.5 miles west along the A4304, in Lutterworth, on the left. (Whitbread) Meals served; Mon-Sat; 12.00-14.00hrs & 18.30-21.30hrs, Sun; 12.00-14.30hrs & 19.00-21.30hrs

---

5   The Denbigh Arms

Tel: (01455) 553537

0.7 miles west along the A4304, in Lutterworth, on the left. (Free House) Open all day. The Players Bar; Bar Snacks served 12.00-14.00hrs & 19.00-21.00hrs daily. Lambert's Restaurant; Breakfast; Mon-Fri; 07.00-09.30hrs, Sat & Sun; 08.00-10.00hrs, Lunch; Sun-Fri; 12.00-14.00hrs, Dinner; 19.00-21.00hrs daily.

---

6 The Hind Hotel

🍴 🍽 ♿ 🐾 🛏B Tel: (01455) 552341
0.8 miles west along the A4304, in Lutterworth, on the left. (Free House) Meals served; 12.00-21.00hrs daily.

NB. Currently being refurbished and due to re-open late 2000.

7 The Shambles

🍴 🍽 🐾 🛏B Tel: (01455) 552620
0.8 miles west along the A4304, in Lutterworth, on the right. (Banks's) Open all day. Meals served; Mon-Fri; 11.00-14.00hrs & 17.00-19.00hrs, Sat & Sun; 12.00-14.00hrs.

8 The Greyhound Hotel

🍴 🍽 🐾 ♿ 🛏B Tel: (01455) 553307
0.9 miles west along the A4304, in Lutterworth, on the left. (Free House) Open all day. Restaurant Open; Lunch; Mon-Fri; 12.00-14.00hrs, Evening Meals; 19.00-21.30hrs daily. The Vaults Bistro; Open; Tues-Sat; 19.00-21.30hrs. Bar Snacks available; Mon-Sat; 12.00-14.00hrs & Sun-Thurs; 19.00-21.30hrs

9 The Cavalier Inn

🍴 🍽 🐾 ♿ Tel: (01455) 552402
1 mile west along the A4304, in Lutterworth, on the left. (Free House) Open all day. Meals served; Mon-Sat; 12.00-21.30hrs, Sun; 12.00-21.00hrs

10 St Mary's Filling Station (Shell)

⛽ ♿ WC 24 Tel: (01455) 553060
1 mile west along the A4304, in Lutterworth, on the left. Access, Visa, Overdrive, All Star, Switch, Dial Card, Mastercard, Amex, Diners Club, Delta, AA Paytrak, Shell Cards, BP Supercharge, Smartcard

PLACES OF INTEREST

Lutterworth Museum

Churchgate, Lutterworth LE17 4AN
Tel: (01455) 284733
Follow the A426 west into Lutterworth (1 Mile)

Devoted to local history, this museum has a wealth of exhibits from Roman times to World War II.

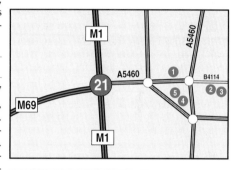

M1 JUNCTION 21

NEAREST A&E HOSPITAL

Leicester Royal Infirmary, Infirmary Square, Leicester LE1 5WW Tel: (0116) 254 1414

Take the exit to Leicester. Follow the A5460 into the city and turn right along Upperton Road. Turn left at the end and the hospital is on the left. (Distance Approx 3.8 miles)

FACILITIES

1 Stakis Leicester

🛏H 🍴 🍽 ♿ 🐾 Tel: (0116) 263 0066
0.4 miles east along the A5460, on the left. Season's Restaurant; Open; Breakfast; 07.00-10.00hrs, Lunch; 12.00-14.00hrs, Dinner; 19.00-22.00hrs daily.

2 Asda Petrol Station

⛽ Tel: (0116) 289 8174
0.9 miles east along the B4114 (Fosse Park Road), on the right. Access, Visa, Overdrive, All Star, Switch, Dial Card, Mastercard, Amex,

Diners Club, Delta, AA Paytrak, Asda Fuelcard. Toilets and cash machine available in adjacent store. Open; 06.00-00.00hrs daily. 24hr Credit card operated pump available Mon-Fri.

3 McDonald's

  Tel: (0116) 263 0563

1 mile east along the B4114 (Fosse Park Road), on the right. Open; Mon-Thurs; 07.00-23.30hrs, Fri & Sat; 07.00-0.00hrs, Sun; 07.00-22.30hrs (Drive-Thru open until 23.30hrs)

4 Sainsbury's Petrol Station

  Tel: (0116) 263 1153

0.4 miles south along the B4114, at the Grove Park Triangle, on the right. Access, Visa, Overdrive, All Star, Switch, Dial Card, Mastercard, Amex, Delta, AA Paytrak, Sainsburys Petrol Card. Toilets and cash machine available in adjacent store.

5 Pizza Hut

  Tel: (0116) 289 2990

0.4 miles south along the B4114, at the Grove Park Triangle, on the right. Open; 12.00-23.00hrs daily.

PLACES OF INTEREST

Leicester

Leicester Tourist Information Centre,
7/9 Every Street, Town Hall Square,
Leicester LE1 6AG
Tel: (0116) 299 8888
e-mail; tic@leicesterpromotions.org.uk
Follow the A5460 north (Signposted 4.2 Miles)

Standing on the River Soar, the origin of Leicester dates back over 2000 years. The Romans built Fosse Way through it and walled the city, naming it Leirceastre, and there are still some remains of their occupation, Peacock Pavement; Roman pavements and Jewry Wall; A 70ft long portion of the city wall, being some of the more notable. Amongst the many fine buildings still to be seen are the 15thC Guild Hall (Tel: 0116-253 2569), 17thC Town Library and the Museum and Art Gallery.

Within the city can be found ....

Jewry Wall Museum

St.Nicholas Circle, Leicester LE1 4JJ
Tel: (0116) 247 3021

Featuring early archaeology from pre-historic times to the Middle Ages the museum is adjacent to the Jewry Wall.

Belgrave Hall & Gardens

Church Road, Leicester LE4 5PE
Tel:(0116) 266 6590

A delightful Queen Anne House dating from 1709 and displaying Edwardian elegance and Victorian cosiness. Set in beautiful gardens.

Castle Park

Tel: (0116) 265 0555

The historic core of Leicester where the Romans, Saxons and Normans settled. Full of fascinating buildings and delightful walks around the area in which the mediaeval town flourished and the industrial city was born. Museums and Specialist Shops.

SERVICE AREA

BETWEEN JUNCTIONS 21 AND 21A

Leicester Forest East Services (Welcome Break)
Tel: (0116) 238 6801 Welcome Lodge & BP Fuel (Northbound) BP Fuel (Southbound)
Granary Restaurant, Red Hen Restaurant, Burger King & KFC on Footbridge connecting sites

M1 JUNCTION 21A

THIS IS A RESTRICTED ACCESS JUNCTION

Vehicles can only exit from the northbound lanes.
Vehicles can only enter the motorway along the southbound lanes

THERE ARE NO FACILITIES WITHIN ONE MILE OF THIS JUNCTION

NEAREST NORTHBOUND A&E HOSPITAL

Leicester Royal Infirmary, Infirmary Square, Leicester LE1 5WW Tel: (0116) 254 1414

Take the A46 north and turn right along the A50 to Leicester. Bear right along the A594 and the hospital is on the right hand side of Infirmary Road in the city centre. (Distance Approx 6.1 miles)

NEAREST SOUTHBOUND A&E HOSPITAL

Leicester Royal Infirmary, Infirmary Square, Leicester LE1 5WW Tel: (0116) 254 1414

Proceed to Junction 21 and take the exit to Leicester. Follow the A5460 into the city and turn right along Upperton Road. Turn left at the end and the hospital is on the left. (Distance Approx 6.5 miles)

M1 JUNCTION 22

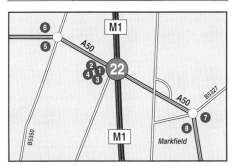

NEAREST A&E HOSPITAL

Leicester Royal Infirmary, Infirmary Square, Leicester LE1 5WW Tel: (0116) 254 1414

Take the exit to Leicester, follow the A50 through the city centre and the hospital is on the right. (Distance Approx 8.6 miles)

FACILITIES

1 Granada Services, Markfield (BP)

🍴 ♿ WC 24   Tel: (01530) 244777

On the west side of the roundabout. Access, Visa, Overdrive, All Star, Switch, Dial Card, Mastercard, Amex, Diners Club, Delta, Routex, AA Paytrak, Shell Agency, BP Cards

2 Little Chef

🍴 ☕ ♿   Tel: (01530) 244777

On the west side of the roundabout. Open; 07.00-22.00hrs daily

3 Burger King

🍴 ☕ ♿   Tel: (01530) 244777

On the west side of the roundabout. Open; 11.00-21.00hrs daily

4 Travelodge Markfield

🛏H ♿   Tel: (01530) 244777

On the west side of the roundabout.

5 Flying Horse Garage (BP)

⛽ WC   Tel: (01530) 244361

0.6 miles west along the A511, on the left. Access, Visa, Overdrive, All Star, Switch, Dial Card, Mastercard, Amex, Diners Club, Delta, Routex, Shell Agency, BP Cards. Open; 05.30-22.30hrs daily.

6 Browns Blue Filling Station (BP)

⛽ ♿ WC 24   Tel: (01530) 249849

0.7 miles west along the A511, on the right. Access, Visa, Overdrive, All Star, Switch, Dial Card, Mastercard, Amex, AA Paytrak, Diners Club, Delta, Routex, Shell Agency, BP Cards.

7 The Field Head Hotel

🍴 🍴 ☕ ♿ 🛏B   Tel: (01530) 245454

1 mile east along the A511, in Markfield, on the left. (Free House) Meals served; Breakfast; Mon-Fri; 07.00-09.30hrs, Sat & Sun; 08.00-10.00hrs, Lunch; Mon-Sat; 12.00-14.00hrs, Dinner; Mon-Sat; 19.00-21.30hrs, Meals served all day Sun; 12.00-21.30hrs.

8 The Coach & Horses

🍴 🍴 ♿ ☕   Tel: (01530) 242312

1 mile east along the A511, in Markfield, on the right. (Everards) Open all day Friday, Saturday & Sunday. Meals served; Mon-Thurs; 12.00-14.00hrs, Sat; 12.00-21.00hrs, Sun; 12.00-19.00hrs

M1

PLACES OF INTEREST

Snibston Discovery Park

Ashby Road, Coalville LE67 3LN
Tel: (01530) 510851 or Tel: (01530) 813256.
Follow the A50 west to Coalville (4.2 Miles)

Based at the former Snibston Colliery the Discovery Park is full of hands-on displays and experiments. There are galleries devoted to Transport, Engineering and Textile & Fashion plus graphic illustrations of 19thC life in a coal mine. There is a surface colliery tour and plenty of activities for children. Gift Shop. Cafe. Disabled facilities.

Bradgate Country Park Visitor Centre

Newtown Linford, Leicestershire LE6 0FA
Tel: (0116) 236 2713
Follow the A50 east and bear left to Newtown Linford (Signposted off A50, 3.2 Miles)

In the centre of Bradgate Park, the Visitor Centre contains displays on Lady Jane Grey and the history of Bradgate Park and Swithland Wood. Within the grounds can be found the ruins of Bradgate House and the Old John Tower. Gift Shop. Some disabled access.

M1　JUNCTION 23

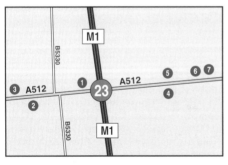

NEAREST NORTHBOUND A&E HOSPITAL

Derby Royal Infirmary, London Road,
Derby DE1 2XX Tel: (01332) 347141

Proceed to Junction 24 and take the A6 west to Derby. The hospital is in the city centre, on the left hand side of the road. (Distance Approx 16 miles)

NEAREST SOUTHBOUND A&E HOSPITAL

Leicester Royal Infirmary, Infirmary Square,
Leicester LE1 5WW Tel: (0116) 254 1414

Proceed to Junction 22 and take the exit to Leicester. Follow the A50 through the city centre and the hospital is on the right. (Distance Approx 13.2 miles)

FACILITIES

1　Junction 23 Lorry Park (Keyfuels)

Tel: (01509) 600108
0.3 miles west along the A512, on the right. Diesel Fuel for commercial vehicles only. Access, Visa, Overdrive, All Star, Mastercard, AA Paytrak, UK Fuelcard, Securicor Fuelserv, IDS, Keyfuels. Cafe Open; Mon-Fri; 06.00-21.00hrs, Sat; 06.00-11.00hrs. Lounge Bar Restaurant Open; Mon-Fri; 12.00-21.00hrs.

2　The Delisle Arms

Tel: (01509) 650170
0.6 miles west along the A512, on the left. (Inn Partnership) Open all day. Lunch; Mon-Sat; 12.00-14.30hrs, Sun (Carvery); 12.00-14.15hrs, Evening Meals; Wed-Sat; 18.00-21.00hrs.

3　Elf Service Station

Tel: (01509) 650214
0.8 miles west along the A512, on the right. Access, Visa, Overdrive, All Star, Switch, Dial Card, Mastercard, Amex, Diners Club, Delta, AA Paytrak, Elf Cards, Total Fina Cards.

4　Charnwood Filling Station (BP)

Tel: (01509) 212071
0.6 miles east along the A512, on the right. Access, Visa, Overdrive, All Star, Switch, Dial Card, Mastercard, Amex, Diners Club, Delta, Routex, AA Paytrak, Shell Agency, BP Cards

5　Shell Temple Filling Station

Tel: (01509) 643770
0.6 miles east along the A512, on the left. Access, Visa, Overdrive, All Star, Switch, Dial Card, Mastercard, Amex, Diners Club, Delta,

 Large Hotel　 Small Hotel　 Motel　 Guest House/ Bed & Breakfast　 Disabled Facilities　 Childrens Facilities

Shell Cards, BP Supercharge, BP Agency, Smartcard

6 Quality Hotel

🛏️🍴🍽️♿🐾 Tel: (01509) 211800

0.9 miles east along the A512, on the left. The Beaumont Restaurant; Breakfast; Mon-Fri; 07.00-09.30hrs, Sat & Sun; 07.30-09.30hrs, Lunch; Sun; 12.30-14.30hrs, Dinner; 19.00-21.30hrs daily

7 The Wheatsheaf Inn

🍴🍽️🔄♿ Tel: (01509) 214165

1 mile east along the A512, on the left. (Harvester) Meals served; 12.00-14.30hrs & 17.00-21.30hrs daily

PLACES OF INTEREST

John Taylor Bellfoundry Museum

Freehold Street, Nottingham Road, Loughborough LE11 1AR Tel: (01509) 233414
Follow the A512 east into Loughborough and continue along the A60 (3.8 Miles)

A unique museum, part of the world's largest working bellfoundry, relating to all aspects of bellfounding from early times. Demonstrating the craft techniques of moulding, casting, tuning and the fitting up of bells. Museum Shop. Foundry tours. Some disabled access.

Great Central Railway

Loughborough Central Station, Great Central Road, Loughborough LE11 1RW Tel: (01509) 230726
Follow the A512 east into Loughborough and turn south along the A6 (Signposted along A6, 3.4 Miles)

The only preserved steam railway on a former main line, the Great Central Railway runs between Loughborough and Rothley. Museum, working signal box and historic loco collection at Loughborough Station. Gift Shop.

Charnwood Museum

Queens Hall, Granby Street, Loughborough LE11 3DZ Tel: (01509) 218113
Follow the A512 east into Loughborough (3 Miles)

The museum for the Borough of Charnwood, the displays include natural history, local history and exhibitions of industry and farming.

M1 JUNCTION 23A

THIS IS A RESTRICTED ACCESS JUNCTION

There is no access from the A453 to the northbound carriageway

There is no exit to the A453 from the southbound carriageway

SERVICE AREA

ACCESSED FROM THIS JUNCTION

Donington Park Services (Granada)

Tel: (01509) 672220 Little Chef, Burger King, Fresh Express Self-Serve Restaurant, Harry Ramsden's, Travelodge & BP Fuel

NEAREST A&E HOSPITAL

Derby Royal Infirmary, London Road, Derby DE1 2XX Tel: (01332) 347141

Proceed to Junction 24 and take the A6 exit west to Derby. The hospital is in the city centre, on the left hand side. (Distance Approx 11.9 miles)

M1 JUNCTION 24

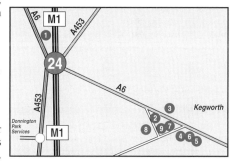

NEAREST A&E HOSPITAL

Derby Royal Infirmary, London Road, Derby DE1 2XX Tel: (01332) 347141

Take the A50 exit west to Derby. The hospital is in the city centre, on the left hand side. (Distance Approx 9.8 miles)

 Pets Welcome   Cash Dispenser   24 Hour Facilities   Toilets   Alcoholic Drinks   Food and Drink   Fuel

M1

FACILITIES

SERVICE AREA

Donington Park Services (Granada)
Tel: (01509) 672220 Little Chef, Burger King, Fresh Express Self-Serve Restaurant, Harry Ramsden's, Travelodge & BP Fuel

1 Hilton National East Midlands Airport Hotel

Tel: (01509) 674000
0.1 miles west along the A6, on the right. Restaurant Open; Breakfast; Mon-Fri; 07.00-09.30hrs, Sat & Sun; 07.30-10.00hrs, Lunch; Sun-Fri; 12.30-14.00hrs, Dinner; Mon-Sat; 19.00-22.00hrs, Sun; 19.00-21.30hrs

2 Kegworth Hotel

Tel: (01509) 672427
0.5 miles east along the A6, on the right. Restaurant Open; Breakfast; Mon-Fri; 07.00-09.00hrs, Sat & Sun; 08.00-10.00hrs, Lunch; 12.00-14.00hrs daily, Dinner; 19.00-21.30hrs daily. Bar Meals served; 08.00-23.00hrs daily

3 Kegworth Service Station (Spot)

Tel: (01509) 673435
0.7 miles east along the A6, on the left. Access, Visa, Overdrive, All Star, Switch, Dial Card, Mastercard, Amex, Delta. Open; Mon-Sat; 07.30-20.30hrs, Sun; 08.00-20.00hrs.

4 Ye Olde Flying Horse

Tel: (01509) 672253
0.9 miles east along the A6, in Kegworth, on the right. (Burghley Taverns) Open all day. Meals served; Mon-Fri; 12.00-15.00hrs & 18.00-21.00hrs, Sat; 12.00-21.00hrs, Sun; 12.00-16.00hrs.

5 The Kegworth Lantern

Tel: (01509) 673989
0.9 miles east along the A6, in Kegworth, on the right. (Free House) Open all day. Lunch; 12.00-14.00hrs daily, Evening Meals; 18.00-21.00hrs daily

6 Public Toilets

WC
0.9 miles east along the A6, in Kegworth, on the right

7 Cottage Restaurant

Tel: (01509) 672449
1 mile along the High Street, in Kegworth, on the right. Open; Lunch; Tues-Fri; 12.00-14.00hrs, Sun; 12.00-14.30hrs, Dinner; Tues-Fri; 19.00-22.00hrs, Sat; 19.00-22.30hrs, Sun; 19.00-21.30hrs. Closed all day on Monday

8 Yew Lodge Hotel

Tel: (01509) 672518
0.8 miles east, on Packington Hill, on the right. (Best Western) Restaurant Open; Breakfast; Mon-Sat; 07.00-09.30hrs, Sun; 07.30-10.30hrs, Lunch; 12.00-14.00hrs daily, Dinner; Mon-Sat; 18.30-21.30hrs, Sun; 19.00-21.00hrs

9 Oddfellows Arms

Tel: (01509) 672552
0.8 miles east, on Packington Hill, on the left. (Banks's) Open all day.

PLACES OF INTEREST

Donington Grand Prix Collection

Donington Park, Castle Donington, Derby DE74 2RP Tel: (01332) 811027
Follow the A6 west (Signposted 3.7 Miles)

The world's largest collection of Grand Prix racing cars with over 120 exhibits in 5 halls depicting the history of the sport from 1901 to the present day. Restaurant, Souvenir Shop.

Kegworth Museum

52 High Street, Kegworth DE74 2DA
Tel: (01509) 672886
Follow the A6 east into Kegworth (0.7 Miles)

The displays include a Victorian parlour, saddlers, local school, knitting industry, air transport, photography and war memorabilia. Some disabled access.

| Large Hotel | Small Hotel | Motel | Guest House/ Bed & Breakfast | Disabled Facilities | Childrens Facilities |

M1 — JUNCTION 24A

THIS IS A RESTRICTED ACCESS JUNCTION

There is no access to the southbound carriageway
There is no exit from the northbound carriageway

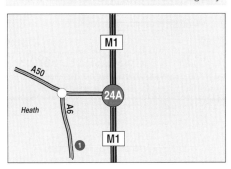

NEAREST A&E HOSPITAL

Derby Royal Infirmary, London Road, Derby DE1 2XX Tel: (01332) 347141

Take the A50 exit west to Derby. The hospital is in the city centre, on the left hand side. (Distance Approx 8.2 miles)

FACILITIES

1 Hilton National East Midlands Airport Hotel

Tel: (01509) 674000
0.5 miles south along the A6, on the left. Restaurant Open; Breakfast; Mon-Fri; 07.00-09.30hrs, Sat & Sun; 07.30-10.00hrs, Lunch; Sun-Fri; 12.30-14.00hrs, Dinner; Mon-Sat; 19.00-22.00hrs, Sun; 19.00-21.30hrs

M1 — JUNCTION 25

NEAREST A&E HOSPITAL

Queens Medical Centre, University Hospital, Derby Road, Nottingham NG7 2UH Tel: (0115) 924 9924

Take the A52 east exit to Nottingham and the hospital is on this route. (Distance Approx 5.6 miles)

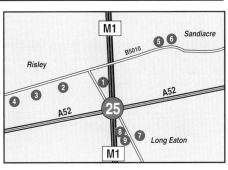

FACILITIES

1 Posthouse Nottingham/Derby

Tel: 0870 400 9062
0.4 miles north along the Risley Road, on the right. Traders Restaurant & Bar; Open all day. Breakfast; Mon-Fri; 06.30-09.30hrs, Sat & Sun; 07.00-10.30hrs. Meals served all day; Mon-Sat; 12.00-22.30hrs, Sun; 12.00-22.00

2 Risley Garage Ltd (Jet)

Tel: (0115) 949 0838
0.7 miles west along the B5010, on the left. Access, Visa, Overdrive, All Star, Switch, Dial Card, Mastercard, Amex, Diners Club, Delta, Jet Cards. Attended Service for disabled drivers. Open; Mon-Sat; 07.00-21.00hrs, Sun & Bank Holidays; 08.30-21.00hrs.

3 Risley Hall Country House Hotel

Tel: (0115) 939 9000
0.8 miles west along the B5010, on the left. The Garden Room Restaurant; Open; Mon-Sat; 12.00-14.00hrs & 19.00-21.30hrs, Sun; 12.00-15.00hrs

4 The Risley Park

Tel: (0115) 939 2313
1 mile west along the B5010, on the left. (Vanguard) Open all day. Meals served; Mon-Thurs; 11.30-21.30hrs, Fri & Sat; 11.30-22.00hrs, Sun; 12.00-21.00hrs

Pets Welcome	£ Cash Dispenser	24 Hour Facilities	WC Toilets	Alcoholic Drinks	Food and Drink	Fuel

M1

5   Sandiacre Service Station (Texaco)

🏪 WC   Tel: (0115) 949 1398

1 mile east along the B5010, in Sandiacre, on the left. Access, Visa, Overdrive, All Star, Switch, Dial Card, Mastercard, Amex, Diners Club, Delta, AA Paytrak, BP Supercharge, Texaco Cards. Open; Mon-Sat; 07.00-21.00hrs, Sun; 09.00-21.00hrs.

6   White Lion

🍴 🍽 🛏 ♿   Tel: (0115) 939 7123

1 mile east along the B5010, in Sandiacre, on the left. (Scottish & Newcastle) Open all day on Fri, Sat & Sun. Meals served; Mon-Sat; 12.00-21.30hrs, Sun; 12.00-15.00hrs.

7   Novotel Hotel Nottingham/Derby

🏨 🍴 🍽 ♿ 🚗   Tel: (0115) 946 5111

0.4 miles south along the Long Eaton Road, on the left. The Garden Brasserie; Meals served daily throughout day from 06.00hrs to 0.00hrs.

8   Branaghans Restaurant & Bar

🍴 🍽 ♿ 🛏   Tel: (0115) 946 2000

0.4 miles south along the Long Eaton Road, on the right. Open; Mon-Thurs; 12.00-14.30hrs & 17.00-22.00hrs, Fri & Sat; 12.00-14.30hrs & 17.00-22.30hrs, Sun; 12.00-21.30hrs

9   Jarvis Nottingham

🏨 ♿ 🚗   Tel: (0115) 946 0000

0.4 miles south along the Long Eaton Road, on the right

PLACES OF INTEREST

Derby

Derby Tourist Information, Assembly Rooms,
Market Place, Derby DE1 3AH
Tel: (01332) 255802
Follow the A52 west (Signposted 7.2 Miles)

A Roman station named Derventio and established on the opposite bank of the River Derwent to today's town centre led to the rise of Derby. The Danes made it an important centre and it was here in 1717 that the first successful silk weaving in England was established. Amongst the many buildings that reflect the town's long history are the Bridge Chapel of 1330, the All Saints Cathedral which originated in 1509 and Derby School, founded in 1160 and re-instituted by Queen Mary in 1554. Derby was an important railway junction as long ago as 1839 and the Midland Railway established extensive workshop premises here with the production of both locomotives and carriages undertaken on the site. Sadly, today they have all but disappeared but the city is still particularly renowned for two of its manufacturers; Rolls-Royce and Royal Crown Derby.

Within the city can be found ...

Derby Industrial Museum

Silk Mill Lane, Off Full Street, Derby DE1 3AR
Tel: (01332) 255308

Housed in a rebuilt version of Britain's first factory, the Silk Mill of 1717-1721, the museum contains galleries featuring Derbyshire industries, railway engineering, stationary power sources and the world's largest collection of Rolls Royce aero-engines.

Royal Crown Derby Visitor Centre

194 Osmaston Road, Derby DE23 8JZ
Tel: (01332) 712800
Follow the A52 west into Derby and take the A514
south towards Melbourne. (8 Miles)

A factory tour and demonstration area are utilized to show visitors how the world famous table and giftware is manufactured from the raw materials of bone ash, clay, stone and water. The traditional skills of hand gilding and flower making, handed down through the generations, are demonstrated by experienced craftsmen and there is a museum which traces the development of the company from 1748 to the present day. Restaurant. Factory Shop. Limited disabled access.

Nottingham

Follow the A52 east (Signposted 8.3 Miles)

For details please see Junction 26 information

| | Large Hotel | | Small Hotel | | Motel | | Guest House/ Bed & Breakfast | | Disabled Facilities | | Childrens Facilities |

Within the city can be found ...

The Caves of Nottingham

Drury Walk, Broad Marsh Shopping Centre,
Nottingham NG1 7LS

For details please see Junction 26 information

Galleries of Justice

Shire Hall, High Pavement, Lace Market,
Nottingham NG1 1HN

For details please see Junction 26 information

The Tales of Robin Hood

30-38 Maid Marian Way, Nottingham NG1 6GF

For details please see Junction 26 information

SERVICE AREA

BETWEEN JUNCTIONS 25 AND 26

Trowell Services (Northbound) (Granada)

Tel: (0115) 932 0291Fresh Express Self Service
Restaurant, Little Chef, Travelodge & Esso Fuel

Trowell Services (Southbound) (Granada)

Tel: (0115) 932 0291Fresh Express Self Service
Restaurant, Burger King & Esso Fuel

Footbridge connection between sites

M1   JUNCTION 26

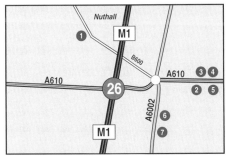

NEAREST NORTHBOUND A&E HOSPITAL

**Kings Mill Hospital, Mansfield Road, Sutton in
Ashfield NG17 4JL Tel: (01623) 622515**

Proceed north to Junction 28 and take the A38 east
towards Mansfield. At the junction with the B6023
bear left and the hospital is along this route.
(Distance Approx 13.5 miles)

NEAREST SOUTHBOUND A&E HOSPITAL

**Queens Medical Centre, University Hospital,
Derby Road, Nottingham NG7 2UH
Tel: (0115) 924 9924**

Take the A610 towards Nottingham and turn right
along the A6514 (A52) and the hospital is along
this route. (Distance Approx 4.6 miles)

FACILITIES

1   The Three Ponds

Tel: (0115) 938 3170

1 mile north along the B600, in Nuthall, on
the left. (Kimberley Ales) Open all day. Lunch;
Mon-Sat; 12.00-14.00hrs, Sun; 12.00-15.00hrs,
Evening meals; Mon-Fri; 18.00-20.00hrs daily.

2   Broxtowe Inn

Tel: (0115) 927 8210

0.9 miles east along the A610, on the right.
(Scottish & Newcastle) Open all day. Meals
served; Mon-Sat; 12.00-21.30hrs, Sun; 12.00-
20.30hrs.

3   The Millers Barn

Tel: (0115) 951 9971

1 mile east along the A610, on the left.
(Whitbread) Open all day. Meals served all day
from 12.00-22.30hrs daily.

4   Travel Inn

Tel: (0115) 951 9971

1 mile east along the A610, on the left.

5   St Mary's Service Station (BP)

Tel: (0115) 927 2707

1 mile east along the A610, on the right. Access,
Visa, Overdrive, All Star, Switch, Dial Card,
Mastercard, Amex, Diners Club, Delta, Routex,
Shell Agency, BP Cards

6   Old Moor Lodge

Tel: (0115) 976 2200

0.9 miles south along the A6002, on the left.
(Whitbread) Open all day. Meals served; Mon-

| Pets Welcome | £ Cash Dispenser | 24 24 Hour Facilities | WC Toilets | Alcoholic Drinks | Food and Drink | Fuel |

Sat; 11.30-22.00hrs, Sun; 12.00-21.30hrs

---

7   Woodhouse Shell Filling Station

🛈 ♿ WC 24   Tel: (0115) 977 7900

1 mile south along the A6002, on the left. Access, Visa, Overdrive, All Star, Switch, Dial Card, Mastercard, Amex, Diners Club, Delta, AA Paytrak, Shell Cards, Smartcard

PLACES OF INTEREST

Nottingham

Nottingham Tourist Information Centre,
1-4 Smithy Row, Nottingham NG1 2BY
Tel: (01636) 678962
Follow the A610 south (Signposted 4.2 Miles)

The city of Nottingham was founded by Snot, the chief of a 6thC Anglo-Saxon tribe who carved out dwellings in the local sandstone and established the settlement of Snottingaham, "home of the followers of Snot". A castle was sited on top of the rock formation by the Normans and this was dismantled by Oliver Cromwell. The edifice seen today is just a 17thC mansion that was destroyed during a riot in 1831 and taken over and converted into a museum by the Corporation in 1875. Beneath the castle is a honeycomb of caves, some utilized as dungeons, and passages the most famous of which is Mortimer's Hole used, it is said, by Edward III to enter the castle in 1330 and arrest Queen Isabella, his mother. At the base of Castle Rock, and set into the sandstone is "The Trip to Jerusalem" dating from 1189 and reputed to be the oldest pub in England. Nottingham, as it became known, is, of course, not only associated with the legend of Robin Hood but also renowned for the manufacture of lace.

Within the city can be found ...

The Caves of Nottingham

Drury Walk, Broad Marsh Shopping Centre,
Nottingham NG1 7LS Tel: (0115) 924 1424

A unique audio tour through the 700 year old man made caves. On view are a mediaeval tannery, an air raid shelter, the pub cellars and the remains of Drury Hill, Nottingham's most historic street. No disabled access.

Galleries of Justice

Shire Hall, High Pavement, Lace Market,
Nottingham NG1 1HN Tel: (0115) 952 0555

A fascinating journey through time revealing the history of justice in England. Set in a Victorian courtroom, county gaol and early 20thC Police Station visitors can witness how crime and punishment methods have changed over the years, from public hangings to present day forensic science detection methods. Disabled Access.

The Tales of Robin Hood

30-38 Maid Marian Way, Nottingham NG1 6GF
Tel: (0115) 948 3284

An indoor family entertainment centre that relates the story of Robin Hood through an exciting 3-dimensional ride. There are exhibitions of mediaeval life and Robin Hood plus opportunities for visitors to try brass rubbings or archery or have a photograph taken in period costume. There is a children's play area, Gift Shop and Cafe. Full disabled access.

Midland Railway Centre

Butterley Station, Ripley, Derbyshire DE5 3QZ
Follow the A610 west (Signposted 9.8 Miles)

For details please see Junction 28 information.

DH Lawrence Birthplace Museum,

8a Victoria Street, Eastwood,
Nottingham NG16 3AW Tel: (01773) 717353.
Follow the A510 west into Eastwood and turn right
along the A608 (5 Miles)

This tiny terraced house is now a museum devoted to the writer's life and reflections on his early life spent in Nottinghamshire. The countryside around Eastwood provided inspiration for novels such as "Sons & Lovers", "Women in Love" and "The Rainbow", as well as the controversial "Lady Chatterley's Lover". (NB The Durban House Heritage Centre, nearby in Mansfield Road has further information on DH Lawrence as well as exhibits on local history). Shop. No disabled access.

The American Adventure

Ilkeston, Derbyshire DE7 5SX Tel: (01773) 531521
Follow the A610 west (Signposted 6.7 Miles)

Set in 390 acres of Derbyshire parkland it is the home to Europe's tallest SkyCoaster, a 200ft freefall ride. Amongst over 100 themed rides,

Nightmare Niagara is the world's highest and wettest triple drop log flume. Restaurant. Gift Shop. Disabled Access.

Routex, AA Paytrak, UK Fuelcard, Shell Cards, BP Cards, Texaco Cards, Esso Cards, Elf Cards, Total Fina Cards. Open; Mon-Sat; 08.00-18.30hrs, Sun; Closed.

M1 JUNCTION 27

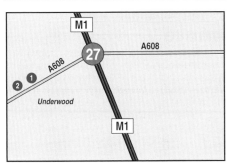

NEAREST NORTHBOUND A&E HOSPITAL

Kings Mill Hospital, Mansfield Road, Sutton in Ashfield NG17 4JL Tel: (01623) 622515

Proceed north to Junction 28 and take the A38 east towards Mansfield. At the junction with the B6023 bear left and the hospital is along this route. (Distance Approx 8 miles)

NEAREST SOUTHBOUND A&E HOSPITAL

Queens Medical Centre, University Hospital, Derby Road, Nottingham NG7 2UH Tel: (0115) 924 9924

Proceed to Junction 26, take the A610 towards Nottingham and turn right along the A6514 (A52). The hospital is along this route. (Distance Approx 10.1 miles)

FACILITIES

1 Sand Hills Tavern

🔲 🔲 🔲 🔲 Tel: (01773) 780330
0.9 miles west along the A608, on the right. (Scottish & Newcastle) Open all day. Bar Snacks available daily. Sunday Carvery; 12.00-14.30hrs.

2 Underwood Garage

🔲 **WC** Tel: (01773) 712554
1 mile west along the A608, on the right. Access, Visa, Overdrive, All Star, Switch, Dial Card, Mastercard, Amex, Diners Club, Delta,

PLACES OF INTEREST

Newstead Abbey & Park

Ravenshead, Nottinghamshire NG15 8GE
Tel: (01623) 455900 Follow the A608 east, signposted "Sherwood Forest" and then signposted to "Newstead Abbey" (6.6 Miles)

The original priory was founded in c1170 and converted into a house, known as Newstead Abbey, by Sir John Byron in the 1540's. It was sold by Lord Byron in 1816 when he moved abroad and the building was refurbished in Gothic style to designs by John Shaw. The building contains many Byron relics and there are 30 beautifully preserved period rooms. It is surrounded by superb grounds including the Japanese Gardens and two lakes. Gift Shop. Cafe. Limited disabled access.

M1 JUNCTION 28

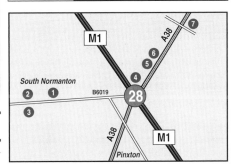

NEAREST NORTHBOUND A&E HOSPITAL

Kings Mill Hospital, Mansfield Road, Sutton in Ashfield NG17 4JL Tel: (01623) 622515

Take the A38 east towards Mansfield and at the junction with the B6023 bear left. The hospital is along this route. (Distance Approx 4.5 miles)

NEAREST SOUTHBOUND A&E HOSPITAL

Queens Medical Centre, University Hospital, Derby Road, Nottingham NG7 2UH Tel: (0115) 924 9924

Proceed to Junction 26, take the A610 towards

M1

Nottingham and turn right along the A6514 (A52). The hospital is along this route. (Distance Approx 13.5 miles)

FACILITIES

1 Meteor Service Station (BP)

Tel: (01773) 811268

0.8 miles west along the B6019, in South Normanton, on the right. Access, Visa, Overdrive, All Star, Switch, Dial Card, Mastercard, Amex, Diners Club, Delta, Routex, AA Paytrak, Shell Agency, BP Cards. Open; Mon-Sat; 06.00-22.00hrs, Sun; 08.00-22.00hrs.

2 The Hawthorns

Tel: (01773) 811328

0.9 miles west along the B6019, in South Normanton, on the right. (Greenalls) Open all day. Meals served all day Mon-Sat; 12.00-21.00hrs, Sun; 12.00-16.30hrs.

3 Carnfield Service Station (BP)

Tel: (01773) 811251

1 mile west along the B6019, in South Normanton, on the left. Access, Visa, Overdrive, All Star, Switch, Dial Card, Mastercard, Amex, Diners Club, Delta, Routex, AA Paytrak, Shell Agency, BP Cards

4 Swallow Hotel Derby/Notts

Tel: (01773) 812000

0.1 miles east along the A38, on the left. Pavilion Restaurant Open; Sunday Lunch; 12.30-14.00hrs, Dinner; Mon-Sat; 19.00-21.45 hrs Chatterley's; Open; Breakfast; 07.00-10.00hrs daily, Lunch; 12.30-14.00hrs daily, Dinner; 19.00-22.00hrs daily

5 The Castlewood

Tel: (01773) 862899

0.2 miles east along the A38, on the left. (Whitbread) Open all day. Meals served all day Mon-Sat; 11.30-22.00hrs, Sun; 12.00-22.00hrs.

6 Travel Inn

Tel: (01773) 862899

0.2 miles east along the A38, on the left.

7 McArthur Glen Designer Outlet Derbyshire

Tel: (01773) 545000

0.7 miles east along the A38, on the right. There are numerous cafes and restaurants, including Madisons, Singapore Sam Chinese Restaurant, Burger King, Arkwright's Fish & Chips, Bradwell's English Restaurant and Spud-U-Like within the Food Court. Open; Mon-Wed & Fri; 10.00-18.00hrs, Thurs; 10.00-20.00hrs, Sat; 09.00-17.00hrs, Sun; 11.00-17.00hrs.

PLACES OF INTEREST

Midland Railway Centre

Butterley Station, Ripley, Derbyshire DE5 3QZ
Tel: (01773) 747674. Follow the A38 west and turn south along the B6179 (Signposted 5.2 Miles)

More than 50 steam, diesel and electric locomotives, and more than 100 items of historic rolling stock are on display. There is a steam hauled passenger service and other attractions include narrow gauge, miniature and model railways, museum, country park and farm park and a demonstration signal box. Gift Shop. Some disabled access.

The National Tramway Museum

Crich, Matlock, Derbyshire DE4 5DP
Tel: (01773) 852565. Follow the A38 west (Signposted 7.7 Miles)

A collection of more than 70 restored horse, steam and electric trams from all over the world. Working trams take visitors along a mile of scenic track and exhibits include depots, power stations workshops and a period street. Gift Shop. Cafe. Playgrounds. Picnic Areas.

 Large Hotel   Small Hotel   Motel   Guest House/ Bed & Breakfast   Disabled Facilities   Childrens Facilities

SERVICE AREA

BETWEEN JUNCTIONS 28 AND 29

Tibshelf Services (Northbound) (RoadChef)
Tel: (01773) 594810 Food Fayre Self-Service Restaurant, Cafe Continental Coffee Shop, Wimpy Bar, RoadChef Lodge & Texaco Fuel

Tibshelf Services (Southbound) (RoadChef)
Tel: (01773) 594810 Food Fayre Self-Service Restaurant, Cafe Continental Coffee Shop, Wimpy Bar & Texaco Fuel

M1　JUNCTION 29

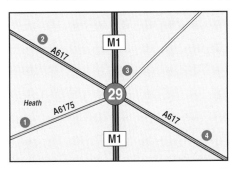

NEAREST NORTHBOUND A&E HOSPITAL

Chesterfield & North Derbyshire Royal, Calow, Chesterfield S44 5BL Tel: (01246) 277271

Take the A617 west to Chesterfield and after 4.8 miles bear right at the roundabout along the A61(T). At the junction with the A619 and A632 turn right along the A632 and the hospital is along this road on the left hand side. (Distance Approx 6.1 miles)

NEAREST SOUTHBOUND A&E HOSPITAL

Kings Mill Hospital, Mansfield Road, Sutton in Ashfield NG17 4JL Tel: (01623) 622515

Take the A617 towards Mansfield and after 5.7 miles turn right along the A6075. The hospital is on the left hand side of this road. (Distance Approx 8.2 miles)

FACILITIES

1　Red House Service Station (Shell)

🛢 ♿ WC　　　Tel: (01246) 850329
0.9 miles south along the A6010, on the right. Access, Visa, Overdrive, All Star, Switch, Dial Card, Mastercard, Amex, Diners Club, Delta, AA Paytrak, Shell Cards, BP Agency, Smartcard. Open; Mon-Fri; 07.00-21.30hrs, Sat & Sun; 08.00-20.00hrs.

2　Heath Service Station (North) (Esso)

🛢 WC ♿ 24 🍴　　　Tel: (01246) 850525
0.6 miles west along the A617, on the right (Actual distance 2.2 miles). Access, Visa, Overdrive, All Star, Switch, Dial Card, Mastercard, Amex, Diners Club, Delta, AA Paytrak, Shell Gold, Shell Europe, Shell Agency, BP Supercharge, Esso Cards. Esso Bakery Shop Open; 06.00-18.00hrs daily.

3　Twin Oaks Motel

🏨 🍴 🍴 ♿　　　Tel: (01246) 855455
0.2 miles along the Palterton Road, on the left. Restaurant Open; Mon-Sat; 12.00-14.00hrs & 19.00-21.00hrs, Sun; 12.00-14.00. Bar Snacks available on Sun;18.30-21.00hrs

4　Ma Hubbard's

🍴 🍴 ♿ 🐾　　　Tel: (01246) 857236
1 mile east along the A617, in Glapwell, on the left. (Mansfield) Meals served; Lunch; 12.00-15.00hrs daily, Evening Meals; 17.00-21.00hrs daily

PLACES OF INTEREST

St Mary and All Saints Church

Church Way, Chesterfield S40 1SF
Tel: (01246) 206506. Follow the A617 west
(Signposted 5.3 Miles)

The symbol of Chesterfield, the famous crooked spire, sits on top of the tower of this church, the largest in Derbyshire. There is much folklore devoted to how this curious shape came about but the answer would appear to be closely connected to the year when it was being constructed - 1349. Precisely at this time, the Black Death was raging through the country and it is not inconceivable that many of the skilled craftsmen engaged in the construction work may have died, leaving less knowledgeable workers to carry on. Certainly no cross braces, utterly essential for the 32tonf

| Pets Welcome | £ Cash Dispenser | 24 24 Hour Facilities | WC Toilets | Alcoholic Drinks | Food and Drink | Fuel |

of tiles that it was to carry, were installed and green timber could easily have been utilized. These two elements, allied to the decaying of the timbers at the base of the 228ft spire would easily account for the deformation that is visible today. The church is open throughout the year and guided trips up the tower are available from time to time. Disabled access to church.

Hardwick Hall (NT)

Doe Lane, Chesterfield, Derbyshire S44 5QJ
Tel: (01246) 850430
Signposted from Motorway (2.7 Miles)

A fine Elizabethan house designed for Bess of Hardwick, the Countess of Shrewsbury. Contains outstanding 16thC furniture, tapestries and needlework, many items known to have been in the house prior to 1601. Walled courtyards enclose fine gardens, orchards and a herb garden and a 300 acre country park contains many rare breeds of sheep and cattle. Some disabled access.

Bolsover Castle

Castle Street, Bolsover S44 6PR
Tel: (01246) 823349 and Tel: (01246) 822844.
Take the unclassified road east to Palterton and turn left to Bolsover. (3.7 Miles)

A castle has stood on this site since Norman times but nothing of the original edifice remains. The building on view today, more of a mansion than a castle, was constructed by Sir Charles Cavendish, during the reign of James I, and added to by his son the 1st Duke of Newcastle in c1660. It contains elaborate fireplaces, panelling, wall paintings and an early indoor riding school. There is an audio tour and a Visitor Centre. Limited disabled access.

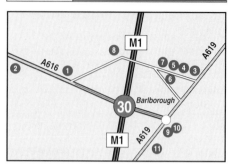

M1 JUNCTION 30

NEAREST NORTHBOUND A&E HOSPITAL

**Northern General Hospital, Herries Road
Sheffield S5 7AU Tel: (0114) 243 4343**

Take the A616 north for 10 miles into Sheffield and at the roundabout follow the A61 north to its junction with the A6178. Turn right along the A6178 and bear left along the A6135. The hospital is on the left hand side of this road. (Distance Approx 12.2 miles)

NEAREST SOUTHBOUND A&E HOSPITAL

**Chesterfield & North Derbyshire Royal, Calow,
Chesterfield S44 5BL Tel: (01246) 277271**

Take the A6135 east and at the first roundabout turn right along the A619 to Chesterfield. After about 6.5 miles turn left at the roundabout along the A61(T) and at the junction with the A619 and A632 turn left along the A632. The hospital is along this road on the left hand side. (Distance Approx 9 miles)

FACILITIES

1   Bridgehouse Service Station (Total)

🅿️ WC   Tel: (01246) 810600
0.6 miles west along the A616, on the right. Access, Visa, Overdrive, All Star, Switch, Dial Card, Mastercard, Amex, Diners Club, Delta, AA Paytrak, Total Cards. Open; Mon-Sat; 07.00-21.00hrs, Sun; 08.00-21.00hrs.

2   The Prince of Wales

🍴 🍽️ ♿ 🔥 B   Tel: (01246) 432108
1 mile west along the A616, on the left. (Free House) Meals served; Lunch; 12.00-14.30hrs

daily, Evening Meals; Mon-Sat; 18.00-20.30hrs.

3   De Rodes Arms

Tel: (01246) 810345
0.6 miles east along the A619, on the left.
(Whitbread) Open all day. Meals served; Mon-
Sat; 11.30-22.00hrs, Sun; 12.00-22.00hrs.

4   Stonecroft

Tel: (01246) 810974
0.7 miles west off the A619, in Barlborough,
on the right

5   The Rose & Crown

Tel: (01246) 810364
0.8 miles west off the A619, in Barlborough,
on the right. (Kimberley Ales) Meals served;
Tues; 12.00-14.00hrs, Wed-Sat; 12.00-14.00hrs
& 18.30-21.00hrs, Sun; 12.00-14.30hrs

6   The Apollo "2 Meals for a Fiver"

Tel: (01246) 810346
0.9 miles west off the A619, in Barlborough,
on the left. (Pubmaster) Open all day. Meals
served; 12.00-15.00hrs daily

7   The Royal Oak

Tel: (01246) 573020
0.9 miles west off the A619, in Barlborough,
on the right. (Bass) Open all day. Meals served;
Mon-Fri; 12.00-14.30hrs & 18.00-21.00hrs, Sat;
12.00-14.00hrs & 18.30-21.30hrs, Sun; 12.00-
17.00hrs

8   Dusty Miller

Tel: (01246) 810507
1 mile west off the A619, in Barlborough, on
the right. (Freehouse) Meals served; Tues-Sat;
12.00-14.00hrs & 17.30-20.00hrs, Sun; 12.00-
14.00hrs

9   Treble Bob

Tel: (01246) 813005
0.4 miles south along the A619, on the left.
(Tom Cobleigh) Open all day. Meals served;

Mon-Sat; 12.00-21.30hrs, Sun; 12.00-21.00hrs.

10 Holiday Inn Express, Barlborough

Tel: (01246) 813222
0.4 miles south along the A619, on the left

11 McDonald's

Tel: (01246) 819520
0.5 miles south along the A619, on the left.
Open; 07.30-23.00hrs daily

PLACES OF INTEREST

Mr Straw's House (NT)

7 Blyth Grove, Worksop S81 0JG
Tel: (01909) 482380 (NB. No access without pre-
booked timed admission ticket)
Follow the A619 east into Worksop and northeast
along the B6045 (8.9 Miles)

A semi-detached house built in c1901 which
belonged to William Straw and his brother,
Walter, has been internally preserved since the
death of their mother in the 1930's. Contains
1920's wallpaper, furnishings and locally made
furniture. There are also museum rooms and a
period suburban garden.

Renishaw Gardens and Museum

Renishaw, Derbyshire S21 3WB
Tel: (01777) 860755
Follow the A616 west (2.9 Miles)

The beautiful formal Italian gardens and
wooded park of Renishaw Hall are open to
visitors, along with a nature trail, a Sitwell
family museum, and art gallery and a display
of Fiori De Henriques sculptures located in the
Georgian stables. Cafe.

Barrow Hill Roundhouse Engine Shed

Campbell Drive, Barrow Hill, Staveley, Chesterfield
S43 2PN Tel: (01246) 472450.
website: www.shu.ac.uk/city/community/bhess
Take the A619 west, continue through Staveley and
at Hollingwood turn right and follow the signposts
to Barrow Hill. (5 miles)

A unique building housing steam and diesel
locomotives and demonstrating the operations
of an engine shed. Casual visitors are welcome
to this industrial museum, with the regular

| Pets Welcome | £ Cash Dispenser | 24 24 Hour Facilities | WC Toilets | Alcoholic Drinks | Food and Drink | Fuel |

M1

special open weekends, featuring a large range of visiting locomotives and train rides well worthy of a visit. The site is generally flat and disabled visitors can be accommodated. Closed on Wednesdays

SERVICE AREA

BETWEEN JUNCTIONS 30 AND 31

Woodall Services (Northbound) (Welcome Break)
Tel: (01142) 486434 The Granary Restaurant, KFC, McDonald's, & Shell Fuel

Woodall Services (Southbound) (Welcome Break)
Tel: (01142) 486434 Red Hen, The Granary Restaurant, KFC, Burger King, La Brioche Doree, Welcome Lodge & Shell Fuel

Footbridge connection between sites

M1　JUNCTION 31

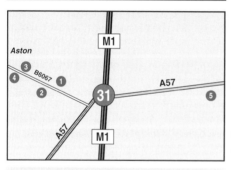

NEAREST NORTHBOUND A&E HOSPITAL

Northern General Hospital, Herries Road Sheffield S5 7AU Tel: (0114) 243 4343

Proceed north to Junction 34 and take the A6178 south towards Sheffield. At the junction with the A6102 turn right and the hospital is on the right hand side along this road. (Distance Approx 10.4 miles)

NEAREST SOUTHBOUND A&E HOSPITAL

Chesterfield & North Derbyshire Royal, Calow, Chesterfield S44 5BL Tel: (01246) 277271

Proceed south to Junction 30 and take the A6135 east. At the first roundabout turn right along the A619 to Chesterfield. After about 6.5 miles turn left at the roundabout along the A61(T) and at the junction with the A619 and A632 turn left along the A632 . The hospital is along this road on the left hand side. (Distance Approx 15.5 miles)

FACILITIES

1　Yellow Lion

Tel: (0114) 287 2283
0.4 miles west along the B6067, on the right. (Whitbread) Open all day. Meals served; Mon-Sat; Lunch; 12.00-15.00hrs & 17.30-21.00hrs, Sun; 12.00-14.30hrs & 15.00-18.00hrs

2　Aston Hall Hotel

Tel: (0114) 287 2309
0.5 miles west along the B6067, on the left. Restaurant Open; Breakfast; Mon-Fri; 07.00-09.00hrs, Sat & Sun; 08.00-10.00hrs, Lunch; 12.00-14.15hrs daily, Dinner; 19.00-21.30hrs daily

3　The Blue Bell

Tel: (0114) 287 1031
0.7 miles west along the B6067, on the right. (Free House) Open all day. Meals served; Mon-Fri; 12.00-14.00hrs & 18.00-20.00hrs, Sat; 12.00-14.00hrs, Sun; 12.00-15.00hrs.

4　Aston Service Station (Jet)

Tel: (0114) 287 4167
0.9 miles west along the B6067, on the left. Access, Visa, Overdrive, All Star, Switch, Dial Card, Mastercard, Amex, Diners Club, Delta, AA Paytrak, Jet Cards

5　The Red Lion Hotel

Tel: (01909) 771654
1 mile east along the A57, on the right. (Whitbread) Open all day. Meals served; Mon-Sat;12.00-22.00hrs, Sun; 12.00-21.30hrs.

PLACES OF INTEREST

Tropical Butterfly House, Falcony & Wildlife Centre

Woodsetts Road, North Anston S25 4EQ
Tel: (01909) 569416. Follow the A57 east
(Signposted along A57, 3.6 Miles)

| Large Hotel | Small Hotel | Motel | Guest House/ Bed & Breakfast | Disabled Facilities | Childrens Facilities |

A Wildlife Centre consisting of a butterfly house containing thousands of free flying butterflies, an Animal Nursery, Bird of Prey Centre, Farm Corner and Nature Trail. Plenty of "hands-on" opportunities. Gift Shop. Cafe. Picnic Area. Disabled access.

Rother Valley Country Park

Mansfield Road, Wales Bar S26 5PQ
Tel: (0114) 247 1452 (General Enquiries)
Follow the A57 west and turn left along the A618
(Signposted 3 Miles)

A landscaped leisure park with excellent facilities for water sports including sailing, windsurfing and canoeing and a 6,600 yards golf course with a driving range. There is a Visitor Centre, Craft Centre and Ranger Service. Equipment and cycles may be hired. Gift Shop. Cafe.

M1 | JUNCTION 32

JUNCTION 32 IS A MOTORWAY INTERCHANGE WITH THE M18 ONLY AND THERE IS NO ACCESS TO ANY FACILITIES

M1 | JUNCTION 33

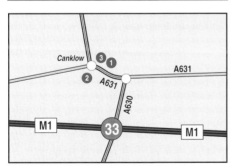

NEAREST A&E HOSPITAL

Northern General Hospital, Herries Road
Sheffield S5 7AU Tel: (0114) 243 4343
Proceed north to Junction 34 and take the A6178 south towards Sheffield. At the junction with the A6102 turn right and the hospital is on the right hand side along this road. (Distance Approx 5.3 miles)

FACILITIES

1 Rotherham Swallow Hotel

Tel: (01709) 830630
0.7 miles west along the A631, on the right. Restaurant Open; Breakfast; Mon-Fri; 07.00-09.30hrs, Sat & Sun; 07.30-10.00hrs, Lunch; Sun-Fri; 12.00-14.00hrs, Dinner; 18.30-21.30hrs daily.

2 Canklow Bridge (Total)

Tel: (01709) 726900
1 mile west along the A631, on the left. Access, Visa, Overdrive, All Star, Switch, Dial Card, Mastercard, Amex, Diners Club, Delta, UK Fuelcard, BP Supercharge, Total Cards.

3 Canklow Service Station

Tel: (01709) 382875
1 mile west along the A631, on the right. Access, Visa, Overdrive, All Star, Switch, Dial Card, Mastercard, Delta. Open; Mon-Fri; 07.00-20.00hrs, Sat; 08.00-17.00hrs, Sun; 10.00-17.00hrs.

PLACES OF INTEREST

Sheffield

Sheffield Tourist Information Centre
Tel: (0114) 273 4671 e-mail; tidestshef@aol.com
website; http://www.sheffieldcity.co.uk
Follow the A630 southwest (Signposted 6 Miles)

From Saxon times, Sheffield was the capital of Hallamshire and by the days of Henry II a castle had been constructed. It was within this fortress, demolished during the Civil War, that Mary, Queen of Scots was imprisoned for fourteen years from 1570. Standing on the River Don with its confluence with the River Sheaf, from which it gained its name, the city found world wide fame from its expertise in the manufacture of steel. Cutlery was made as early as the 14thC and the Company of Cutlers gained their charter in 1624. The present Cutlers Hall (Tel: 0114-272 8456) in Church Street, is in Grecian style and dates from 1832, the third such building to be erected on the site. The Cathedral Church of St Peter and St Paul (Tel: 0114-275 3434), also in Church Street, was built in c1430 with parts added or altered until 1805. There are many fine Victorian structures including the Town Hall in Pinstone Street. Following the contraction of the steel industry in the 1970's some of the steel works sites have been revitalized with new projects including the Don Valley Stadium and the Meadowhall Shopping Centre.

Within the city can be found ...

The National Centre for Popular Music

Paternoster Row, Sheffield S1 2QQ
Tel; (0114) 296 6060 (24hr Information Line)
website; www.ncpm.co.uk

This exhibition tells the fascinating story of music through the ages and from many cultures using interactive exhibits, real and virtual instruments, audio-visual simulations and items of rare memorabilia, The centre's striking building houses the world's first 360° surround-sound auditorium and visitors can have a go at playing an instrument, mixing a record or being a DJ. Cafe. Licensed Bar. Music themed shop.

Kelham Island Museum

Alma Street (Off Corporation Street)
Sheffield S3 8RY Tel: (0114) 272 2106.

This living museum tells the story of Sheffield, its industry and life. The exhibits include the River Don Engine, the most powerful working steam engine in Europe, the Melting Shop, an interactive steelmaking display (only open on Sundays and School holidays) and workshops demonstrating cutlery manufacture.

Graves Art Gallery

Surrey Street, Sheffield S1 1XZ
Tel: (0114) 273 5158

A wide-ranging collection of British art from the 16thC to the present, European paintings, a fine collection of watercolours, drawings and prints and the Grice Collection of Chinese Ivories forms the centrepiece to the collection of non-European artefacts.

Rotherham

Rotherham Tourist Information Centre,
Central Library, Walker Place, Rotherham S65 1JH
Tel: (01709) 835904
website: www.rotherham.gov.uk
e-mail; tic@rotherham.gov.uk
Follow the A630 North (Signposted 3.2 Miles)

Where the rivers Rother and Don meet, the Romans worked iron here and built a fort on the south bank of the River Don at Templeborough and finds from excavations of the area can be seen at Clifton Park Museum. In 1161 the monks of Rufford Abbey were given the right to prospect for, and to smelt, iron and plant an orchard, leading to the tradition in Rotherham of industry and agriculture existing side by side. Heavy industry put the town on the map when, in the early days of the Industrial Revolution, the Walker family established themselves here in 1746. The Walker Company manufactured cannons, some of which featured in such famous conflicts as the American War of Independence and the Battle of Trafalgar, and bridges, with those at Southwark and Sunderland being better known examples. The town's connection with bridges does not end there either, as Sir Donald Coleman Bailey, designer and inventor of the Bailey Bridge was born here in 1901. In total contrast to heavy engineering, Rockingham Pottery was produced here in the late 18thC and early 19thC.

Within the town can be found ...

The Chapel of our Lady

Rotherham Bridge, Rotherham S60 1QJ
Tel: (01709) 364737

One of only a few surviving complete bridge chapels, construction commenced in 1483 following a legacy of 3/4d (17p) from John Bokyng, master of the grammar school, and was probably completed, and paid for, by Thomas Rotherham. It was sumptuously furnished but only lasted until the reign of King Edward VI (1547-1553) when it was closed and then saw a variety of commercial uses until reconsecrated in 1924 by the Bishop of Sheffield.

 Large Hotel   Small Hotel   Motel   Guest House/ Bed & Breakfast   Disabled Facilities   Childrens Facilities

Clifton Park Museum

Clifton Lane, Rotherham S60 2AA
Tel: (01709) 382121

Built in 1783 for the Rotherham ironmaster, Joshua Walker, the interior has little changed and now houses a fine collection of Rockingham Pottery including the Rhinoceros Vase which stands almost 4ft high. In addition there is also a collection of Yorkshire pottery, English glass, silver and British oil paintings and water colours. Gift Shop. Limited disabled access.

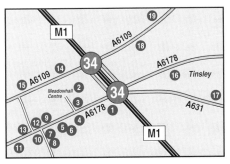

M1 JUNCTION 34

NEAREST NORTHBOUND A&E HOSPITAL

**Northern General Hospital, Herries Road
Sheffield S5 7AU Tel: (0114) 243 4343**

Take the A6178 south and at the junction with the A6102 turn right. The hospital is on the right along this road. (Distance Approx 2.6 miles)

NEAREST SOUTHBOUND A&E HOSPITAL

**Northern General Hospital, Herries Road
Sheffield S5 7AU Tel: (0114) 243 4343**

Take the A6109 south and at the junction with the A6102 turn right. The hospital is on the right along this road. (Distance Approx 2.7 miles)

FACILITIES

1 Tinsley Transport Cafe

Tel: (0114) 244 2046

Adjacent to the south side of the roundabout. Open; Mon-Fri; 06.00-14.30hrs, Sat & Sun; 06.00-10.30hrs

2 Meadowhall Shopping Centre

Tel: 0345 573 618

Adjacent to the west side of the junction. There are numerous cafes and restaurants within the shopping centre. Open; Mon-Wed; 10.00-20.00hrs, Thurs & Fri; 10.00-21.00hrs, Sat; 09.00-19.00hrs, Sun; 11.00-17.00hrs

3 Sainsbury's Petrol South

Tel: (0114) 235 4200

0.2 miles west along the A6178, on the right. Access, Visa, Overdrive, All Star, Switch, Dial Card, Mastercard, Amex, Delta, AA Paytrak, UK Fuelcard, Sainsbury's Fuelcard. Open; Mon-Wed; 07.00-21.00hrs, Thurs; 07.00-22.00hrs, Fri; 07.00-22.30hrs, Sat; 07.00-21.00hrs, Sun; 08.00-19.00hrs. Toilets available in adjacent store.

4 Tinsley Filling Station (Jet)

Tel: (0114) 256 2343

0.3 miles west along the A6178, on the left. Visa, Jet Card, Diners Club, Overdrive, All Star, Mastercard, Amex, Switch, Dial Card.

5 Pizza Hut

Tel: (0114) 256 2211

0.6 miles west along the A6178, on the left. Open; Sun-Thurs; 12.00-22.00hrs, Fri & Sat; 12.00-23.00hrs.

6 McDonald's

Tel: (0114) 261 9269

0.6 miles west along the A6178, on the left. 'Drive-Thru' & Restaurant. Open; 07.00-0.00hrs Daily.

7 KFC

Tel: (0114) 249 5210

0.8 miles west along the A6178, on the left in Broughton Lane. Open; Sun-Thurs; 11.00-0.00hrs, Fri/Sat & Sat/Sun; 11.00-02.45hrs

Pets Welcome	Cash Dispenser	24 Hour Facilities	Toilets	Alcoholic Drinks	Food and Drink	Fuel

M1

8   Burger King

Tel: (0114) 242 6197
0.8 miles west along the A6178, on the left in Broughton Lane. Open; Sun-Thurs; 10.00-0.00hrs, Fri/Sat & Sat/Sun; 10.00-03.30hrs

9   Carbrook Hall

Tel: (0114) 244 0117
0.8 miles west along the A6178, on the right. (Punch Taverns) Open all day on Friday. Meals served; 12.00-14.00hrs & 18.00-20.00hrs daily

10   The Stumble Inn

Tel: (0114) 244 1530
0.9 miles west along the A6178, on the left. (Free House) Open all day on Friday. Lunch; Mon-Fri; 12.00-14.00hrs.

11   The Players Sports Cafe

Tel: (0114) 296 0007
1 mile west along the A6178, on the left. Open; Mon-Sat; 12.00-23.00hrs, Sun; 12.00-22.30hrs

12   Arena Square

Tel: (0114) 243 2320
1 mile west along the A6178, on the right. (Brewsters) Open all day. Meals served; Mon-Sat; 11.30-22.00hrs, Sun; 12.00-22.00hrs

13   Sheffield Travel Inn

Tel: (0114) 242 2802
1 mile west along the A6178, on the right. The Potter's Bar & Restaurant; Breakfast; Mon-Fri; 07.00-09.00hrs, Sat & Sun; 08.00-10.00hrs, Dinner; 17.30-22.30hrs daily

14   Sainsbury's Petrol North

Tel: (0114) 235 4200
0.4 miles west along the A6109, on the right. Access, Visa, Overdrive, All Star, Switch, Dial Card, Mastercard, Amex, Delta, AA Paytrak, UK Fuelcard, Sainsbury's Fuelcard. Open; Mon-Wed; 07.00-21.00hrs, Thurs; 07.00-22.00hrs, Fri; 07.00-22.30hrs, Sat; 07.00-21.00hrs, Sun; 08.00-19.00hrs. Toilets available in adjacent store.

15   The Crown Pub and B&B

Tel: (0114) 243 1319
0.9 miles west along the A6109, on the right. (Free House)

16   The Fox & Duck

Tel: (0114) 244 1938
0.4 miles east along the A6178, on the right

17   The Fairways Restaurant

Tel: (01709) 838111
1 mile south along the A631, on the left. (Millhouse Inns) Open all day. Meals served; Mon-Sat; 12.00-20.00hrs, Sun; 12.00-14.30hrs (Bookings only on Sundays)

18   The Pudding & Pint at the Meadowbank Hotel

Tel; (01709) 512114
0.9 miles east along the A6109, on the right. (Hop & Bean Co) Open all day Thurs-Sun. Meals served; Mon-Wed; 12.00-15.00hrs & 17.00-21.00hrs, Thurs-Sat; 12.00-21.00hrs, Sun; 12.00-15.00hrs

19   Meadowbank Filling Station (Jet)

Tel: (01709) 740440
1 mile east along the A6109, on the left. Access, Visa, Overdrive, All Star, Switch, Dial Card, Mastercard, Amex, Diners Club, Delta, BP Supercharge, Jet Card.

PLACES OF INTEREST

Sheffield

Follow the A631 west. (Signposted 4.2 Miles)

For details please see Junction 33 information

Within the city can be found ...

The National Centre for Popular Music

Paternoster Row, Sheffield S1 2QQ

For details please see Junction 33 information

| Large Hotel | Small Hotel | Motel | Guest House/ Bed & Breakfast | Disabled Facilities | Childrens Facilities |

Kelham Island Museum

Alma Street (Off Corporation Street)
Sheffield S3 8RY

For details please see Junction 33 information

Graves Art Gallery

Surrey Street, Sheffield S1 1XZ

For details please see Junction 33 information

Meadowhall

Adjacent to the west side of Junction 34
Tel: (0345) 573618
website; www.meadowhall.co.uk

A large shopping centre with more than 270 shops and an 11 screen Warner Village cinema. Other features include the Oasis, a Mediterranean-style food court with a giant video wall, and The Lanes, an avenue of specialist and craft shops.

Rotherham

Follow the A6178 east (Signposted 2.9 Miles)

For details please see Junction 33 information.

Within the town can be found ...

The Chapel of our Lady

Rotherham Bridge, Rotherham S60 1QJ

For details please see Junction 33 information.

Clifton Park Museum

Clifton Lane, Rotherham S60 2AA

For details please see Junction 33 information.

Blackburn Meadows

Steel Street, Holmes, Rotherham S61 1DF
(Admin: Recreation Offices, Grove Road,
Rotherham S60 2ER)
Tel: (01709) 822041
Follow the A6109 east and turn right down
Psalters Lane. (1.4 Miles)

An unusual haven for wildlife within an urban area, this conservation site has an extraordinary history. Part of the floodplain of the River Don, Holmes Farm was established here and then the land was taken over by the Tinsley Sewage Farm. Surprisingly, wildlife and especially migrating birds were attracted to the huge, liquid sewage lagoons but modern changes in waste management led them to start drying up and the local birdwatchers, aided by the local councils and Yorkshire Water, stepped in to save the site. The green heart of a projected large scale development of the Lower Don Valley it is planned to improve the facilities and build a Visitor Centre.

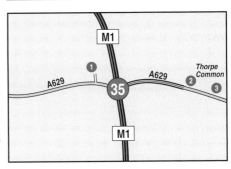

M1 JUNCTION 35

NEAREST NORTHBOUND A&E HOSPITAL

Barnsley & District General Hospital, Gawber Road, Barnsley S75 2PW Tel: (01226) 730000

Proceed north to Junction 37, take the A628 east towards Barnsley, and after about 0.5 miles turn left at the traffic lights along Pogmoor Road. Turn left at the end of this road into Gawber Road and the hospital is immediately on the left. (Distance Approx 8.4 miles)

NEAREST SOUTHBOUND A&E HOSPITAL

Northern General Hospital, Herries Road Sheffield S5 7AU Tel: (0114) 243 4343

Take the A629 west and after about 0.5 miles turn left along Nether Lane. At the end of this lane turn left along the A6135 and the hospital is on the right hand side of this road. (Distance Approx 4 miles)

FACILITIES

1 The Travellers

Tel: (0114) 246 7870

0.2 miles west along the A629, on the right in Smithy Wood Road. (Free House) Lunch & Evening Meals

2 Star Scholes (Texaco)

Tel: (0114) 246 7124

0.8 miles east along the A629, on the left.

M1

Access, Visa, Overdrive, All Star, Switch, Dial Card, Mastercard, Diners Club, Delta, UK Fuelcard, BP Cards, Texaco Cards.

3 The Sportsman's

Tel: (0114) 257 2011

1 mile east along the A629, on the left. (Bass) Open all day. Lunch; Mon-Sat; 12.00-14.30hrs, Evening Meals; Mon-Sat;17.00-21.00hrs, Meals served Sun; 12.00-18.00hrs.

M1 JUNCTION 35A

THIS IS A RESTRICTED ACCESS JUNCTION

Vehicles can only exit from the northbound lanes

Vehicles can only enter the motorway along the southbound lanes

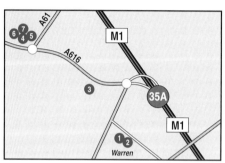

NEAREST NORTHBOUND A&E HOSPITAL

Barnsley & District General Hospital, Gawber Road, Barnsley S75 2PW Tel: (01226) 730000

Proceed north to Junction 37, take the A628 east towards Barnsley, and after about 0.5 miles turn left at the traffic lights along Pogmoor Road. Turn left at the end of this road into Gawber Road and the hospital is immediately on the left. (Distance Approx 6.9 miles)

NEAREST SOUTHBOUND A&E HOSPITAL

Northern General Hospital, Herries Road Sheffield S5 7AU Tel: (0114) 243 4343

Proceed south to Junction 35 and take the A629 west. After about 0.5 miles turn left along Nether Lane and at the end of this lane turn left along the A6135. The hospital is on the right hand side of this road. (Distance Approx 5.5 miles)

FACILITIES

1 Thorncliffe Arms

Tel: (0114) 245 8942

0.3 miles south along Warren Lane, on the right. (Swallow) Open all day Fri & Sat, Lunch; Mon-Sat; 12.00-14.00hrs, Sun; 12.00-15.00hrs, Evening Meals; Mon-Fri; 18.00-21.00hrs, Sat; 18.00-21.30hrs, Sun; Closed.

2 The Miners Arms

Tel: (0114) 257 0092

0.3 miles south along Warren Lane, on the right. (The Nice Pub Co.) Open all day Fri-Sun. Lunch; 12.00-14.30hrs, Evening Meals; 17.00-20.30hrs Daily.

3 Truckers Cafe (Transport)

0.4 miles north along the A616, on the left

4 McDonald's

Tel: (01226) 740025

1 mile north along the A616, on the right. Open; 07.00-23.00hrs ["Drive Thru" open until 0.00hrs] Daily.

5 Wentworth Park Service Station (Save)

Tel: (01226) 350479

1 mile north along the A616, on the right. Access, Visa, Overdrive, All Star, Switch, Dial Card, Mastercard, Delta, Routex, Keyfuels, AA Paytrak, UK Fuelcard.

6 Travel Inn

Tel: (01226) 350035

1 mile north along the A616, on the right

7 The Wentworth

Tel: (01226) 350035

1 mile north along the A616, on the right. Open all day. Meals served all day Mon-Sat; 11.30-22.00hrs, Sun; 12.00-22.00hrs.

| Large Hotel | Small Hotel | Motel | Guest House/ Bed & Breakfast | Disabled Facilities | Childrens Facilities |

M1 JUNCTION 36

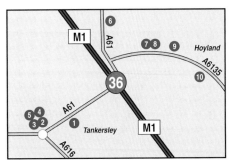

NEAREST NORTHBOUND A&E HOSPITAL

Barnsley & District General Hospital, Gawber Road, Barnsley S75 2PW Tel: (01226) 730000

Proceed north to Junction 37, take the A628 east towards Barnsley, and after about 0.5 miles turn left at the traffic lights along Pogmoor Road. Turn left at the end of this road into Gawber Road and the hospital is immediately on the left. (Distance Approx 5.3 miles)

NEAREST SOUTHBOUND A&E HOSPITAL

Northern General Hospital, Herries Road Sheffield S5 7AU Tel: (0114) 243 4343

Proceed south to Junction 35 and take the A629 west. After about 0.5 miles turn left along Nether Lane and at the end of this lane turn left along the A6135. The hospital is on the right hand side of this road. (Distance Approx 7.1 miles)

FACILITIES

1 Tankersley Manor Hotel & Pub

Tel: (01226) 744700
0.7 miles west along the A61, on the left. Bar Meals; Open All Day. Meals served; 12.00-21.30hrs daily. Restaurant; Breakfast; Mon-Fri; 07.00-09.30hrs, Sat & Sun; 08.00-10.00hrs, Lunch; Mon-Fri; 12.00-14.00hrs, Sat; Closed; Sun; 12.00-14.30hrs, Dinner; Mon-Sat; 19.00-21.30hrs, Sun; Closed.

2 McDonald's

Tel: (01226) 740025
1 mile west along the A61, on the right. Open;

07.00-23.00hrs ["Drive Thru" open until 0.00hrs] Daily.

3 Wentworth Park Service Station (Save)

Tel: (01226) 350479
1 mile west along the A61, on the right. Access, Visa, Overdrive, All Star, Switch, Dial Card, Mastercard, Delta, Routex, Keyfuels, AA Paytrak, UK Fuelcard.

4 Travel Inn

Tel: (01226) 350035
1 mile west along the A61, on the right

5 The Wentworth

Tel: (01226) 350035
1 mile west along the A61, on the right. Open all day. Meals served all day Mon-Sat; 11.30-22.00hrs, Sun; 12.00-22.00hrs.

6 Hilltop Service Station (Shell)

Tel: (01226) 284412
1 mile north along the A61, on the right. Access, Visa, Overdrive, All Star, Switch, Dial Card, Mastercard, Amex, Diners Club, Delta, Routex, Keyfuels, AA Paytrak, Shell Cards, BP Agency, Esso Cards. Open; Mon-Sat; 07.00-21.00hrs, Sun; 08.00-20.30hrs.

7 The Cross Keys

Tel: (01226) 742277
0.4 miles south along the A6135, on the left. (Whitbread) Open all day on Sunday. Lunch; Mon-Sat; 12.00-14.00hrs, Evening Meals, Mon-Sat; 17.30-21.00hrs, Meals served all day Sundays; 12.00-21.00hrs.

8 Cross Keys Garage (Jet)

Tel: (01226) 743331
0.4 miles south along the A6135, on the left. Access, Visa, Overdrive, All Star, Switch, Dial Card, Mastercard, Delta, UK Fuelcard, Jet Card. Open; Mon-Sat; 07.00-22.00hrs, Sun; 08.00-22.00hrs.

9   Hare & Hounds

Tel: (01226) 742283

0.5 miles south along the A6135, on the left. (John Smith's) Open all day. Bar Meals; Mon-Sat; 12.00-14.00hrs and Mon-Fri; 17.30-19.00hrs. Restaurant; Mon-Sat; 19.00-21.30hrs, Sun 12.00-14.30hrs.

10   Hoylake Common Save Petrol Station

Tel: (01226) 746475

0.7 miles south along the A6135, on the right. Access, Visa, Overdrive, All Star, Switch, Dial Card, Mastercard, Delta, AA Paytrak, Save Business Card. Open; 06.00-23.00hrs Daily.

PLACES OF INTEREST

Worsbrough Mill Museum & Country Park

Off Park Road, Worsbrough Bridge, Barnsley S70 5LJ Tel: (01226) 774527
Follow the A61 north (Signposted 2.1 Miles)

A water powered corn mill dating from c1625 and a 19thC steam mill have been restored to full working order and form the centrepiece of this industrial museum which is surrounded by a 200 acre park.

Elsecar Heritage Centre

Wath Road, Elsecar, Barnsley S74 8HJ
Tel: (01226) 740203. Follow the A61 northeast (Signposted 2.9 Miles)

Situated in the South Yorkshire countryside, this science and history centre features "hands-on" science in the Power House and trips on the Elsecar Steam Railway. Displays include the interactive "Living History Exhibition" and interactive multi-media in the Newcomen Beam Engine Centre. Gift Shop. Tea Room. Disabled access.

Wigfield Farm

Haverlands Lane, Worsbrough Bridge, Barnsley S70 5NQ Tel: (01226) 733702
Follow the A61 northeast (Signposted along A61, 2.4 Miles)

An opportunity to view rare and traditional breeds of farm animals with milking demon-strations being held daily. There is also an Exotic Animal House featuring reptiles and snakes. Gift Shop. Cafe. Picnic and play areas. Disabled access.

M1   JUNCTION 37

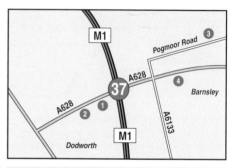

NEAREST A&E HOSPITAL

Barnsley & District General Hospital, Gawber Road, Barnsley S75 2PW Tel: (01226) 730000

Take the A628 east and after about 0.5 miles turn left at the traffic lights along Pogmoor Road. Turn left at the end of this road into Gawber Road and the hospital is immediately on the left. (Signposted Distance Approx 1.3 miles)

FACILITIES

1   The Gate Inn

Tel: (01226) 282705

0.3 miles west along the A628, on the left. (Pubmaster) Lunch; Sun-Fri; 12.00-14.30hrs, Evening Meals; Mon-Fri; 17.00-19.30hrs. No food served on Saturdays.

2   Brooklands Restaurant & Motel

Tel: (01226) 299571

0.5 miles west along the A628, on the left. Restaurant Open; 18.30-22.00hrs Daily

3   Intake 1 Jet Service Station

Tel: (01226) 286424

1 mile north along Pogmoor Road, on the left. Access, Visa, Overdrive, All Star, Switch, Dial Card, Mastercard, Amex, Diners Club, Delta, Jet Card.

4   Shell Barnsley

▢ ♿ WC 24   Tel: (01226) 737100

0.6 miles east along the A628, on the right.
Access, Visa, Overdrive, All Star, Switch, Dial
Card, Mastercard, Amex, Diners Club, Shell
Cards, BP Agency, Esso Card, Smart Card.

PLACES OF INTEREST

Metrodome Leisure Complex

Queen's Ground, Queen's Road, Barnsley S71 1AN
Tel: (01226) 730060. Follow the A628 into
Barnsley, continue along Peel Street and Kendray
Street (Signposted 1.8 Miles)

Britain's most exciting and imaginative indoor
water based theme attraction, featuring the
Terrorship 3000, Black Hole, Red River Cruiser
and Lunar Run, the ultimate in aqua
excitement challenge and daring! Dry facilities
include squash, badminton and bowls. Cafe.
Disabled access.

Cannon Hall Museum, Park & Gardens

Cawthorne, Barnsley S75 4AT
Tel: (01226) 790270
Follow the A628 west and turn right towards
Barugh Green. Turn left along the A635 to
Cawthorne and it is signposted locally. (4.3 Miles)

The country house, remodelled by John Carr
of York in the 18thC, houses a collection of
pottery, furniture, paintings and glassware
from the 18th to 20thC and is the Regimental
Museum of the 13th/18th Royal Hussars. It is
surrounded by 70 acres of parkland landscaped
by Richard Woods in the 1760's, with lawns, a
walled garden full of historic pear trees and
lakes. Shop. Victorian Kitchen Cafe.

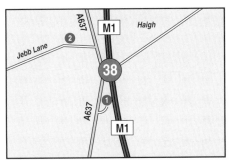

M1   JUNCTION 38

NEAREST NORTHBOUND A&E HOSPITAL

**Pinderfield Hospital, Aberford Road,
Wakefield WF1 4DG Tel: (01924) 201688**

Proceed north to Junction 39. Take the A636 east
exit and after 2.5 miles turn right along the A638.
At the junction of the A642 and A61 turn right along
the A642 and the hospital is along this road on the
left. (Distance Approx 7.7 miles)

NEAREST SOUTHBOUND A&E HOSPITAL

**Barnsley & District General Hospital, Gawber
Road, Barnsley S75 2PW Tel: (01226) 730000**

Take the A637 south towards Barnsley and at the
roundabout carry straight on along the A635.
Almost immediately turn first right into Redbrook
Road and continue into Gawber Road. The hospital
is on the right hand side. (Distance Approx 4.4
miles)

FACILITIES

1   The Old Post Office

▮▮ ▮▮ ♿ ⤴   Tel: (01226) 387619

0.1 miles south along the A637, on the left.
(Free House) Open all day on Saturday. Lunch;
12.00-14.00hrs Daily, Evening Meals; Mon-
Thurs; 17.30-19.30hrs, Fri-Sat; 17.30-21.00hrs.

2   The Old Haigh Cottage

🛏B   Tel: (01924) 830456

0.2 miles west along Jebb Lane, on the right

PLACES OF INTEREST

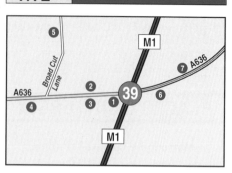

M1   JUNCTION 39

National Coal Mining Museum for England

Caphouse Colliery, New Road, Overton,
Wakefield WF4 4RH Tel: (01924) 848 806
Follow the A637 north (Signposted 5.4 Miles)

Displays and audio-visual presentations show mining conditions from the early days to the present whilst there is a one hour underground tour (warm clothes and sensible shoes recommended) of a real coal mine with an experienced local miner as a guide.

Yorkshire Sculpture Park

Bretton Hall, West Bretton, Wakefield WF4 4LG
Tel: (01924) 830 302. Follow the A637 north
(Signposted 1 mile)

One of Europe's leading open air galleries, set in 100 acres of 18thC parkland, with a programme of international exhibitions of contemporary sculpture. There are indoor galleries, a craft shop, cafe and sculpture bookshop. Electric scooters (pre-booked by telephone) are available and there are audio guides and braille information. The neighbouring Bretton Country Park contains a number of works by Henry Moore

SERVICE AREA

BETWEEN JUNCTIONS 38 AND 39

Woolley Edge Services (Northbound) (Granada)
Tel: (01924) 830371 Self Service Restaurant, Burger King, Travelodge & Esso Fuel
Woolley Edge Services (Southbound) (Granada)
Tel: (01924) 830371 Self Service Restaurant, Burger King & Esso Fuel

NEAREST NORTHBOUND A&E HOSPITAL

Pinderfield Hospital, Aberford Road, Wakefield WF1 4DG Tel: (01924) 201688
Take the A636 east exit and after 2.5 miles turn right along the A638. At the junction of the A642 and A61 turn right along the A642 and the hospital is along this road on the left. (Distance Approx 4.3 miles)

NEAREST SOUTHBOUND A&E HOSPITAL

Barnsley & District General Hospital, Gawber Road, Barnsley S75 2PW Tel: (01226) 730000
Proceed south to Junction 38. Take the A637 south towards Barnsley and at the roundabout carry straight on along the A635. Almost immediately turn first right into Redbrook Road and continue into Gawber Road. The hospital is on the right hand side. (Distance Approx 7.9 miles)

FACILITIES

1 Cedar Court Hotel

Tel: (01924) 276310
On the west side of the roundabout. Restaurant Open; Breakfast; Mon-Fri; 07.00-10.00hrs, Sat; 07.30-10.30hrs, Sun; 08.00-10.30hrs, Lunch; Mon-Fri;12.00-14.00hrs [Sat; Closed], Sun; 12.00-15.00hrs, Dinner; 19.00-22.00hrs daily.

2 Knight of Wakefield Total Filling Station

Tel: (01924) 274756
0.2 miles west along the A636, on the right. Access, Visa, Overdrive, All Star, Switch, Dial Card, Mastercard, Delta, Routex, BP Supercharge, Total Cards, Energy Plus,

	Large Hotel		Small Hotel		Motel		Guest House/ Bed & Breakfast		Disabled Facilities		Childrens Facilities

Fourcourt Business Card, Total Fina Card. Open; Mon-Fri; 07.00-21.00hrs, Sat; 08.00-19.00hrs, Sun; 09.00-19.00hrs.

3   King Fishers Fish & Chip Restaurant

Tel: (01924) 274994
0.2 miles west along the A636, on the left. Open; 11.30-23.30hrs Daily.

4   The British Oak

Tel: (01924) 275286
0.8 miles west along the A636, on the left. (Tetley) Open all day. Meals served; 12.00-21.30hrs daily.

5   The Navigation Inn

Tel: (01924) 274361
0.5 miles along Broad Cut Lane, on the left. (Free House) Open all day. Lunch; Mon-Sat; 12.00-14.00hrs, Sun; 12.00-15.00hrs, Evening Meals; Mon-Sat; 17.30-20.00hrs.

6   The New Inn

Tel: (01924) 255897
0.2 miles east along the A636, on the right. (Allied Domecq) Open all day. Lunch; Mon-Sat; 11.30-14.00hrs.

7   Grange Service Station (Esso)

Tel: (01924) 371209
0.7 miles east along the A636, on the left. Access, Visa, Overdrive, All Star, Switch, Dial Card, Mastercard, Amex, Diners Club, Delta, AA Paytrak. Shell Gold, BP Supercharge, Style. Open; Mon-Sat; 07.00-22.00hrs, Sun; 08.00-22.00hrs.

PLACES OF INTEREST

Wakefield

Follow the A 636 north (Signposted 3.1 Miles)
For details please see Junction 40 information

Within the city can be found ...

The Cathedral Church of All Saints

Northgate, Wakefield WF1 1HG
For details please see Junction 40 information

The Chantry Chapel of St Mary on Wakefield Bridge

For details please see Junction 40 information

Sandal Castle

Off Manygates Lane, Sandal, Wakefield.
Follow the A636 east towards Wakefield, turn south along the A6186 and continue north along the A61. (Signposted along A61, 2.7 Miles)
For details please see Junction 40 information

M1   JUNCTION 40

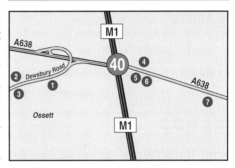

NEAREST A&E HOSPITAL

Pinderfield Hospital, Aberford Road, Wakefield WF1 4DG Tel: (01924) 201688
Take the A638 east exit and at the junction of the A642 and A61 turn right along the A642. The hospital is along this road on the left. (Distance Approx 4.2 miles)

FACILITIES

1   The Commercial Inn

Tel: (01924) 274197
0.8 miles west along the Dewsbury Road, in Ossett, on the left. (Bass Yorkshire) Open; Mon-Wed; 15.00-23.00hrs, Thurs-Sat; 12.00-23.00hrs, Sun; 12.00-22.30hrs. Food Served 17.00-21.00hrs daily

| Pets Welcome | £ Cash Dispenser | 24 Hour Facilities | WC Toilets | Alcoholic Drinks | Food and Drink | Fuel |

M1

2 Mercury Service Station (UK Fuels)

📶 WC   Tel: (01924) 275680

1 mile west along the Dewsbury Road, in Ossett, on the right. Access, Visa, Overdrive, All Star, Switch, Dial Card, Mastercard, Amex, Diners Club, Delta, Keyfuels, AA Paytrak, BP Supercharge. Open; Mon-Sat; 07.30-20.00hrs, Sun; 10.00-20.00hrs.

3 The Red Lion

🍴 🛏 ♿ ☂   Tel: (01924) 273487

1 mile west along the Dewsbury Road, in Ossett, on the left. (Free House) Open all day. Lunch; Mon-Sat; 12.00-14.00hrs, Sun; 12.00-18.00hrs, Evening Meals; Tues-Sat; 19.00-21.00hrs.

4 Shell Ossett

📶 ♿ WC 24   Tel: (01924) 282570

0.3 miles east along the A638, on the left. Access, Visa, Overdrive, All Star, Switch, Dial Card, Mastercard, Amex, Diners Club, Delta, UK Fuelcard, Shell Cards, BP Agency, Smart.

5 Posthouse Wakefield

🏨H 🍴 🛏 🍽 ☂   Tel: (01924) 276388

0.3 miles east along the A638, on the right. Restaurant Open; Mon-Fri; 06.30-22.30hrs, Sat & Sun; 06.30-22.00hrs

6 Total Fina Ossett Service Station

📶 WC   Tel: (01924) 265947

0.3 miles east along the A638, on the right. Access, Visa, Overdrive, All Star, Switch, Dial Card, Mastercard, Amex, Diners Club, Delta, BP Supercharge, Elf Cards, Total Fina Cards. Open; Mon-Fri; 06.00-23.00hrs, Sat & Sun; 07.00-23.00hrs.

7 The Old Malt Shovel

🍴 🛏 ☂ ♿   Tel: (01924) 201561

0.6 miles east along the A638, on the right. (Free House) Open all day. Carvery; Sun Lunch; 12.00-16.00hrs. Bar Meals; Mon-Sat; 12.00-14.00hrs. Pizzas served 11.00-23.00hrs Daily.

PLACES OF INTEREST

Wakefield

Wakefield Tourist Information Centre, Town Hall,
Wood Street, Wakefield WF1 2HQ
Tel: (01924) 305000/1
Follow the A 638 east (Signposted 2.4 Miles)

This ancient Cathedral City dates back to at least Roman times and was once an inland grain and cloth port. The famous Battle of Wakefield took place on Wakefield Green in 1460 with defeat for the Royalists. Amongst the fine buildings are the parish church, dating from the 14thC and with the tallest spire in Yorkshire, which became a cathedral in 1888 and St Mary's Chapel.

Within the town can be found ...

The Cathedral Church of All Saints

Northgate, Wakefield WF1 1HG
Tel: (01924) 373923

The Cathedral offers a quiet and tranquil respite from the lively and bustling pedestrian precinct that surrounds it. There are magnificent wood carvings throughout the building, notably the choir and screen, and the stained glass windows form a comprehensive collection of glass by Kempe who contributed work to the cathedral throughout his life. Book Shop. Disabled access.

The Chantry Chapel of St Mary on Wakefield Bridge

Chairman: John Gilbey, 2 Westfield Terrace,
Wakefield WF1 3RD. Tel: (01924) 373847.

Standing on Wakefield's Mediaeval bridge, this chapel is a very rare survival of the once common practice in the Middle Ages of siting places of worship on bridges where priests could minister to travellers and hold services. The stone bridge upon which it stands was built in the 1340's and the tolls charged to use it were of considerable economic importance to the townspeople. The chapel was extensively restored in 1847.

Sandal Castle

Off Manygates Lane, Sandal, Wakefield
Follow the A631 into Wakefield and turn south along the A61 (Signposted along A61, 4 Miles)

This motte and bailey castle dates from the 12thC and the later stone castle, owned by Richard, Duke of York, overlooks the site of the Battle of Wakefield (1460). The structure was demolished upon the orders of Parliament after the siege of 1645 and the ruins are now part of a park.

NB. Artefacts and further information about Sandal Castle can be seen at Wakefield Museum, Wood Street, Wakefield WF1 2HQ. Tel: (01924) 305351

M1 JUNCTION 41

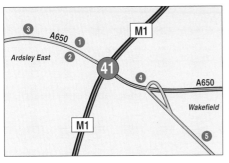

NEAREST NORTHBOUND A&E HOSPITAL

Leeds General Infirmary, Great George Street, Leeds LS1 3EX Tel: (0113) 243 2799

Proceed north to Junction 7 (M621) and take the A61 north into Leeds city centre. The hospital is signposted in the city. (Distance Approx 8 miles)

NEAREST SOUTHBOUND A&E HOSPITAL

Pinderfield Hospital, Aberford Road, Wakefield WF1 4DG Tel: (01924) 201688

Take the A650 exit east and continue along the A61. At the junction of the A642 and A61 turn left along the A642 and the hospital is along this road on the left. (Distance Approx 3.4 miles)

FACILITIES

1   The Bay Horse

Tel: (01924) 825926

0.3 miles west along the A650, on the right. (Allied Domecq) Open all day. Lunch; Mon-Fri; 12.00-15.00hrs, Evening Meals, Mon-Fri; 18.00-22.00hrs, Meals served all day on Saturdays; 12.00-22.00hrs and Sundays; 12.00-19.00hrs.

2   Manor Service Station (BP)

Tel: (01924) 822260

0.3 miles west along the A650, on the left. Access, Visa, Overdrive, All Star, Switch, Dial Card, Mastercard, Amex, Diners Club, Delta, Routex, UK Fuelcard, Shell Cards, BP Cards. Open; Mon-Fri; 06.00-22.00hrs, Sat & Sun; 07.00-22.00hrs.

3   Ardsley Esso Service Station

Tel: (01924) 823192

0.6 miles west along the A650, on the right. Access, Visa, Overdrive, All Star, Switch, Dial Card, Mastercard, Amex, Diners Club, Delta, UK Fuelcard, Shell Gold, Esso Cards.

4   The Malt Shovel

Tel: (01924) 875521

0.4 miles east along the A650, on the left. (Bass) Open all day. Lunch; 12.00-14.00hrs daily, Evening Meals [Fri only]; 17.30-19.30hrs.

5   The Poplars

Tel: (01924) 375682

0.8 miles east along the road to Wrenthorpe, on the left.

PLACES OF INTEREST

Wakefield

Follow the A 650 south (Signposted 3.1 Miles)

For details please see Junction 40 information

Within the city can be found ...

The Cathedral Church of All Saints

Northgate, Wakefield WF1 1HG

For details please see Junction 40 information

The Chantry Chapel of St Mary on Wakefield Bridge

For details please see Junction 40 information

Sandal Castle

Off Manygates Lane, Sandal, Wakefield. Follow the A650 south into Wakefield and continue south along the A61.(Signposted along A61, 4.8 Miles)

For details please see Junction 40 information

M1   JUNCTION 42

THIS JUNCTION IS A MOTORWAY INTERCHANGE ONLY AND THERE IS NO ACCESS TO ANY FACILITIES

NEAREST NORTHBOUND A&E HOSPITAL

Leeds General Infirmary, Great George Street, Leeds LS1 3EX Tel: (0113) 243 2799

Proceed north to Junction 7 (M621) and take the A61 north into Leeds city centre. The hospital is signposted in the city. (Distance Approx 6.8 miles)

NEAREST SOUTHBOUND A&E HOSPITAL

Pinderfield Hospital, Aberford Road, Wakefield WF1 4DG Tel: (01924) 201688

Proceed south to Junction 41, take the A650 exit east and continue along the A61. At the junction of the A642 and A61 turn left along the A642 and the hospital is along this road. (Distance Approx 4.6 miles)

M1   JUNCTION 43

THIS JUNCTION IS A MOTORWAY INTERCHANGE ONLY AND THERE IS NO ACCESS TO ANY FACILITIES

NEAREST NORTHBOUND A&E HOSPITAL

Leeds General Infirmary, Great George Street, Leeds LS1 3EX Tel: (0113) 243 2799

Proceed north to Junction 7 (M621) and take the A61 north into Leeds city centre. The hospital is signposted in the city. (Distance Approx 4.2 miles)

NEAREST SOUTHBOUND A&E HOSPITAL

Pinderfield Hospital, Aberford Road, Wakefield WF1 4DG Tel: (01924) 201688

Proceed south to Junction 41, take the A650 exit east and continue along the A61. At the junction of the A642 and A61 turn left along the A642 and the hospital is along this road. (Distance Approx 6.4 miles)

M1   JUNCTION 44

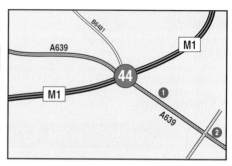

NEAREST NORTHBOUND A&E HOSPITAL

Leeds General Infirmary, Great George Street, Leeds LS1 3EX Tel: (0113) 243 2799

Follow the B6481 north and continue along the A61 into Leeds city centre.The hospital is signposted within the city. (Distance Approx 4.1 miles)

NEAREST SOUTHBOUND A&E HOSPITAL

Pinderfield Hospital, Aberford Road, Wakefield WF1 4DG Tel: (01924) 201688

Proceed south to Junction 41, take the A650 exit east and continue along the A61. At the junction of the A642 and A61 turn left along the A642 and the hospital is along this road. (Distance Approx 7.6 miles)

FACILITIES

1   John O'Gaunts

Tel: (0113) 282 2243

0.6 miles south along the A639, on the left. (Whitbread) Lunch; Mon-Fri; 11.30-14.30hrs, Evening Meals; Mon-Sat; 17.30-22.00hrs. Meals served all day on Sundays; 12.00-21.00hrs.

2   Rothwell BP Petrol Station

Tel: (0113) 282 1489

1 mile south along the A639, on the left. Access, Visa, Overdrive, All Star, Switch, Dial Card, Mastercard, Amex, Diners Club, Delta, BP Cards, Shell Agency, Mobil Cards.

 Large Hotel   Small Hotel   Motel   Guest House/ Bed & Breakfast   Disabled Facilities   Childrens Facilities

PLACES OF INTEREST

Leeds

Follow the B6481 north and continue along the A61.(Signposted 4 Miles)

For details please see Junction 27 (M62) information

Within the city centre ...

Royal Armouries

Armouries Drive, Leeds LS10 1LT

For details please see Junction 27 (M62) information

Thackray's Medical Museum

Beckett Street, Leeds LS9 7LN
Follow the B6481 north, continue along the A61 and then the A58 (3.8Miles)

For details please see Junction 27 (M62) information

M1   JUNCTION 45

THIS JUNCTION IS NOT CURRENTLY IN SERVICE AND HAS BEEN BUILT TO FACILITATE ACCESS TO THE EAST LEEDS RADIAL ROUTE WHICH HAS BEEN PROJECTED FOR CONSTRUCTION BY DECEMBER 2002

NEAREST NORTHBOUND A&E HOSPITAL

Leeds General Infirmary, Great George Street, Leeds LS1 3EX Tel: (0113) 243 2799

Proceed north to Junction 46, take the A63 exit north and continue along the A64 and M64 into the city centre. The hospital is signposted within the city. (Distance Approx 8.6 miles)

NEAREST SOUTHBOUND A&E HOSPITAL

Pinderfield Hospital, Aberford Road, Wakefield WF1 4DG Tel: (01924) 201688

Proceed south to Junction 41, take the A650 exit east and continue along the A61. At the junction of the A642 and A61 turn left along the A642 and the hospital is along this road. (Distance Approx 8.9 miles)

M1   JUNCTION 46

IT IS PROPOSED TO ENLARGE THIS JUNCTION TO INCORPORATE A GYRATORY ROAD SYSTEM

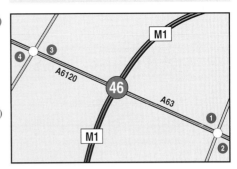

NEAREST A&E HOSPITAL

Leeds General Infirmary, Great George Street, Leeds LS1 3EX Tel: (0113) 243 2799

Take the A6120 exit north and continue along the A64 and M64 into the city centre. The hospital is signposted within the city. (Distance Approx 6 miles)

FACILITIES

1   The Old George

Tel: (0113) 286 2100
1 mile south along the A63, on the left. (Beefeater) Meals served 12.00-22.30hrs daily.

2   Hilton National, Garforth

Tel: (0113) 286 6556
1 mile south along the A63, on the right. Breakfast; Mon-Fri; 07.00-09.30hrs, Sat & Sun; 08.00-10.00hrs, Lunch; Sun-Fri; 12.30-14.00hrs, Dinner; 19.00-22.00hrs daily. Bar meals are available 11.00-23.00hrs daily.

3   Austhorpe Filling Station (Texaco)

Tel: (0113) 284 0100
0.2 miles north along the A6120, on the right. Access, Visa, Overdrive, All Star, Switch, Dial Card, Mastercard, Amex, Diners Club, Delta, BP Chargecard, Fast Fuel, Texaco Cards.

 Pets Welcome   Cash Dispenser   24 Hour Facilities   Toilets   Alcoholic Drinks   Food and Drink   Fuel

M1

4   Sainsburys Petrol Station

🏪 24   Tel: (0113) 232 8154

0.3 miles north along the A6120, on the left. Access, Visa, Overdrive, All Star, Switch, Dial Card, Mastercard, Amex, Delta, AA Paytrak. Toilets available in adjacent store during store opening hours.

PLACES OF INTEREST

Harewood House

Harewood, Leeds LS17 9LQ Tel: (0113) 288 6331
website; www.harewood.org
e-mail; business@harewood.org
Follow the A63 north, continue along the A6120
and turn north along the A61(Signposted 10 miles)

Designed by John Carr of York in the neo-classical style and completed in 1772, Harewood House is one of the great treasure houses of England with interiors and plaster work ceilings by Robert Adam and State Rooms furnished by Thomas Chippendale. Outstanding art collections include works by JMW Turner, Girtin, Reynolds, Gainsborough and Picasso, whilst The Gallery and the China Room display, amongst other things, examples of fine porcelain. The magnificent gardens were designed in the 1840's by Sir Charles Barry and the Terrace contains an Italian-style garden with ornate fountains. There are extensive walks around the grounds, boat trips around the lake, a Bird Garden and a children's adventure play ground. Cafe. Gift Shops. Disabled access.

Temple Newsam House

Leeds LS15 0AE
Tel: (0113) 264 7321 or (0113) 264 1358.
Follow the A63 north (Signposted 2.1 miles)

Dubbed the "Hampton Court of the North", Temple Newsam is a magnificent Tudor-Jacobean country house set in 1200 acres of parkland with 30 rooms recently restored to their original splendour and representing many of the different styles employed during its existence. The Home Farm has the largest collection of rare breeds in the country. Cafe.

M1   JUNCTION 47

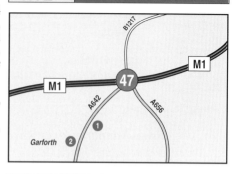

NEAREST A&E HOSPITAL

Leeds General Infirmary, Great George Street, Leeds LS1 3EX Tel: (0113) 243 2799

Proceed south to Junction 46, take the A6120 exit north and continue along the A64 and M64 into the city centre. The hospital is signposted within the city. (Distance Approx 8.9 miles)

FACILITIES

1   Aagrar Restaurant

🍴 🍴 🔌 ♿   Tel: (0113) 287 6606

0.6 miles south along the A642, on the left. Open; Sun-Thurs; 18.00-23.30hrs, Sat; 17.30-23.30hrs

2   Toll Bar Garage (UK Fuel)

🏪 ♿ 🚻   Tel: (0113) 286 2926

1 mile south along the A642, in Garforth, on the right. Access, Visa, Mastercard, Amex, Delta. Open; Mon-Fri; 08.00-18.00hrs, Sat; 08.00-13.00hrs, Sun; Closed.

PLACES OF INTEREST

Lotherton Hall

Aberford, Yorkshire LS25 3EB
Tel: (0113) 281 3259
Signposted from junction. (2.6 miles)

A small Edwardian country house with period gardens, housing the Gascoigne Collection and displays of fashions up to recent times. The

| H Large Hotel | h Small Hotel | M Motel | B Guest House/ Bed & Breakfast | Disabled Facilities | Childrens Facilities |

Gardens, created by Mrs Gwendolen Gascoigne, retain their Edwardian character, whilst the Bird Garden is home to many rare and endangered species. Cafe.

M1   JUNCTION 48

NEAREST A&E HOSPITAL

Leeds General Infirmary, Great George Street, Leeds LS1 3EX Tel: (0113) 243 2799

Proceed south to Junction 46, take the A6120 exit north and continue along the A64 and M64 into the city centre. The hospital is signposted within the city. (Distance Approx 10.2 miles)

MOTORWAY ENDS

(TOTAL LENGTH OF MOTORWAY 191.1 MILES)

 Pets Welcome   Cash Dispenser   24 Hour Facilities   Toilets   Alcoholic Drinks   Food and Drink   Fuel

M1

OFF THE MOTORWAY

M4

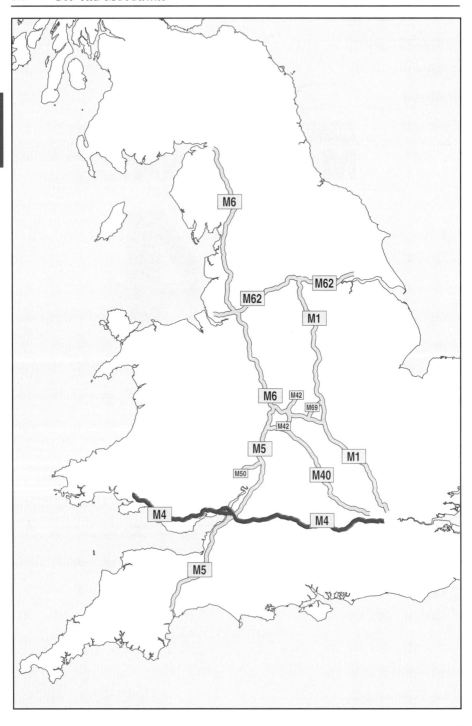

The M4 (& M48) Motorway

Linking London and South West Wales, the motorway commences in Chiswick and heads due west. To the south, planes can be seen circling whilst waiting to land at Heathrow Airport and, after crossing the M25, Windsor Castle dominates the skyline as it overlooks the town. Slough and Maidenhead are passed before Reading, the County Town of Berkshire, is reached and here the modern lines of the Madejski Stadium, which opened as recently as August 1998 can be seen to the north.

The motorway continues westwards through Wiltshire, passes the port city of Bristol, to the south and links up with the M5 before reaching the east bank of the River Severn. The river is crossed in spectacular fashion by one of two bridges; the first, a suspension bridge which opened in 1961 and is now on the M48 Loop, and the later Second Severn Crossing a cable stayed bridge which opened in 1996.

Proceeding westwards, the important crossing point of Newport and Cardiff, once the richest coal-exporting city in the world, are passed to the south and, at Tongwynlais, the fairy tale Castell Coch which was built for the 3rd Marquess of Bute can be seen nestling in the trees on the north side. Bridgend is passed to the south before the motorway turns north and follows the coastline, slipping between the steel town of Port Talbot to the west and Margam Country Park to the east before crossing the River Neath.

The M4 skirts around the north side of Swansea before heading north west to terminate at Pont Abraham, Junction 49, where it forms an end-on junction with the A48 to Carmarthen and Fishguard and the A483 to Llandovery and Central Wales.

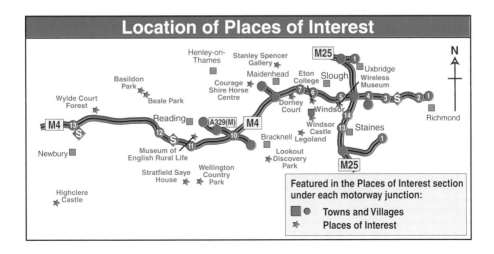

Location of Places of Interest

Featured in the Places of Interest section under each motorway junction:

■ ● Towns and Villages
✷ Places of Interest

M4

Location of Places of Interest

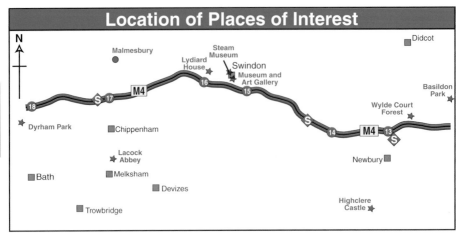

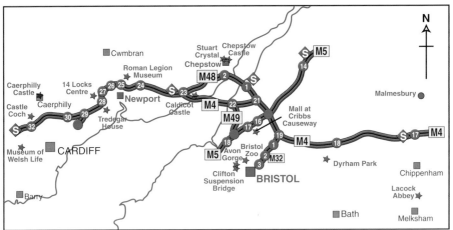

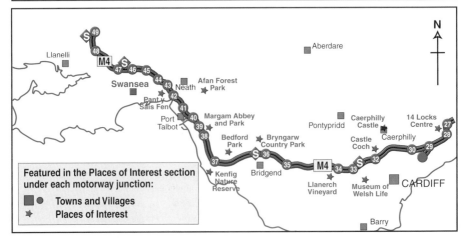

Featured in the Places of Interest section under each motorway junction:

■ ● Towns and Villages

✳ Places of Interest

M4 | JUNCTION 2

THIS IS AN URBAN MOTORWAY AND A VARIETY OF FACILITIES ARE WITHIN EASY REACH OF THIS JUNCTION

NEAREST A&E HOSPITAL

The Ealing Hospital, Uxbridge Road, Southall UB1 3HW Tel: (020) 8574 2444

Follow the A4 west, turn right along the A3002 and continue along the A4020. The hospital is on the left. (Distance Approx 3.6 miles)

SERVICE AREA

BETWEEN JUNCTIONS 2 AND 3

Heston Services (Westbound) (Granada)

Tel: (020) 8580 2000 Fresh Express Self Service Restaurant, Burger King, Travelodge & BP Fuel

Heston Services (Eastbound) (Granada)

Tel: (020) 8580 2000 Little Chef, Burger King, Travelodge & BP Fuel

M4 | JUNCTION 3

THIS IS AN URBAN MOTORWAY AND A VARIETY OF FACILITIES ARE WITHIN EASY REACH OF THIS JUNCTION

NEAREST A&E HOSPITAL

The Ealing Hospital, Uxbridge Road, Southall UB1 3HW Tel: (020) 8574 2444

Follow the A312 north and turn right along the A4020. The hospital is on the right. (Distance Approx 4.1 miles)

M4 | JUNCTION 4

THIS IS AN URBAN MOTORWAY AND A VARIETY OF FACILITIES ARE WITHIN EASY REACH OF THIS JUNCTION

NEAREST A&E HOSPITAL

Hillingdon Hospital, Pield Heath Road, Hillingdon UB8 3NN Tel: (01895) 238282

Follow the A408 north, turn right along the B465 and left along the A437. (Distance Approx 2.7 Miles)

M4 | JUNCTION 4B

THIS JUNCTION IS A MOTORWAY INTERCHANGE WITH THE M25 ONLY AND THERE IS NO ACCESS TO ANY FACILITIES

NEAREST A&E HOSPITAL

Wexham Park Hospital, Wexham Road, Slough SL2 4HL Tel: (01753) 633000

Proceed west to Junction 5 and take the A4 exit north. Turn right along the A412 and the hospital is signposted. (Distance Approx 6.4 Miles)

M4 | JUNCTION 5

NEAREST A&E HOSPITAL

Wexham Park Hospital, Wexham Road, Slough SL2 4HL Tel: (01753) 633000

Take the A4 exit north, turn right along the A412 and the hospital is signposted. (Distance Approx 4.3 Miles)

FACILITIES

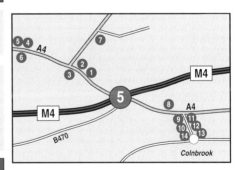

1   Toby Carvery, Langley

Tel: (01753) 591212
0.2 miles north along the A4, on the right. (Bass) Open all day. Bar Meals served; 12.00-22.00hrs daily. Carvery; Open; Mon-Fri; 12.00-14.00hrs & 17.00-22.00hrs, Sat & Sun; 12.00-22.00hrs.

| Pets Welcome | Cash Dispenser | 24 Hour Facilities | WC Toilets | Alcoholic Drinks | Food and Drink | Fuel |

M4

2 Toby Lodge, Langley

🛏B ♿ Tel: (01753) 444123

0.2 miles north along the A4, on the right. (Bass)

3 The Montagu Arms

🍴 🍽 ♿ 🧒 Tel: (01753) 543009

0.3 miles north along the A4, on the left. (Harvester) Open all day. Meals served; Sun-Thurs; 12.00-22.00hrs, Fri & Sat; 12.00-23.00hrs

4 St Francis House

🛏B Tel: (01753) 578888

1 mile north along the A4, on the right

5 Slough East Service Station (Total Fina)

⛽ ♿ WC 🍽 Tel: (01753) 574041

1 mile north along the A4, on the right. Keyfuels, Access, Visa, Overdrive, All Star, Switch, Dial Card, Mastercard, Amex, AA Paytrak, Diners Club, Delta, BP Supercharge, Total Fina Cards. Open; 06.00-23.00hrs daily. Wimpy Hot Snacks available; 09.00-22.00hrs daily.

6 Upton Court Service Station (Esso)

⛽ WC 24 Tel: (01753) 671679

1 mile north along the A4, on the left. Access, Visa, Overdrive, All Star, Switch, Dial Card, Mastercard, Amex, Diners Club, Delta, Shell Gold, Esso Cards.

7 BP Langley

⛽ ♿ WC 🍽 £ Tel: (01753) 583058

0.7 miles east along Parlaunt Road, on the right. Access, Visa, Overdrive, All Star, Switch, Dial Card, Mastercard, Amex, AA Paytrak, Diners Club, Delta, Routex, BP Cards, Shell Agency. Open; Mon-Fri; 06.00-23.00hrs, Sat & Sun; 07.00-23.00hrs. Delice de France Hot Snacks available; Mon-Fri; 06.00-17.00hrs, Sat & Sun; 07.00-17.00hrs

8 Brands Hill Filling Station (BP)

⛽ ♿ WC 🍽 24 £ Tel: (01753) 545531

0.3 miles south along the A4, on the left.

Access, Visa, Overdrive, All Star, Switch, Dial Card, Mastercard, Amex, AA Paytrak, Diners Club, Delta, Routex, BP Cards, Shell Agency. Delice de France Hot Snacks available; Mon-Fri; 06.00-17.00hrs, Sat & Sun; 07.00-17.00hrs

9 Quality Hotel Heathrow

🛏H 🍽 🍴 Tel: (01753) 684001

0.4 miles south along the A4, on the right. The Windsor Brasserie; Breakfast; Mon-Fri; 06.30-09.30hrs, Sat; 07.00-09.30hrs, Sun; 08.00-10.00hrs, Dinner; 19.00-22.00hrs daily. Bar Meals served 10.30-22.30hrs daily

10 Brands Hill Lodge

🛏B Tel: (01753) 680377

0.6 miles south along the Colnbrook Road, on the right

11 Regent House

🛏B ♿ 🧒 Tel: (01753) 683093

0.6 miles south along the Colnbrook Road, on the left

12 Gibtel Cafe

🍴 ♿ Tel: (01753) 683093

0.6 miles south along the Colnbrook Road, on the left. Open; Mon-Sat; 06.30-15.00hrs & 19.00-22.00hrs

13 The Crown

🍴 🍽 ♿ 🧒 🛏B Tel: (01753) 682026

0.8 miles south along the Colnbrook Road, on the left. (Punch Taverns) Meals served; Mon-Sat; 12.00-15.00hrs & 19.00-21.00hrs

14 Golden Cross Service Station (BP)

⛽ Tel: (01753) 686321

1 mile south along the Colnbrook Road, on the right. Access, Visa, Overdrive, All Star, Switch, Dial Card, Mastercard, Amex, AA Paytrak, Diners Club, Delta, Routex, BP Cards, Shell Agency. Open; Mon-Sat; 06.00-21.00hrs, Sun; 08.00-21.00hrs.

PLACES OF INTEREST

Museum of Ancient Wireless & Historic Gramophone Apparatus

Tel: (01753) 542242 Follow the B470 west (2 Miles) Open by Appointment Only.

Located in Datchet, this small private museum features a unique and fascinating collection of vintage wireless, gramophone, phonograph, telephone, wire and tape-recording apparatus. The enthusiastic innovator Capt. Maurice Seddon, also renowned as the inventor of medical heated clothing, is on hand to share his considerable expertise and opinions (in any one of five languages) with visitors. The extensive collection includes a large stock of redundant, but serviceable, spare parts many of which are available for sale.

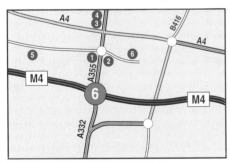

M4 JUNCTION 6

NEAREST A&E HOSPITAL

Wexham Park Hospital, Wexham Road, Slough SL2 4HL Tel: (01753) 633000

Follow the A355 north, turn right along the A4 and left along the B416. The hospital is signposted. (Distance Approx 3.8 Miles)

FACILITIES

1 Copthorne Slough/Windsor Hotel

Tel: (01753) 516222
0.3 miles north along the A355, on the left. The Verandah Restaurant; Breakfast; Mon-Fri;

06.30-10.00hrs, Sat & Sun; 07.30-11.00hrs, Lunch; Mon-Fri; 12.00-14.00hrs, Dinner; 18.30-22.00hrs daily. Bar Meals served 10.00-22.00hrs daily.

2 Courtyard Slough/Windsor Marriott Hotel

Tel: (01753) 551551
0.3 miles north along the A355, on the right. The No.1 Restaurant; Breakfast; Mon-Fri; 07.00-09.30hrs, Sat & Sun; 07.30-11.00hrs, Lunch; Mon-Fri; 12.30-14.00hrs, Dinner; Mon-Sat; 19.00-22.00hrs, Sun; 19.00-21.30hrs. Bar Meals served 10.00-22.00hrs daily.

3 The Three Tuns

Tel: (01753) 521911
0.7 miles north along the A355, on the left. (Scottish & Newcastle) Open all day. Meals served; 11.00-21.30hrs daily

4 Farnham Road Service Station (Elf)

Tel: (01753) 534848
0.9 miles north along the A355, on the left. Access, Visa, Overdrive, All Star, Switch, Dial Card, Mastercard, Amex, Diners Club, Delta, Elf Cards, Total Fina Cards.

5 The Earl of Cornwall

Tel: (01753) 578333
1 mile west along Cippenham Lane, on the left. (Whitbread) Open all day. Meals served; 12.00-14.00hrs daily. Prior booking required for evening meals

6 The Foresters Arms

Tel: (01753) 643340
0.7 miles east along Church Street on the left. (Courage) Open all day Sat & Sun; Meals served; Mon-Sat; 12.00-14.30hrs & 18.30-22.00hrs, Sun; 12.30-15.00hrs & 19.00-21.30hrs

PLACES OF INTEREST

Eton College & Museum of Eton Life

High Street, Eton SL4 6DW Tel: (01753) 671177
website; http://www.etoncollege.com
e-mail; visits@etoncollege.org.uk.
Follow the A332 south. (Signposted 1.4 Miles)

Founded in 1440 by Henry VI, Eton College is one of the oldest schools in the country. Although the architecture is mainly of the 16thC with additions in 1889, earlier buildings such as the Lower School and Long Chamber of 1500 and the chapel, constructed between 1476 and 1482, are on view. The library, which stands on the site of the south cloisters and was erected in 1730, contains many rare manuscripts, including the original of Gray's Elegy. Still a bastion of privilege and a foundation stone of the British Establishment many political and commercial leaders can be counted amongst its former pupils.

Windsor Castle

Windsor SL4 1NJ Visitor Office
Tel: (01753) 868286
website; http://www.royal.gov.uk.
Follow the A332 south (Signposted 2 Miles)

Following the Norman Conquest of England in 1066 a strategic ring of defences was established around London, each of one days march apart and one days march from the Tower of London, at the centre. A stronghold of earth and timber was constructed on a steep chalk hill overlooking the River Thames at Windsor and it was quickly rebuilt in stone to a plan much as it is seen today. Improved and enlarged over the ages it was following the restoration of the monarchy and under Charles II that the role of the castle changed from that of a fortification to a royal palace. The state apartments contain some of the finest works of art, armour, pictures and interiors in the world and it was part of this section that was destroyed by a fire in November 1992. Ironically, it was this fire that signalled a shift in the relationship between the Royal family and the public. The Government's immediate response to underwrite all the costs of repairs was met with a nationwide disbelief as to why taxpayers should have to bail out one of the world's richest women, and this feeling was endorsed when a national appeal raised very little. Today, partially as a result of the proceeds raised from the opening of Buckingham Palace to visitors, the £37m restoration has been completed.

Legoland Windsor

Winkfield Road, Windsor SL4 4AY Tel: 0990 040404
website; www.legoland.co.uk.
Follow the A332 south (Signposted 3.3 Miles)

An 150 acre major children's adventure park packed with exciting rides, hands-on features, displays and exhibitions all with the Lego theme and designed to enhance creativity, fun, development, play and learning. Restaurants. Gift Shops. Disabled access.

M4   JUNCTION 7

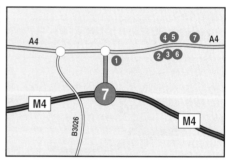

NEAREST A&E HOSPITAL

Wexham Park Hospital, Wexham Road, Slough SL2 4HL Tel: (01753) 633000

Proceed east to Junction 6, follow the A355 north, turn right along the A4 and left along the B416. The hospital is signposted. (Distance Approx 5.5 Miles)

FACILITIES

1   The Huntercombe

Tel: (01628) 663177
Adjacent to south east side of roundabout. (Unique Pub Co) Open all day. Meals served; Sun; 12.00-16.00hrs. Bar snacks served all day daily.

2   KFC

🍴 ♿   Tel: (01628) 603020

0.4 miles east along the A4, on the right. Open: Sun-Thurs; 11.30-0.00hrs, Fri & Sat; 11.30-01.00hrs

3   Slough West Service Station (Esso)

🛢 WC 24 £   Tel: (01628) 666877

0.5 miles east along the A4, on the right. Access, Visa, Overdrive, All Star, Switch, Dial Card, Mastercard, Amex, AA Paytrak, Diners Club, Delta, BP Supercharge, Shell Gold, Esso Cards, Style Card.

4   PMA Service Station (Total)

🛢 ♿ WC 24   Tel: (01628) 662013

0.6 miles east along the A4, on the left. Access, Visa, Overdrive, All Star, Switch, Dial Card, Mastercard, Amex, AA Paytrak, Diners Club, Delta, Total Fina Cards.

5   Shell Cippenham

🛢 ♿ WC   Tel: (01628) 607900

0.7 miles east along the A4, on the left. Access, Visa, Overdrive, All Star, Switch, Dial Card, Mastercard, Amex, Diners Club, Delta, BP Supercharge, Shell Cards, Smartcard.

6   The White Horse

🍴 🍴 ♿ 🐾   Tel: (01628) 603666

0.7 miles east along the A4, on the right. (Bass) Open all day. Meals served; 12.00-22.00hrs daily.

7   McDonald's

🍴 ♿   Tel: (01753) 531816

0.8 miles east along the A4, on the left. Open; 07.00-23.00hrs daily

PLACES OF INTEREST

Dorney Court

Dorney, Nr Windsor SL4 6QP Tel: (01628) 604638
Follow the A4 west and turn left along the B3026
(1.4 Miles)

Built in c1440, Dorney Court is one of the finest Tudor Manor Houses in England. The rooms contain 15th & 16thC oak, beautiful 17thC lacquer furniture and 400 years of history of the family who still live here. The very first pineapple to be raised in England was grown here and presented to Charles II in 1661. Visitors can enjoy lunch or afternoon tea in the walled garden at the Plant Centre and, during June to September, there is the opportunity to pick-your-own fruit and vegetables.

M4   JUNCTION 8/9

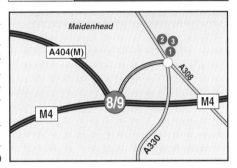

NEAREST EASTBOUND A&E HOSPITAL

Wexham Park Hospital, Wexham Road, Slough SL2 4HL Tel: (01753) 633000

Proceed to Junction 6, follow the A355 north, turn right along the A4 and left along the B416. The hospital is signposted. (Distance Approx 8.7 Miles)

NEAREST WESTBOUND A&E HOSPITAL

Royal Berkshire Hospital, London Road, Reading RG1 5AN Tel: (0118) 987 5111

Proceed to Junction 10, follow the A329(M) north and turn left along London Road (Signposted from end of A329[M]. Distance Approx 12.5 Miles)

FACILITIES

1   Brayswick Service Station (Esso)

🛢 ♿ WC 24 £   Tel: (01628) 623315

0.8 miles north along the A308, on the right. Access, Visa, Overdrive, All Star, Switch, Dial Card, Mastercard, Amex, Diners Club, Delta, Shell Gold, Esso Cards.

| Pets Welcome | £ Cash Dispenser | 24 Hour Facilities | WC Toilets | Alcoholic Drinks | Food and Drink | Fuel |

M4

2 The Stag at Bray

🏨 🍴 ♿ ⚓   Tel: (01628) 624100

0.8 miles north along the A308, on the right. (Whitbread) Open all day on Sun. Meals served; Mon-Sat; 12.00-14.30hrs & 19.00-22.00hrs, Sun; 12.00-22.00hrs.

3 The Old Coach House

📮B   Tel: (01628) 671244

0.8 miles north along the A308, on the right

PLACES OF INTEREST

Stanley Spencer Gallery

King's Hall, High Street, Cookham on Thames, Berkshire SL6 9SJ Tel: (01628) 520890
Follow the A308(M) and A308 north, turn right along the A4 and left along the B4442 to Cookham. (Signposted in Cookham area, 5 Miles)

The gallery is exclusively devoted to the works of Sir Stanley Spencer, who died in 1959, and is unique in Britain inasmuch as it is in the village of his birthplace. Located in the former Victorian Methodist Chapel that he used to attend as a child, there are also displays, memorabilia and artefacts devoted to his life. The gallery contains a permanent collection of his work and over a thousand pictures have been shown since it opened in 1962. Sales Desk. Disabled access.

Courage Shire Horse Centre

Cherry Garden Lane, Maidenhead Thicket, Maidenhead SL6 3QD Tel: (01628) 824848
Follow the A404(M) north to Junction 9b and take the A4 west. (2.5 Miles)

A small centre devoted to the Shire horse and the drays that were hauled by them. A number of horses are on view and there is a display of old agricultural implements and drays, a childrens play area and a small farm animals section. Picnic area. Tea Room. Disabled access.

M4   JUNCTION 10

THIS JUNCTION IS A MOTORWAY INTERCHANGE WITH THE A329(M) ONLY AND THERE IS NO ACCESS TO ANY FACILITIES

NEAREST A&E HOSPITAL

Royal Berkshire Hospital, London Road, Reading RG1 5AN Tel: (0118) 9875111

Follow the A329(M) north and turn left along London Road (Signposted from end of A329[M]. Distance Approx 5.2 Miles)

PLACES OF INTEREST

Look Out Discovery Park

Nine Mile Ride, Bracknell RG12 7QW
Tel: (01344) 354400 Follow the A329(M) south (Signposted from motorway 5.9 Miles)

A hands-on science and nature exhibition featuring over 70 exhibits, including a hot air balloon. Set within 2600 acres of woodland there are nature walks and mountain bike trails, with bikes available for hire. There is also a children's adventure playground, Coffee Shop and Gift Shop.

M4   JUNCTION 11

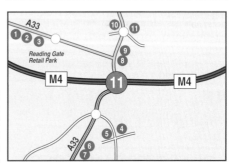

NEAREST A&E HOSPITAL

Royal Berkshire Hospital, London Road, Reading RG1 5AN Tel: (0118) 987 5111

Take the A33 north and turn right along the A4. (Distance Approx 2.9 Miles)

FACILITIES

1 Pizza Hut

[icons] Tel: (0118) 945 1100
0.8 miles north along the A33, on the left in Reading Gate Retail Park. Open; 12.00-23.00hrs daily

2 McDonald's

[icons] Tel: (0118) 975 2343
0.8 miles north along the A33, on the left in Reading Gate Retail Park. Open; 08.00-23.00hrs daily

3 KFC

[icons] Tel: (0118) 957 3541
0.8 miles north along the A33, on the left in Reading Gate Retail Park. Open; Sun-Thurs; 11.30-0.00hrs, Fri & Sat; 11.30-01.00hrs

4 The Swan

[icons] Tel: (0118) 988 3674
0.5 miles south along the Basingstoke Road, on the left in Three Mile Cross. (Freehouse) Open all day. Meals served; 12.00-15.00hrs & 18.00-22.00hrs daily

5 Roman Road Service Station (BP)

[icons] Tel: (0118) 988 2334
0.5 miles south along the Basingstoke Road, on the right in Three Mile Cross. Access, Visa, Overdrive, All Star, Switch, Dial Card, Mastercard, Amex, AA Paytrak, Diners Club, Delta, Routex, BP Cards, Shell Agency. Open; 07.00-22.00hrs daily

6 M's Family Diner

[icons] Tel: (0118) 988 4026
1 mile south along the A33, on the left. Open; Mon-Thurs; 07.00-18.00hrs, Fri-Sun; 07.00-22.00hrs

7 Swallowfields Service Station (Repsol)

[icons] Tel: (0118) 988 6065
1 mile south along the A33, on the left. Keyfuels, Access, Visa, Overdrive, All Star, Switch, Dial Card, Mastercard, Amex, AA Paytrak, Diners Club, Delta, BP Supercharge, Repsol Card. Toilets available between 07.00-20.00hrs only.

8 Little Chef

[icons] Tel: (0118) 931 3465
0.4 miles north along the B3031, on the right. Open; 07.00-22.00hrs daily

9 Shell Fairfield

[icons] Tel: (0118) 922 4500
0.5 miles north along the B3031, on the right. Access, Visa, Overdrive, All Star, Switch, Dial Card, Mastercard, Amex, Diners Club, Delta, BP Agency, Shell Cards, Smartcard.

10 Posthouse Reading

[icons] Tel: 0870 400 9067
0.8 miles north along the B3031, on the left. Restaurant Open; Breakfast; Mon-Fri; 07.00-09.30 hrs, Sat & Sun; 07.30-10.00hrs, Lunch; Mon-Fri; 12.30-14.30hrs, Sun; 12.30-15.00hrs, Dinner; Mon-Sat; 18.30-22.30hrs, Sun; 19.00-22.00hrs.

11 Whitley Wood Garage (BP)

[icons] Tel: (0118) 987 1278
0.9 miles north along the B3031, on the right. Access, Visa, Overdrive, All Star, Switch, Dial Card, Mastercard, Amex, AA Paytrak, Diners Club, Delta, Routex, BP Cards, Shell Agency

PLACES OF INTEREST

Museum of English Rural Life

University of Reading, Whiteknights, Reading RG6 6AG Tel: (0118) 931 8663
website; http://www.reading.ac.uk/Instits/im/
Follow the B3350 east and turn left along the A327 (2.6 Miles)

Established in 1951 by Reading University as a centre of information and research on all aspects of country living, the museum contains farm implements, tools and domestic

| Pets Welcome | Cash Dispenser | 24 Hour Facilities | Toilets | Alcoholic Drinks | Food and Drink | Fuel |

equipment. There is also a records resource including manuscripts, photographs, prints and drawings.

Wellington Country Park

Riseley, Nr Reading RG7 1SP Tel: (0118) 932 6444
Follow the A33 south (Signposted 4.7 Miles)

A 350 acre park surrounding a 35 acre lake and containing many activities both for adults and children. There are several nature trails, a deer park, an animal farm, an adventure playground, a miniature railway and a crazy golf course. Activities at the lake include coarse fishing, from the banks, and boating and there are picnic areas and barbecue sites. Gift shop. Cafe. Some disabled access.

Stratfield Saye House

Stratfield Saye, Reading RG7 2BZ
Tel: (01256) 882882
Follow the A33 south (Signposted 7.7 Miles)

Presented to the Duke of Wellington in 1817, this elegant and stylish house has been the family home ever since. Within the beautiful grounds, which include many rare trees, is the grave of Copenhagen, the Iron Duke's famous charger. The Wellington Exhibition graphically illustrates his life and the exhibits include his magnificent funeral carriage. Gift Shop. Tea Room.

SERVICE AREA

BETWEEN JUNCTIONS 11 AND 12

Reading Services (Westbound) (Granada)
Tel: (0118) 956 6966 Fresh Express Self Service Restaurant, Burger King, Little Chef, Travelodge & BP Fuel

Reading Services (Eastbound) (Granada)
Tel: (0118) 956 6966 Fresh Express Self Service Restaurant, Burger King, Travelodge & BP Fuel

M4 JUNCTION 12

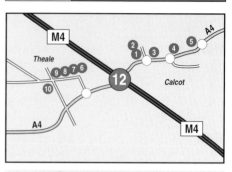

NEAREST A&E HOSPITAL

Royal Berkshire Hospital, London Road, Reading RG1 5AN. Tel: (0118) 987 5111

Follow the A4 east into Reading (Distance Approx 4.7 Miles)

FACILITIES

1 Pincent Manor Hotel

Tel: (0118) 932 3511
0.2 miles east along the A4, on the left. Restaurant Open; Breakfast; 07.00-10.00hrs, Lunch; 12.00-14.00hrs, Dinner; 19.00-21.30hrs daily. Bar Meals available; 10.00-22.00hrs daily.

2 McDonald's

Tel: (0118) 941 5744
0.2 miles east along the A4, on the left. Open; 07.00-23.00hrs daily

3 Sainsburys Petrol

Tel: (0118) 938 2200
0.2 miles east along the A4, on the left. Access, Visa, Overdrive, All Star, Switch, Dial Card, Mastercard, Amex, Delta, AA Paytrak, UK Fuelcard, Sainsburys Fuel Card

4 BP Express Calcot

Tel: (0118) 942 7912
0.8 miles east along the A4, on the left. Access, Visa, Overdrive, All Star, Switch, Dial Card, Mastercard, Amex, AA Paytrak, Diners Club,

Delta, Routex, BP Cards, Shell Agency. Open; 06.00-22.00hrs daily. Hot Food snacks available; Mon-Fri; 07.00-19.00hrs.

---

5 Calcot Hotel

Tel: (0118) 941 6423
0.9 miles east along the A4, on the left. Restaurant Open; Breakfast; Mon-Fri; 07.00-09.00hrs, Sat & Sun; 08.00-10.00hrs, Lunch; Sun-Fri; 12.00-14.00hrs, Dinner; 19.00-22.00hrs daily.

---

6 The Bull

Tel: (0118) 930 3478
0.7 miles on the right, in Theale village. (Wadworth) Open all day. Meals served; Mon-Sat; 12.00-15.00hrs. Prior booking required for evening meals

---

7 The Falcon

Tel: (0118) 930 2523
0.7 miles on the right, in Theale village. (Freehouse) Open all day. Meals served; Mon-Sat; 12.00-14.30hrs & 18.00-21.00hrs, Sun; 12.00-14.30hrs.

---

8 The Village Tea Room

Tel: (0118) 932 3433
0.8 miles on the right, in Theale village. Open; Mon-Sat; 07.30-16.30hrs.

---

9 The Crown Inn

Tel: (0118) 930 2310
0.9 miles on the right, in Theale village. (Whitbread) Open all day. Meals served; Mon-Fri; 12.00-14.00hrs.

---

10 The Red Lion

Tel: (0118) 930 2394
1 mile on the left, in Theale village. (Pubmaster) Open all day Tues-Sun. Meals served; Mon-Sat; 12.00-14.00hrs & 19.00-21.00hrs, Sun; 12.00-16.00hrs

---

PLACES OF INTEREST

Basildon Park (NT)

Lower Basildon, Reading RG8 9NR
Tel: (0118) 984 3040 Follow the A4 south, turn right along the A340 and continue along the A340
(Signposted "Beale Park" 7.8 Miles)

A classical Georgian 18thC house set in a beautiful garden and 400 acres of parkland, Basildon Park was built on a fortune made in the East Indies. The mansion later held the art collection of a wealthy industrialist but by the conclusion of World War II, it had been reduced to a near ruin after being a billet for British and American troops as well as a prisoner of war camp. In the 1950's it was rescued by Lord & Lady Iliffe who lovingly restored it before handing it over to the National Trust in 1978. It contains an unusual octagonal drawing room, fine plasterwork, Graham Sutherland's studies for Coventry Cathedral and a decorative shell room and there are woodland walks. Tea Room. NT Shop.

Just north of Basildon Park ...

Beale Park

Lower Basildon, Reading RG8 9NH
Tel: (0118) 984 5172

The park contains a diverse range of activities centred around an amazing collection of rare birds and animals. The attractions include a variety of activity and play areas for children, a deer park, children's fun fair, a model boat exhibition, picnic areas and gardens and walks. Shop. Cafe.

M4

M4 | **JUNCTION 13**

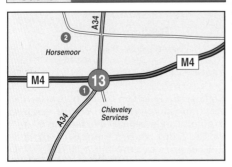

NEAREST EASTBOUND A&E HOSPITAL

Royal Berkshire Hospital, London Road, Reading RG1 5AN Tel: (0118) 987 5111

Proceed to Junction 12 and take the A4 east to Reading (Distance Approx 16.2 Miles)

NEAREST WESTBOUND A&E HOSPITAL

Princess Margaret Hospital, Okus Road, Swindon SN1 4JU Tel: (01793) 536231

Proceed to Junction 15 and follow the A419 north. (Signposted within the town. Distance Approx 24.3 Miles)

FACILITIES

SERVICE AREA

ACCESSED FROM THIS JUNCTION

Chieveley Services (Granada)

Tel: (01635) 248024 Fresh Express Self Service Restaurant, Burger King, Travelodge & BP Fuel

1   Stakis Newbury Hotel

Tel: (01635) 247010

0.1 miles south along the A34, on the right (2.3 miles detour required to gain access). The Saddlers Restaurant; Breakfast; Mon-Fri; 07.00-10.00hrs, Sat & Sun; 08.00-10.00hrs, Lunch; Sun-Fri; 12.30-14.00hrs, Dinner; Sun-Thurs; 19.00-22.00hrs, Fri & Sat; 18.30-22.00hrs.

2   Ye Olde Red Lion

Tel: (01635) 248379

0.9 miles north along the Chieveley Road, on the left, in Chieveley. (JC Taverns) Meals served; Mon-Sat; 11.15-14.30hrs & 18.00-21.30hrs, Sun; 12.00-15.00hrs & 19.00-21.30hrs.

PLACES OF INTEREST

Highclere Castle

Near Newbury, Berkshire RG20 9RN
24hr Information Line, Tel: (01635) 253204
website; www.highclerecastle.co.uk
e-mail; theoffice@highclerecastle.co.uk
Follow the A34 south (10.2 Miles)

Designed by Sir Charles Berry, Highclere Castle is considered to be the finest Victorian house still in existence. The castle is the family home of Lord & Lady Carnarvon and the ornate architecture is supplemented by fine period furniture and an art collection including works by Gainsborough, Reynolds and Van Dyck. Exhibitions include displays on the Tomb of Tutankhamun, discovered by the 5th Earl of Carnarvon and Howard Carter, and horse racing, the present Earl being engaged as the Racing Manager to the monarch. The magnificent parkland surrounding the castle is dominated by the Cedars of Lebanon and the sweeping vistas, inset with follies, present breathtaking views and relaxing walks. Gift Shop. Restaurant & Tea Rooms.

Wyld Court Rainforest

Hampstead Norreys, Thatcham. Berkshire
RG18 0TN Tel: (01635) 200221
e-mail; wyldcourt.rainforest@lineone.net
Follow the A34 north, take the first exit and turn right towards Hermitage and, at the end, turn left along the B4009 towards Hampstead Norris. (5.5 miles)

Experience the sheer beauty of this unique rainforest conservation project created under $20,000ft^2$ of glass and containing a stunning collection of dramatic and endangered rare plant species and rainforest creatures, including the tamarin monkeys, that thrive in tropical temperatures. Wyld Court Rainforest is the World Land Trust's UK Conservation and Education Centre and qualified staff are on hand to give guided tours. Picnic Area. Cafe. Gift Shop. Disabled access.

M4 — JUNCTION 14

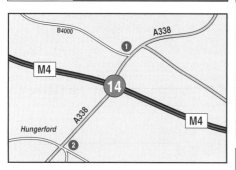

NEAREST EASTBOUND A&E HOSPITAL

Royal Berkshire Hospital, London Road, Reading RG1 5AN Tel: (0118) 987 5111

Proceed to Junction 12 and take the A4 east to Reading (Distance Approx 23.9 Miles)

NEAREST WESTBOUND A&E HOSPITAL

Princess Margaret Hospital, Okus Road, Swindon SN1 4JU Tel: (01793) 536231

Proceed to Junction 15 and follow the A419 north. (Signposted within the town. Distance Approx 16.6 Miles)

FACILITIES

1   The Pheasant Inn

Tel: (01488) 648284
0.4 miles north along the B4000, on the right. (Freehouse) Meals served; Mon-Sat; 12.00-14.15hrs & 18.30-21.15hrs, Sun; 12.00-14.15hrs & 19.00-21.15hrs.

2   Tally Ho!

Tel: (01488) 682312
0.9 miles south along the A338, on the left. (Wadworth) Open all day Sat & Sun. Meals served; Mon-Fri; 12.00-14.30hrs & 18.00-21.30hrs, Sat; 12.00-21.30hrs, Sun; 12.00-20.00hrs

SERVICE AREA

BETWEEN JUNCTIONS 14 AND 15

Membury Services (Westbound) (Welcome Break)

Tel: (01488) 72336 La Brioche Doree French Cafe, Granary Restaurant, Burger King, KFC, Red Hen Restaurant, Days Inn & BP Fuel

Membury Services (Eastbound) (Welcome Break)

Tel: (01488) 72336 Granary Restaurant, Burger King & BP Fuel

Footbridge connection between sites.

M4 — JUNCTION 15

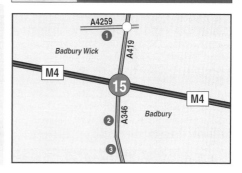

NEAREST A&E HOSPITAL

Princess Margaret Hospital, Okus Road, Swindon SN1 4JU Tel: (01793) 536231

Follow the A419 north. (Signposted within the town. Distance Approx 4.6 Miles)

FACILITIES

1   Kingsbridge House

Tel: (01793) 522861
1 mile north along the A4259, on the left

2   The Plough Inn

Tel: (01793) 740342
0.4 miles south along the A346, on the right. (Arkells) Meals served; Mon-Sat; 12.00-14.00hrs & 19.00-22.30hrs, Sun; 12.00-14.00hrs & 19.00-22.00hrs

Pets Welcome	Cash Dispenser	24 Hour Facilities	Toilets	Alcoholic Drinks	Food and Drink	Fuel	

3   Chiseldon Camp Service Station (Esso)

🍴 ♿ **WC** **24**   Tel: (01793) 740251

1 mile south along the A346, on the right. Access, Visa, Overdrive, All Star, Switch, Dial Card, Mastercard, Amex, AA Paytrak, Diners Club, Delta, BP Supercharge, Shell Gold, Esso Cards.

PLACES OF INTEREST

Swindon Museum & Art Gallery

Bath Road, Swindon SN1 4BA Tel: (01793) 466556
Follow the A419 north into Swindon (1.5 Miles)

The museum contains artefacts dating from Roman times and, amongst other exhibits, a comprehensive local studies section illustrating life in Swindon. The art gallery houses one of the most important collections of British 20thC Art. Established in 1944 the collection ranges from Sir George Clausen of 1896 to 1986 and Gillian Ayres, and embraces works by such major exponents as Lowry, Nicolson, Sutherland and Moore.

M4 JUNCTION 16

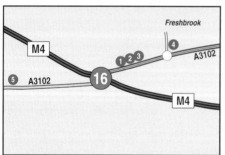

NEAREST A&E HOSPITAL

Princess Margaret Hospital, Okus Road, Swindon SN1 4JU Tel: (01793) 536231

Follow the A3102 east. (Signposted within the town. Distance Approx 2.8 Miles)

FACILITIES

1   Hilton National Swindon

🏨H 🍴 ♿ 🚆   Tel: (01793) 881777

0.1 miles north along the A3102, on the left. Minsky's Restaurant; Breakfast; 06.30-10.00hrs daily, Lunch; Sun-Fri; 12.30-14.00hrs, Dinner; 19.00-22.00hrs daily

2   The Lydiard

🍴 ♿ 🚆   Tel: (01793) 881490

0.1 miles north along the A3102, on the left. (Beefeater) Open all day. Restaurant Open; Breakfast; Mon-Fri; 07.00-09.00hrs, Sat & Sun; 08.00-10.00hrs, Lunch & Evening Meals; Sun-Thurs; 12.00-22.30hrs, Fri & Sat; 12.00-23.00hrs. Light refreshments available 09.00-12.00hrs and Bar Meals available 12.00-23.00hrs daily.

3   Travel Inn

🏨M ♿   Tel: (01793) 881490

0.1 miles north along the A3102, on the left

4   Save Petrol Station

🍴 ♿ **WC**   Tel: (01793) 881654

0.5 miles north along the A3102, on the left. Keyfuels, Access, Visa, Overdrive, All Star, Switch, Dial Card, Mastercard, Delta, Save Card. Open; 06.00-23.00hrs daily.

5   Sally Pusseys Inn

🍴 ♿   Tel: (01793) 852430

0.9 miles west along the A3102, on the right. (Arkells) Meals served; Mon-Sat; 12.00-14.00hrs & 18.00-21.30hrs, Sun; 12.00-14.30hrs & 19.00-21.00hrs

PLACES OF INTEREST

Lydiard House & Park

Hook Street, Lydiard Tregoze, Swindon SN5 9PA Tel: (01793) 770401 Follow the A3102 east and turn left along the B4534 (Signposted 1.7 Miles)

Lydiard Park is the delightful, yet little known,

home of the Viscounts Bolingbroke which was rescued from ruin by Swindon Corporation in 1943 and restored to its former glory. The elegant ground floor apartments contain ornate plasterwork and original family furnishings are preserved alongside portraits of the St John family who lived here from Elizabethan times. Other attractions include the Blue Dressing Room, devoted to the talented 18thC society artist Lady Diana Spencer, and a 17thC painted window by Abraham Van Linge. The Visitor Centre houses countryside displays and a cafe and there are extensive paths and trails through the parkland and gardens as well as adventure playgrounds for children. Gift Shop. Limited disabled access.

Steam-Museum of the Great Western Railway

Great Western Village, Swindon
Tel (01793) 466646 Follow the A3102 east
(Signposted "Great Western" 3.2 Miles)

Formerly sited in an old chapel in Faringdon Road, the Great Western Railway Museum is due to re-open at a new location in Spring 2000. The No.20 Shop in the former GWR Railway Works is being re-furbished and converted into a new museum facility to show the story of the works and the Great Western Railway and their place in railway history. It will feature static and interactive displays as well as famous locomotives, carriages and wagons.

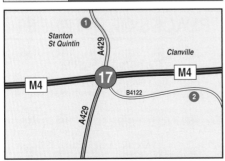

M4   JUNCTION 17

NEAREST EASTBOUND A&E HOSPITAL

Princess Margaret Hospital, Okus Road, Swindon SN1 4JU Tel: (01793) 536231

Proceed to Junction 16 and follow the A3102 east. (Signposted within the town. Distance Approx 15.2 Miles)

NEAREST WESTBOUND A&E HOSPITAL

**Frenchay Hospital, Frenchay Park Road, Frenchay, Bristol BS16 1L
Tel: (0117) 970 1212**

Proceed to Junction 19 and then follow the M32 to Junction 1. Take the A4174 east exit and the hospital is signposted. (Distance Approx 19.1 miles)

FACILITIES

1   Murco Petrol Station

🅿 ♿ WC 24 🍴 £   Tel: (01666) 837161
0.7 miles north along the A429, on the left.. Access, Visa, Overdrive, All Star, Switch, Dial Card, Mastercard, Amex, AA Paytrak, Delta, BP Supercharge, Murco Cards.

2   Silvey Truckstop Chippenham (UK Fuels)

🅿 🍴 🍴 WC 24   Tel: (01249) 750645
0.8 miles east along the B4122, on the right. Diesel Fuel Only. Access, Visa, Switch, Securicor Fuelserv, IDS, Keyfuels, AS24, Silvey Card.

PLACES OF INTEREST

Malmesbury

Malmesbury Tourist Information Centre, Town Hall, Market Lane, Malmesbury SN16 9BZ
Tel: (01666) 823748. Follow the A429 north
(Signposted 5.5 Miles)

England's oldest borough, the hill-top town is dominated by the impressive remains of the Norman Abbey. St Aldhelm, the first abbot, created a place of pilgrimage in the 7thC and Athelstan, the first King of England, chose Malmesbury as his capital and was buried here. In 1010 Elmer, a monk, attempted to fly, making himself a set of wings and launching himself 620ft from the top of the Abbey roof. He survived and, despite breaking both legs, enjoyed a long life. There are delightful walks around the town and beside the River Avon.

Lacock Abbey & The Fox Talbot Museum of Photography

Chippenham, Wiltshire SN15 2LG
Tel: (01249) 730227 Follow the A429 south past Chippenham and continue along the A350 towards Melksham. Lacock is signposted on the left
(7.7 Miles)

A 13thC abbey, it was acquired by Sir William Sharington, one time treasurer at the Bristol Mint, in 1540 and adapted as a Tudor mansion by adding an octagonal tower and twisted chimneys whilst retaining the 13th & 15thC cloisters, chapter house and nuns parlour. The house was altered in neo-Gothic style in 1753 with further modifications being made in 1828. It was here in 1839-41 that William Fox Talbot perfected his talbotype technique which laid the foundations of modern photography and the adjacent museum is devoted to this achievement.

SERVICE AREA

BETWEEN JUNCTIONS 17 AND 18

Leigh Delamere Services (Eastbound) (Granada)

Tel: (01666) 837691Self Service Restaurant, Fresh Express, Harry Ramsdens, Burger King, Little Chef, Travelodge & Esso Fuel

Leigh Delamere Services (Westbound) (Granada)

Tel: (01666) 837691 Self Service Restaurant, Fresh Express, Harry Ramsdens, Burger King, Little Chef, Travelodge & Esso Fuel

Footbridge connection between sites.

M4   JUNCTION 18

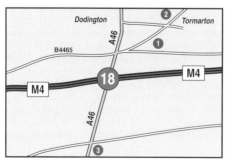

NEAREST A&E HOSPITAL

Frenchay Hospital, Frenchay Park Road, Frenchay, Bristol BS16 1LE
Tel: (0117) 970 1212

Proceed west to Junction 19 and then follow the M32 to Junction 1. Take the A4174 east exit and the hospital is signposted. (Distance Approx 8.7 miles)

FACILITIES

1   The Compass Inn

Tel: (01454) 218242
0.4 miles north along the Tormarton Road, on the right. (Best Western) Bittles Restaurant; Open; 07.00-22.30hrs daily. Avon Room Restaurant; Open; Mon-Sat; 19.00-21.15hrs. Bar Snacks available; 11.00-22.00hrs daily.

2   Chestnut Farm B & B

Tel: (01454) 218563
0.8 miles north along the Tormarton Road, on the left

3   The Crown Inn at Tolldown

Tel: (01225) 891231
1 mile south along the A46, on the left. (Wadworth) Meals served; Sun-Thurs; 12.00-14.15hrs & 19.00-21.30hrs, Fri & Sat; 12.00-14.15hrs & 19.00-22.00hrs

PLACES OF INTEREST

Dyrham Park (NT)

Dyrham, Nr Chippenham, Wiltshire SN14 8E
Tel: (01179) 372501. Follow the A46 south
(Signposted 2.6 Miles)

A country mansion built in c1698 to designs by William Talman for William Blathwayt, poilitician and Secretary of State to William III in Flanders. The house contains portraits and tapestries, furniture used by Pepys and Evelyn, the diarists, and rooms panelled in oak, walnut and cedar. There is a park, garden and orangery and a number of picnic sites. Licensed Restaurant. Gift Shop. Limited disabled access.

 Large Hotel   Small Hotel   Motel   Guest House/ Bed & Breakfast   Disabled Facilities   Childrens Facilities

M4 | JUNCTION 19

THIS JUNCTION IS A MOTORWAY
INTERCHANGE WITH THE M32 ONLY
AND THERE IS NO ACCESS TO ANY
FACILITIES

NEAREST A&E HOSPITAL

Frenchay Hospital, Frenchay Park Road,
Frenchay, Bristol BS16 1LE
Tel: (0117) 970 1212

Follow the M32 to Junction 1 and take the A4174
east exit and the hospital is signposted. (Distance
Approx 1.5 miles)

PLACES OF INTEREST

Bristol

Follow the M32 south (Signposted 5.6 Miles)

For details please see M5 Junction 16 & 17
information

M4 | JUNCTION 20

THIS JUNCTION IS A MOTORWAY
INTERCHANGE WITH THE M5 ONLY AND
THERE IS NO ACCESS TO ANY
FACILITIES

NEAREST A&E HOSPITAL

Frenchay Hospital, Frenchay Park Road,
Frenchay, Bristol BS16 1LE
Tel: (0117) 970 1212

Proceed east to Junction 19 and then follow the
M32 to Junction 1. Take the A4174 east exit and
the hospital is signposted. (Distance Approx 4.9
miles)

M4 | JUNCTION 21

THIS JUNCTION IS A MOTORWAY
INTERCHANGE WITH THE M48 ONLY
AND THERE IS NO ACCESS TO ANY
FACILITIES

M4 | JUNCTION 22

THIS JUNCTION IS A MOTORWAY
INTERCHANGE WITH THE M49 ONLY
AND THERE IS NO ACCESS TO ANY
FACILITIES

NEAREST WESTBOUND A&E HOSPITAL

The Royal Gwent Hospital, Cardiff Road,
Newport NP20 2UB Tel: (01633) 234234.

Proceed to Junction 24 and follow the A48 west
(Signposted along route. Distance Approx 15.9
miles)

NEAREST EASTBOUND A&E HOSPITAL

Frenchay Hospital, Frenchay Park Road,
Frenchay, Bristol BS16 1LE
Tel: (0117) 970 1212

Proceed to Junction 19 and then follow the M32 to
Junction 1. Take the A4174 east exit and the
hospital is signposted. (Distance Approx 10 miles)

START OF M48 LOOP

M48 | JUNCTION 1

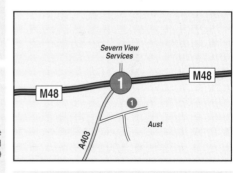

NEAREST WESTBOUND A&E HOSPITAL

The Royal Gwent Hospital, Cardiff Road,
Newport NP20 2UB Tel: (01633) 234234

Proceed to Junction 24 and follow the A48 west
(Signposted along route. Distance Approx 18.4
miles)

NEAREST EASTBOUND A&E HOSPITAL

Frenchay Hospital, Frenchay Park Road,
Frenchay, Bristol BS16 1LE
Tel: (0117) 970 1212

Proceed to Junction 19 and then follow the M32 to

 Pets Welcome   Cash Dispenser   24 Hour Facilities   Toilets   Alcoholic Drinks   Food and Drink   Fuel

Junction 1. Take the A4174 east exit and the hospital is signposted. (Distance Approx 9 miles)

FACILITIES

SERVICE AREA

ACCESSED FROM THIS JUNCTION

Severn View Services (Granada)
Tel: (01454) 633199 Self Service Restaurant, Burger King, Travelodge & BP Fuel

1 Boars Head

Tel: (01454) 632278
0.4 miles south in Aust Village. (Eldridge Pope)
Meals served; Mon-Sat; 12.00-14.00hrs & 19.00-21.30hrs, Sun; 12.00-14.00hrs

M48 JUNCTION 2

THERE ARE NO FACILITIES WITHIN ONE MILE OF THE JUNCTION

NEAREST WESTBOUND A&E HOSPITAL

The Royal Gwent Hospital, Cardiff Road, Newport NP20 2UB Tel: (01633) 234234.

Proceed to Junction 24 and follow the A48 west (Signposted along route. Distance Approx 15.3 miles)

NEAREST EASTBOUND A&E HOSPITAL

Frenchay Hospital, Frenchay Park Road, Frenchay, Bristol BS16 1LE
Tel: (0117) 970 1212

Proceed to Junction 19 and then follow the M32 to Junction 1. Take the A4174 east exit and the hospital is signposted. (Distance Approx 12 miles)

PLACES OF INTEREST

Chepstow Castle

Chepstow, Monmouthshire NP6 5EZ
Tel: (01291) 624065 Follow the A466 north
(Signposted 2.1 Miles)

Standing guard over a strategic crossing point into Wales, Chepstow Castle was amongst the first of the Norman stone-built strongholds. It was constructed by William fitz Osbern to secure his territories along the Welsh borders and was improved and added to over the centuries until it ceased to have a military use in 1690. It is one of the few castles in Britain in which the evolution of mediaeval military architecture can be traced from start to finish. A major exhibition entitled "A Castle at War" reflects the changing role of Chepstow through the Middle Ages and there are displays of life sized models of the mediaeval lords and a dramatic Civil War battle scene.

Sited opposite to the castle ...

Stuart Crystal

Bridge Street, Chepstow, Monmouthshire NP6 5EZ
Tel: (01291) 620135

A Visitor Centre and Factory Shop featuring an exhibition of old glass, a glass engraving display and an audio visual presentation on glass making. There is a glass repair and personalized glassware service, and also a picnic area. Coffee Shop.

END OF M48 LOOP

M4 JUNCTION 23

JUNCTION 23 IS A MOTORWAY INTERCHANGE WITH THE M48 ONLY AND THERE IS NO ACCESS TO ANY FACILITIES

M4 JUNCTION 23A

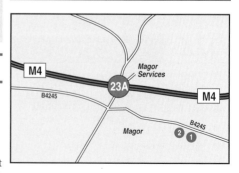

NEAREST WESTBOUND A&E HOSPITAL

The Royal Gwent Hospital, Cardiff Road,
Newport NP20 2UB Tel: (01633) 234234
Proceed to Junction 24 and follow the A48 west
(Signposted along route. Distance Approx 7.5 miles)

NEAREST EASTBOUND A&E HOSPITAL

Frenchay Hospital, Frenchay Park Road,
Frenchay, Bristol BS16 1LE
Tel: (0117) 970 1212
Proceed to Junction 19 and then follow the M32 to
Junction 1. Take the A4174 east exit and the
hospital is signposted. (Distance Approx 18.4 miles
[via M4])

FACILITIES

SERVICE AREA

ACCESSED FROM THIS JUNCTION

Magor Services (First)

Tel: (01633) 881515   Wendys Restaurant, Self
Service Restaurant, Comfort Inns Lodge & Esso Fuel

---

1   Reliance Garage (Esso)

 Tel: (01633) 880229
1 mile along the B4245 east, on the right.
Access, Visa, Overdrive, All Star, Switch, Dial
Card, Mastercard, Amex, AA Paytrak, Diners
Club, Delta, BP Supercharge, Shell Gold, Esso
Cards. Open; Mon-Sat; 07.30-19.30hrs, Sun;
09.00-14.00hrs.

---

2   Wheatsheaf Inn

 Tel: (01633) 880608
1 mile along the B4245 east, in Magor.
(Whitbread) Open all day. Meals served; Mon-
Sat; 12.00-14.00hrs & 19.00-21.30hrs, Sun;
12.00-14.00hrs.

PLACES OF INTEREST

Caldicot Castle & Country Park

Caldicot, Monmouthshire NP6 4HU
Tel: (01291) 420241 Follow the B4245 east
(Signposted 4.7 Miles)

Caldicot Castle's well preserved fortifications
were founded by the Normans and fully
developed by the late 14thC. Restored as a
family home by a wealthy Victorian, the castle
offers the chance to explore mediaeval walls
and towers in a setting of tranquil gardens and
wooded country park. There are personal stereo
sound tours, play areas and picnic and barbecue
sites within the park. Castle Shop, Tea Room.

M4   JUNCTION 24

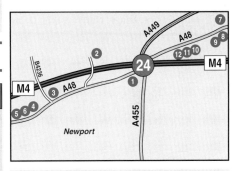

NEAREST A&E HOSPITAL

The Royal Gwent Hospital, Cardiff Road,
Newport NP20 2UB Tel: (01633) 234234.
Follow the A48 west (Signposted along route.
Distance Approx 3.6 miles)

FACILITIES

1   Holiday Inn Newport

Tel: (01633) 412777
0.1 miles west along the A48, on the left.
Harpers Bar & Restaurant; Restaurant Open;
Breakfast; Mon-Fri; 06.30-09.30hrs, Sat & Sun;
07.00-11.00hrs, Lunch; Closed; Dinner; 19.00-
22.00hrs daily. Bar Meals served 09.00-22.30hrs
daily.

---

2   Celtic Manor Hotel

Tel: (01633) 413000
0.2 miles west along the A48, on the right. The
Olive Tree Restaurant; Breakfast; 07.00-10.30hrs
daily, Lunch; 12.30-14.30hrs daily, Dinner;
19.00-22.30hrs daily. Owen's Restaurant; Open;
Tues-Sat; 19.00-23.00hrs.

---

3 The Royal Oak

🚹 🍴 ♿ 🔗   Tel: (01633) 282155
0.6 miles west along the A48, on the right.
(Bass) Open all day. Meals served; Mon-Sat;
12.00-14.00hrs & 17.30-21.00hrs, Sun; 12.00-
21.00hrs

4 Treberth Service Station (Esso)

⛽ ♿ WC 24 £   Tel: (01633) 281241
0.8 miles west along the A48, on the right.
Access, Visa, Overdrive, All Star, Switch, Dial
Card, Mastercard, Amex, Diners Club, Delta,
Esso Cards.

5 Chepstow Road Service Station (Elf)

⛽   Tel: (01633) 271473
1 mile west along the A48, on the right. Access,
Visa, Overdrive, All Star, Switch, Mastercard,
Amex, AA Paytrak, Diners Club, Delta, Elf
Cards. Open; 06.00-22.00hrs daily

6 Man of Gwent

🚹 🍴 ♿ 🔗 🛏B   Tel: (01633) 281263
1 mile west along the A48, on the right.
(Scottish & Newcastle) Open all day. Meals
served; 11.00-21.30hrs daily

7 Stakis Hotel Newport

🛏H 🚹 🍴 ♿ 🔗   Tel: (01633) 413733
0.2 miles east along the A48, on the left.
Seasons Restaurant; Breakfast; Mon-Sat; 07.00-
10.00hrs, Sun; 08.00-10.00hrs, Lunch; Sun;
12.30-15.00hrs, Dinner; Mon-Sat; 19.00-
22.00hrs, Sun; 19.00-21.30hrs. Cafe Chino;
Open; Mon-Fri; 09.00-17.00hrs.

8 Lynwood Garage

⛽ WC   Tel: (01633) 412354
0.2 miles east along the A48, on the right. Most
major credit cards accepted. Open; Mon-Fri;
07.00-19.00hrs, Sat; 07.30-18.00hrs, Sun; 08.00-
18.00hrs.

9 Taylor's Cafe

🍴 ♿   Tel: (01633) 412354
0.2 miles east along the A48, on the right.
Open; Mon-Fri; 07.00-19.00hrs, Sat; 07.30-

13.00hrs, Sun; 08.00-13.00hrs

10 McDonald's

🍴 ♿   Tel: (01633) 412087
0.1 miles east along the A48, on the right.
Open; 07.30-23.00hrs daily. "Drive Thru" Open
until 0.00hrs on Fri & Sat.

11 Travel Inn

🛏M ♿   Tel: (01633) 411390
0.1 miles east along the A48, on the right

12 The Coldra

🚹 🍴 🔗   Tel: (01633) 411390
0.1 miles east along the A48, on the right.
(Beefeater) Open all day. Meals served; Mon-
Fri; 12.00-14.30hrs & 17.30-22.45hrs, Fri; 12.00-
14.30hrs & 17.00-23.00hrs, Sat & Sun; 12.00-
23.00hrs.

PLACES OF INTEREST

Roman Legionary Museum

High Street, Caerleon, Newport NP6 1AE
Tel: (01633) 423134. Follow the A48 south and
turn right along the B4236 (Signposted 1.9 Miles)

The legionary fortress of "Isca" was one of the
principal military bases in Roman Britain. The
museum illustrates the history of Roman
Caerleon and the daily life of its garrison with
a display of many fascinating discoveries,
including a remarkable collection of engraved
gemstones from the Fortress Baths, an
intriguing labyrinth mosaic, tombstones and
arms and equipment of the Roman soldiers.
An amphitheatre, Barracks, Baths and Roman
remains are to be seen nearby. The Capricorn
Centre, sited next door, is an educational
facility for schools and families.

 Large Hotel   Small Hotel   Motel   Guest House/ Bed & Breakfast   Disabled Facilities   Childrens Facilities

M4 JUNCTION 25 & JUNCTION 25A

THESE TWO JUNCTIONS ARE ADJACENT. ALTHOUGH EACH ONE HAS RESTRICTED ACCESS, THE COMBINATION IS SUCH THAT VEHICLES ARE ABLE TO ENTER AND EXIT IN BOTH DIRECTIONS

THERE ARE NO FACILITIES WITHIN ONE MILE FOR VEHICLES EXITING FROM JUNCTION 25A

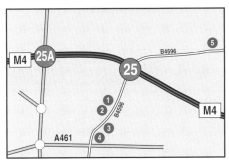

NEAREST A&E HOSPITAL

The Royal Gwent Hospital, Cardiff Road, Newport NP20 2UB Tel: (01633) 234234

Follow the A4042 south and continue along the A48 (Signposted along route. Distance Approx 2.1 miles)

FACILITIES
(FROM JUNCTION 25)

1 The Victoria

Tel: (01633) 258252
0.4 miles south along the B4596, on the right. (Punch Taverns) Open all day. Meals served; Mon-Sat; 12.00-20.00hrs, Sun; 12.00-16.00hrs.

2 Polash Tandoori Restaurant

Tel: (01633) 252891
0.5 miles south along the B4596, on the right. Open; Mon-Thurs; 12.00-14.00hrs & 18.00-0.00hrs, Fri; 18.00-01.30hrs, Sat; 12.00-14.00hrs & 18.00-01.30hrs, Sun; 18.00-0.00hrs.

3 Ashburton House

Tel: (01633) 211140
0.6 miles south along the B4596, on the left

4 Caerleon Service Station (Esso)

Tel: (01633) 266581
0.7 miles south along the B4596, on the left. Access, Visa, Overdrive, All Star, Switch, Dial Card, Mastercard, Amex, Diners Club, Shell Gold, Esso Cards. Open; 07.00-23.00hrs daily

5 St.Julian Inn

Tel: (01633) 258663
1 mile north along the B4596, on the left. (Courage) Open all day. Meals served; Mon-Sat; 11.30-14.45hrs & 18.00-20.45hrs

PLACES OF INTEREST

Newport

Newport Tourist Information Centre, John Frost Square, Newport NP9 IH2. Tel; (01633) 842962. Follow the A4042 south (Signposted 1.6 Miles)

Bronze Age fishermen settled around the estuary of the River Usk and, later the Celtic Silures built hill forts overlooking it. On the western edge of their empire, the Romans built a fortress at Caerleon to defend the river crossing and later the Normans arrived to build a castle on Stow Hill. In the 14thC a new castle was constructed next to a river crossing and around this a new town grew to become Newport. The most famous landmark in the town is the recently renovated Transporter Bridge, which opened in 1906 and was erected to provide a crossing between the east and west banks of the Usk. This is one of only two working in Britain, Middlesbrough has the other, and there are now believed to be only five such examples in the world of this unusual form of transport which allows traffic to cross the river without obstructing shipping.

Pets Welcome	Cash Dispenser	24 Hour Facilities	Toilets	Alcoholic Drinks	Food and Drink	Fuel

M4 JUNCTION 26

NEWPORT TOWN CENTRE IS WITHIN ONE MILE OF THE JUNCTION

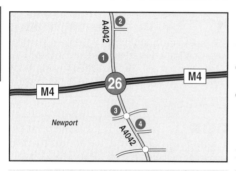

NEAREST A&E HOSPITAL

The Royal Gwent Hospital, Cardiff Road, Newport NP20 2UB Tel: (01633) 234234

Follow the A4042 south and continue along the A48 (Signposted along route. Distance Approx 2.1 miles)

FACILITIES

1   The Borderer

🍴 🏨 ♿ ☕   Tel: (01633) 858667
0.4 miles north along the A4051, on the left. (Harvester) Open all day. Meals served; 12.00-22.00hrs daily.

2   Malpas BP Service Station

⛽ 24   Tel: (01633) 858057
1 mile north along the A4051, on the right. Access, Visa, Overdrive, All Star, Switch, Dial Card, Mastercard, Amex, AA Paytrak, Diners Club, Delta, Routex, BP Cards.

3   Newport Shell

⛽ ♿ WC 24   Tel: (01633) 820000
0.7 miles south along the A4051, on the right. Access, Visa, Overdrive, All Star, Switch, Dial Card, Mastercard, Amex, AA Paytrak, Delta, Routex, Shell Cards, Smartcard.

4   The Old Rising Sun

🍴 🏨 ♿ ☕   Tel: (01633) 223452
0.9 miles south along the A4051, on the left. (Freehouse) Open all day. Meals served; Mon-Wed; 11.00-15.00hrs, Thurs & Fri; 11.00-15.00hrs & 18.00-20.00hrs, Sat; 11.00-15.00hrs, Sun; 12.00-15.00hrs. Meals can be served at other times by prior notification.

PLACES OF INTEREST

Newport

Follow the A4042 south (Signposted 1.6 Miles)
For details please see Junction 25 information..

M4 JUNCTION 27

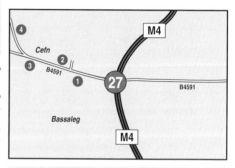

NEAREST A&E HOSPITAL

The Royal Gwent Hospital, Cardiff Road, Newport NP20 2UB Tel: (01633) 234234

Follow the B4591 east, bear right along the B4240 and turn left along the A48. (Signposted along route. Distance Approx 1.8 miles)

FACILITIES

1   High Cross Shell

⛽ ♿ WC   Tel: (01633) 890940
0.2 miles north along the B4591, on the left. Access, Visa, Overdrive, All Star, Switch, Dial Card, Mastercard, Amex, AA Paytrak, Diners Club, Delta, BP Agency, Shell Cards, Smartcard. Open; 06.00-22.00hrs daily

---

| H Large Hotel | h Small Hotel | M Motel | B Guest House/ Bed & Breakfast | Disabled Facilities | Childrens Facilities |

2   Cefn Smithy Service Station (Elf)

🍴 WC 24   Tel: (01633) 894021
0.5 miles north along the B4591, on the right.
Access, Visa, All Star, Switch, Dial Card,
Mastercard, Amex, Diners Club, Elf Cards, Total
Fina Cards.

3   The Rising Sun Hotel & Restaurant

🍴 🍽 ♿ 🐾 ⌁B   Tel: (01633) 895126
0.8 miles north along the B4591, on the left.
(Free House) Meals served; Mon-Sat; 12.00-
14.15hrs & 18.00-21.30hrs, Sun; 12.00-14.15hrs

4   The Olde Oak

🍴 🍽 ♿ 🐾   Tel: (01633) 892883
0.9 miles north along the B4591, on the right
in Ruskin Avenue. (Ushers) Open all day Sat &
Sun. Meals served; Mon-Sat; 12.00-14.30hrs &
18.30-21.30hrs, Sun; 12.00-15.00hrs.

PLACES OF INTEREST

Newport

Follow the B4591 east (Signposted 2.2 Miles)

For details please see Junction 25 information.

Fourteen Locks Canal Centre

Cwm Lane, Rogerstone NP10 9GN
Tel: (01633) 894802. Follow the B4591 north
towards Risca (Signposted along this route 1 Mile)

On the Monmouthshire Canal which opened
in 1796, Fourteen Locks is a complicated system
of locks, ponds, channels, tunnels and weirs
designed to enable barges to be lowered or
raised 168ft in 0.5 miles with the minimum
wastage of water. At the Visitor Centre, an
exhibition traces the growth and decline of the
canal in Gwent and a display explains how the
locks and water storage system worked. There
are numerous walks and picnic sites around
the canal centre.

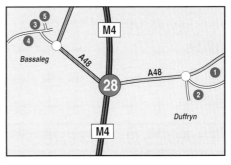

M4 JUNCTION 28

NEAREST A&E HOSPITAL

**The Royal Gwent Hospital, Cardiff Road,
Newport NP20 2UB. Tel: (01633) 234234.**

Follow the A48 east. (Signposted along route.
Distance Approx 2 miles)

FACILITIES

1   Newport BP Petrol Station

🍴 24   Tel: (01633) 815921
0.9 miles east along the A48, on the right.
Access, Visa, Overdrive, All Star, Switch, Dial
Card, Mastercard, Amex, AA Paytrak, Diners
Club, Delta, Routex, BP Cards, Shell Agency

2   The Stonehouse

🍴 🍽 ♿ 🐾   Tel: (01633) 810541
0.9 miles east along the B4239, on the left.
(Whitbread) Open all day. Meals served; Mon-
Sat; 11.30-22.00hrs, Sun; 12.00-22.00hrs.

3   The Tredegar Arms

🍴 🍽 ♿ 🐾   Tel: (01633) 893247
0.9 miles north along the A468, on the right.
(Whitbread) Open all day. Meals served; 12.00-
14.30hrs & 18.00-21.00hrs daily.

4   Bassaleg BP Service Station

🍴 ♿ WC   Tel: (01633) 893321
1 mile north along the A468, on the left. Access,
Visa, Overdrive, All Star, Switch, Dial Card,
Mastercard, Amex, Diners Club, Delta, Routex,

| | Pets Welcome | | £ Cash Dispenser | | 24 24 Hour Facilities | | WC Toilets | | Alcoholic Drinks | | Food and Drink | | Fuel |

BP Cards, Shell Agency. Open; Mon-Fri; 07.30-20.30hrs, Sat; 07.30-20.00hrs, Sun; 09.00-20.00hrs.

5   Junction 28 Restaurant

Tel: (01633) 891891

1 mile north along Forge Lane, on the left, in Station Approach. Open; Mon-Sat; 12.00-14.00hrs & 17.30-21.30hrs, Sun; 12.00-16.00hrs.

PLACES OF INTEREST

Tredegar House & Park

Newport NP1 9YW Tel: (01633) 815880.
Follow the A48 south.(Signposted 0.8 Miles)

Set in a beautiful 90 acre park, Tredegar House is one of the best examples of a 17thC, Charles II mansion in Britain with some parts of the building dating back to the early 1500's. All the rooms have been restored and reflect the rise and fall of the home's former owners, the Morgan family of whom the most notorious member was Sir Henry Morgan, the pirate. The second Lord Tredegar, Godfrey Morgan, survived the Charge of the Light Brigade and his horse is buried in the grounds. The stunning state rooms are adorned with fine paintings, carvings and elaborate ceilings; illustrating the opulence of William Morgan, the builder of the house. There is a children's adventure playground and nature lakeside walks, as well as walled gardens and a beautiful Orangery. Craft Workshops. Gift Shop. Tea Room. Disabled access.

M4   JUNCTION 29

THIS IS A RESTRICTED ACCESS MOTORWAY INTERCHANGE WITH THE A48(M) ONLY

Vehicles can only exit along the A48(M) from the southbound lanes.

Vehicles can only enter the motorway from the A48(M) along the northbound lanes.

THERE ARE NO FACILITIES WITHIN ONE MILE OF THIS JUNCTION

NEAREST WESTBOUND A&E HOSPITAL

University Hospital of Wales, Heath Park, Cardiff CF4 4XW Tel: (029) 2074 7747.

Follow the A48(M) and A48 (Signposted. Distance Approx 7.1 Miles)

NEAREST EASTBOUND A&E HOSPITAL

The Royal Gwent Hospital, Cardiff Road, Newport NP20 2UB Tel: (01633) 234234.

Proceed to Junction 28 and follow the A48 east. (Signposted along route. Distance Approx 4.3 miles)

PLACES OF INTEREST

Cardiff

Cardiff Tourist Information Centre, Central Station, Cardiff CF1 1QY Tel: (029) 2022 7281. Follow the A48 south (Signposted 5 Miles)

Once a walled town with five gates, a Benedictine priory, a Dominican house, a Franciscan and two churches, Cardiff, in 150 years rose grew to become the capital of the Principality of Wales, a major seaport, and a university city. The prosperity of the city was founded upon the establishment of the first Dock, opened in 1839 and upon which the Marquess of Bute risked the whole of his fortune. It became the richest coal exporting city in the world with an extensive port of 5 docks, 12 graving docks, 5 miles of quays and great timber basins and the legacy of this wealth is reflected in the Victorian and Edwardian architecture to be found in the city. Today the docks have all but disappeared and they are now the centre of a 2,700 acre redevelopment to revitalize the area.

Within the city centre...

Cardiff Castle

Castle Street, Cardiff CF1 2RB
Tel: (029) 2087 8100
e-mail; cardiffcastle@cardiff.gov.uk

A motte with wooden buildings was raised in c1093 on the site of a Roman fortification; the present stone keep was erected to replace wooden buildings in the late 12thC and additions were made in the 15th and each subsequent century following. In 1861 the 3rd Marquess of Bute and William Burges designed additions in the Gothic, Arab and Classical Greek idioms. There are also two military

---

 Large Hotel   Small Hotel   Motel   Guest House/ Bed & Breakfast   Disabled Facilities   Childrens Facilities

museums, splendid gardens, gift shop and a tea room. Some disabled access

National Museum & Gallery

Cathays Park, Cardiff CF1 3NP
Tel: (029) 2039 7951
website; http:www.nmgw.ac.uk

No other British museum offers a similar range of art, natural history and science under one roof. There is a spectacular exhibition on the creation of Wales, complete with animated Ice Age creatures and simulated Big Bang, there are natural history galleries with woodland and wildlife displays and the art section includes works from Canaletto to Cezanne. Disabled access.

Cardiff Bay: Millennium Waterfront

Cardiff Bay Visitor Centre Tel: (029) 2046 3833
(Signposted from Motorway)

Europe's largest regeneration project, a Covent Garden on the waterfront with entertainment, arts, places to eat and a range of leisure attractions including; Techniquest (Tel: 01222-475475), a Science Discovery Centre, Centre for Visual Arts (Tel: 01222-388922) with an interactive discovery gallery and the Atlantic Wharf Leisure Village with a 12-screen multiplex cinema, 26 lane Hollywood Bowl, micro brewery, family entertainment complex, nightclubs, cafes and bars..

M4   JUNCTION 30

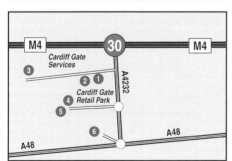

NEAREST A&E HOSPITAL

University Hospital of Wales, Heath Park,
Cardiff CF4 4XW Tel: (029) 2074 7747

Follow the A48 south (Signposted. Distance Approx 5.3 Miles)

FACILITIES

1   Cardiff Gate Services (Elf)

🅿 ♿ WC 24   Tel: (029) 2073 5618

Adjacent to the south west side of roundabout. Access, Visa, Overdrive, All Star, Switch, Dial Card, Mastercard, Amex, Diners Club, Delta, Securicor Fuelserv, Keyfuels, Elf Cards, Total Fina Cards, Maxol.

2   Holiday Inn Express Cardiff

H ♿ 🐾   Tel: (029) 2073 3222

Adjacent to the south west side of roundabout

3   The Cardiff Gate

🍴 🍽 ♿ 🐾   Tel: (029) 2054 1132

Adjacent to the south west side of roundabout. (Bass Retail) Open all day. Meals served; Mon-Sat; 12.00-22.00hrs, Sun; 12.00-21.30hrs

4   McDonald's

🍽 ♿   Tel: (029) 2073 3228

0.5 miles south along the A4232, on the right, in Cardiff Gate Retail Park. Open; Sun-Thurs; 07.30-23.00hrs, Fri & Sat; 07.30-0.00hrs.

5   Asda Petrol Station

🅿   Tel: (029) 2034 0276

0.5 miles south along the A4232, on the right, in Cardiff Gate Retail Park. Access, Visa, Overdrive, All Star, Switch, Dial Card, Mastercard, Amex, Diners Club, Delta, BP Supercharge, Asda Business Card. Open; Mon-Wed; 07.00-22.00hrs, Thurs & Fri; 07.00-23.00hrs, Sat; 07.30-22.00hrs, Sun; 09.00-18.00hrs. Toilets and cash machine available in adjacent store.

6   Hotel Campanile Cardiff

H 🍴 🍽 ♿ 🐾   Tel: (029) 2054 9044

1 mile south along the A4232, on the right. The Bistro Restaurant; Open; 12.00-14.00hrs & 19.00-22.00hrs daily

---

Pets Welcome	£ Cash Dispenser	24 24 Hour Facilities	WC Toilets	Alcoholic Drinks	Food and Drink	Fuel

M4

PLACES OF INTEREST

Cardiff

Follow the A48 south (Signposted 5 Miles)

For details please see Junction 29A information

Within the city centre...

Cardiff Castle

Castle Street, Cardiff CF1 2RB

For details please see Junction 29A information

National Museum & Gallery

Cathays Park, Cardiff CF1 3NP

For details please see Junction 29A information

Cardiff Bay: Millennium Waterfront

For details please see Junction 29A information

M4 JUNCTION 31

THERE IS NO JUNCTION 31

M4 JUNCTION 32

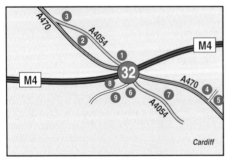

NEAREST A&E HOSPITAL

University Hospital of Wales, Heath Park,
Cardiff CF4 4XW Tel: (029) 2074 7747

Follow the A470 south (Distance Approx 2.4 Miles)

FACILITIES

1 Quality Hotel Cardiff

Tel: (029) 2052 9988

0.1 miles north along the A4054, on the right. The Hillside Restaurant; Breakfast; Mon-Fri; 07.00-09.45hrs, Sat & Sun; 07.30-10.00hrs, Lunch; Sun-Fri; 12.30-13.45hrs, Dinner; Mon-Sat; 19.00-21.45hrs, Sun; 19.00-21.30hrs.

2 Old Tom Inn

Tel: (029) 2081 1865

0.6 miles north along the A4054, on the left, in Tongwynlais. (Whitbread) Open all day. Meals served; Mon-Sat; 11.00-21.00hrs, Sun; 12.00-21.00hrs

3 The Lewis Arms

Tel: (029) 2081 0330

0.6 miles north along the A4054, on the right, in Tongwynlais. (Brain's) Open all day. Meals served; Mon-Sat; 12.00-14.45hrs & 18.00-20.45hrs, Sun; 12.00-14.45hrs

4 Masons Arms Hotel

Tel: (029) 2069 2554

1 mile south along the A470, on the left. (Toby) Open all day on Sun. Restaurant Open; Mon-Thurs; 12.00-14.00hrs & 17.30-22.00hrs, Fri & Sat; 12.00-14.00hrs & 17.30-22.30hrs, Sun; 12.00-21.30hrs. Bistro Open; Mon-Fri; 11.00-15.00hrs & 17.30-23.00hrs, Sat; 12.00-14.30hrs & 17.30-23.00hrs, Sun; 12.00-22.30hrs.

5 Safeway BP Filling Station

Tel: (029) 2061 6414

1 mile south along the A470, on the left. Access, Visa, Overdrive, All Star, Switch, Dial Card, Mastercard, Amex, Diners Club, BP Cards, Shell Agency

6 Village Hotel & Leisure Club

Tel: (029) 2052 4300

0.1 miles south along the A4054, on the right. Salingers Bar & Restaurant; Open; Mon-Sat; 12.00-22.00hrs, Sun; 12.30-22.00hrs. The

Village Pub; Meals served; Mon-Sat; 11.00-23.00hrs, Sun; 12.00-22.30hrs. Cafe Copra; Open; 08.00-21.00hrs daily

7   The Holly Bush

🍴 ⛽ ♿ 🔧   Tel: (029) 2062 5037
0.4 miles south along the A4054, on the left. (Bass) Open all day. Meals served; 12.00-14.00hrs & 17.30-22.00hrs daily.

8   Coryton Services (Esso)

⛽ ♿ 🚻 24   Tel: (029) 2061 6044
Adjacent to the south west side of roundabout. Access, Visa, Overdrive, All Star, Switch, Dial Card, Mastercard, Amex, AA Paytrak, Diners Club, Delta, BP Supercharge, Shell Gold, Esso Cards.

9   McDonald's

🍴 ♿   Tel: (029) 2069 1700
Adjacent to the south west side of roundabout. Open; Sun-Thurs; 07.30-23.00hrs. Fri & Sat; 07.30-0.00hrs

PLACES OF INTEREST

Cardiff

Follow the A470 south (Signposted 4.2 Miles)
For details please see Junction 29A information

Within the city centre...

Cardiff Castle

Castle Street, Cardiff CF1 2RB
For details please see Junction 29A information

National Museum & Gallery

Cathays Park, Cardiff CF1 3NP
For details please see Junction 29A information

Cardiff Bay: Millennium Waterfront

For details please see Junction 29A information

Caerphilly Castle

Caerphilly CF83 1JD Tel: (029) 2088 3143
Follow the A480 north (Signposted 5.5 Miles)
Built by the Anglo-Norman lord, Gilbert de Clare, in the late 13thC to consolidate his

control on the lands he had captured, it sprawls over 30 acres and is one of the largest castles in Britain. The ingenuity of the "walls within walls" fortifications and the scale of its water defences ensured that it was out of the range of military catapults and never taken in battle. Due to immense reconstructions carried out by the former owner, the 4th Marquess of Bute, and since 1947 by Government agencies. Caerphilly Castle can be viewed today as a splendid example of early military architecture. Limited disabled access.

Castell Coch

Tongwynlais, Nr Cardiff CF15 7J
Tel: (029) 2081 0101 Follow the A470 north
(Signposted 0.7 Miles)

Built on the site of a mediaeval castle, Castell Coch was commissioned by the 3rd Marquess of Bute in 1865. The architect, William Burges, was given free reign to create a Victorian dream of the Middle Ages as a companion piece to the patron's home at Cardiff Castle and the result was the ultimate fairytale castle. The conical towers and needle sharp turrets peek out from a wooded slope and present a vision more in tune with the Bavarian countryside, the pages of Sleeping Beauty, or even a Walt Disney creation than a Welsh hillside. The interior is extravagantly decorated in the "anything goes" spirit of the Victorian age and there are fantastic furnishings and fireplaces. Gift Shop. Tea Room. Disabled access (free) to ground floor only.

M4   JUNCTION 33

APART FROM CARDIFF WEST SERVICES
THERE ARE NO OTHER FACILITIES
WITHIN ONE MILE OF THIS JUNCTION

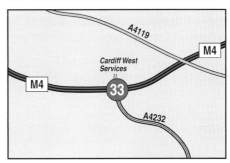

NEAREST WESTBOUND A&E HOSPITAL

The Royal Glamorgan Hospital, Llantrisant
CF72 8XR Tel: (01443) 443443

Proceed to Junction 34 and it is signposted from the Junction. (Distance Approx 5 Miles)

NEAREST EASTBOUND A&E HOSPITAL

University Hospital of Wales, Heath Park, Cardiff CF4 4XW Tel: (029) 2074 7747

Proceed to Junction 32 and follow the A470 south (Distance Approx 5.9 Miles)

SERVICE AREA

ACCESSED FROM THIS JUNCTION

Cardiff West Services (Granada)

Tel: (029) 2089 1141   Self Service Restaurant, Burger King, Travelodge & Esso Fuel

PLACES OF INTEREST

Cardiff

Follow the A4232 south (Signposted 7 Miles)

For details please see Junction 29A information

Within the city centre...

Cardiff Castle

Castle Street, Cardiff CF1 2RB

For details please see Junction 29A information

National Museum & Gallery

Cathays Park, Cardiff CF1 3NP

For details please see Junction 29A information

Cardiff Bay: Millennium Waterfront

For details please see Junction 29A information

Museum of Welsh Life

St Fagans, Cardiff CF5 6XB Tel: (029) 2057 3500
website; www.nmgw.ac.uk
Follow the A4232 south (Signposted 1.8 Miles)

Over forty buildings have been taken down and rebuilt here to recreate Welsh daily life throughout history. Set in 100 acres of beautiful countryside, the open air museum follows the evolution of national life from a Celtic village of 2000 years ago to a miner's cottage of the 1980's. Disabled access.

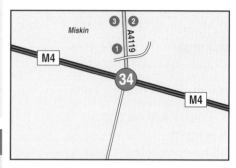

M4   **JUNCTION 34**

NEAREST A&E HOSPITAL

The Royal Glamorgan Hospital, Llantrisant
CF72 8XR Tel: (01443) 443443

Signposted from Junction. (Distance Approx 2.7 Miles)

FACILITIES

1   Miskin Manor Hotel

Tel: (01443) 224204

0.6 miles north along the A4119, on the left. Restaurant Open; Lunch; Sun-Fri; 12.00-14.00hrs, Dinner; 18.30-21.45hrs daily. (Pre-booking advisable).

2   Castell Mynach

Tel: (01443) 220940

0.7 miles north along the A4119, on the right. (Vintage Inns) Open all day. Meals served; Mon-Sat; 11.30-22.00hrs, Sun; 12.00-21.00hrs.

3   Corner Park Garage (Texaco)

Tel: (01443) 224115

1 mile north along the A4119, on the left. Access, Visa, Overdrive, All Star, Switch, Dial Card, Mastercard, Amex, Diners Club, Delta, BP Supercharge, Fast Fuel, Texaco Cards.

PLACES OF INTEREST

Llanerch Vineyard

Hensol, Pendoylan, Vale of Glamorgan CF72 8JU
Tel: (01443) 225877
e-mail; llanerch@cariadwines.demon.co.uk
Signposted "Vineyard" (1 Mile)

The largest vineyard in Wales, producing estate bottled Welsh wines marketed under the "Cariad" label. The extensive grounds include a 6.5 acre vineyard with a vineyard trail, landscaped gardens and a country park of woodland and lakes. There is a Visitor Centre with a coffee shop for light lunches and wine sales. Disabled access.

M4 JUNCTION 35

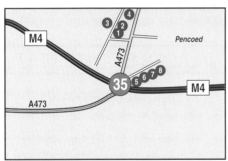

NEAREST A&E HOSPITAL

The Princess of Wales Hospital, Coity Road, Bridgend CF31 1RQ Tel: (01656) 752752

Proceed west to Junction 36 (Signposted from Junction. Distance Approx 4.9 Miles)

FACILITIES

1 Old Kings Head

Tel: (01656) 860203
1 mile north, on the right, in Pencoed. (Punch Taverns) Open all day. Bar Snacks served 11.30-23.00hrs daily

2 Pencoed Service Station (BP)

Tel: (01656) 860432
1 mile north, on the right, in Pencoed. Access,
Visa, Overdrive, All Star, Switch, Dial Card, Mastercard, Amex, AA Paytrak, Diners Club, Delta, Routex, BP Cards, Shell Agency. Open; 06.30-22.00hrs daily

3 Cafe Petite Steak House

Tel: (01656) 862783
1 mile north, on the left, in Pencoed. Open; Sun; 12.00-14.30hrs & 18.00-0.00hrs, Mon-Thurs; 18.00-23.00hrs, Fri & Sat; 18.00-0.00hrs

4 Riverside Filling Station (BP)

Tel: (01656) 860783
1 mile north along the A473, on the left. Access, Visa, Overdrive, All Star, Switch, Dial Card, Mastercard, Amex, AA Paytrak, Diners Club, Delta, Routex, BP Cards, Shell Agency. Open; 06.30-21.00hrs daily

5 Cross Roads Service Station (Texaco)

Tel: (01656) 863686
Adjacent to north east side of roundabout. Keyfuels, Access, Visa, Overdrive, All Star, Switch, Dial Card, Mastercard, Amex, Diners Club, Delta, BP Supercharge, Fast Fuel, Texaco Cards.

6 McDonald's

Tel: (01656) 865484
Adjacent to north east side of roundabout. Open; 07.30-23.00hrs daily

7 Pantruthyn Farm

Tel: (01656) 860133
Adjacent to north east side of roundabout. (Whitbread) Open all day. Meals served; Mon-Sat; 11.30-22.00hrs, Sun; 12.00-22.00hrs

8 Travel Inn

Tel: (01656) 860133
Adjacent to north east side of roundabout

M4 JUNCTION 36

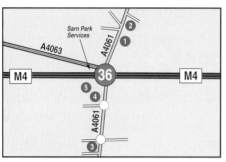

NEAREST A&E HOSPITAL

The Princess of Wales Hospital, Coity Road, Bridgend CF31 1RQ Tel: (01656) 752752

Signposted from Junction along A4061 south (Distance Approx 1.2 Miles)

FACILITIES

SERVICE AREA

ACCESSED FROM THIS JUNCTION

Sarn Park Services (Welcome Break)

Tel: (01656) 655332   Red Hen Restaurant, Granary Restaurant, Welcome Lodge & Shell Fuel

---

1   The Royal Oak

Tel: (01656) 720083
0.7 miles north along the A4061, on the right. (Ushers) Open all day. Meals served; Mon-Sat; 10.15-13.30hrs, Sun; 12.00-15.00hrs. Bar Meals available; 19.00-21.00hrs daily.

---

2   The Masons Arms Hotel

Tel: (01656) 720253
1 mile north along the A4061, on the right. Meals served; Mon-Sat; 12.00-14.30hrs & 18.00-22.00hrs, Sun; 12.00-14.30hrs

---

3   The Red Dragon

Tel: (01656) 654753
0.7 miles south along the Litchard Road, in Litchard. (Scottish & Newcastle) Open all day.

Meals served; 11.00-22.00hrs daily

4   Sainsbury's Petrol Station

Tel: (01656) 648951
Adjacent to south west side of roundabout in Litchard Park. Access, Visa, Overdrive, All Star, Switch, Dial Card, Mastercard, Amex, Diners Club, Delta, Style Card, Sainsbury's Fuel Card. Toilets and Cash Machine are available in adjacent store.

---

5   McArthur Glen Designer Outlet

Tel: (01656) 665700
Adjacent to south west side of roundabout in. There are numerous cafes and restaurants, including Harry Ramsden's, McDonald's, Madisons, Singapore Sam Chinese Restaurant, Fat Jackets, Sidoli's and Costa Coffee, within the Food Court. Open; Mon-Sat; 09.00-21.00hrs, Sun; 10.00-21.00hrs.

PLACES OF INTEREST

Bryngarw Country Park

Brynmenyn, Nr Bridgend CF32 8UU Park
Tel: (01656) 725155, House Tel: (01656) 729009
Follow the A4061 north (Signposted 1 Mile)

Bryngarw House, built in 1834, is set above a delightful ornamental garden, part of which is the exotic Japanese Garden of 1910. It is surrounded by a secluded 113 acre park with meadows, lake and woodland. The facilities include barbecue and picnic areas, a Visitor Centre with a touch screen computer and a children's play area. The Harlequin Restaurant serves cream teas, light refreshments and candle-lit dinners. Gift Shop. Disabled facilities

---

M4 JUNCTION 37

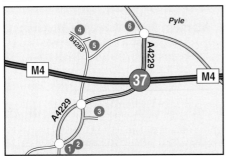

NEAREST EASTBOUND A&E HOSPITAL

The Princess of Wales Hospital, Coity Road, Bridgend CF31 1RQ Tel: (01656) 752752

Proceed to Junction 36 (Signposted from Junction. Distance Approx 6.8 Miles)

NEAREST WESTBOUND A&E HOSPITAL

Neath General Hospital, Penrhiwtyn, Neath SA11 2LQ Tel: (01639) 641161

Proceed to Junction 41, follow the A48 north and turn right along the A474. (Distance Approx 10.8 Miles)

FACILITIES

1   South Cornelly Service Station (Esso)

Tel: (01656) 746658

1 mile south along the A4229, on the left. Access, Visa, Overdrive, All Star, Switch, Dial Card, Mastercard, Amex, AA Paytrak, Diners Club, Delta, BP Supercharge, Shell Gold, Esso Cards.

2   Smokey Cott Drive Thru Chicken Inn

Tel: (01656) 741174

1 mile south along the A4229, on the left. Open; Mon-Sat; 07.30-22.00hrs, Sun; 08.00-22.00hrs

3   The Three Horsehoes

Tel: (01656) 740037

0.7 miles south, on the left, in South Cornelly. (Courage) Meals served; Mon-Sat; 12.00-14.15hrs & 18.30-21.30hrs, Sun; 12.00-14.30hrs & 19.00-20.30hrs

4   Mike Poacher & Sons Ltd (UK Fuels)

Tel: (01656) 740305

1 mile north, on the right, in North Cornelly. Access, Visa, Overdrive, All Star, Switch, Dial Card, Mastercard, Diners Club, Delta. Open; Mon-Fri; 07.30-18.00hrs, Sat; 08.00-16.00hrs, Sun; Closed.

5   The New House Inn

Tel: (01656) 747911

1 mile west, on the right in North Cornelly. (Bass) Open all day. Mon-Sat; 12.00-14.00hrs & 17.30-21.30hrs, Sun; 12.00-14.00hrs

6   The Crown Inn

Tel: (01656) 740021

0.7 miles north along the A48, on the left. (Pubmaster) Open all day. Meals served daily

PLACES OF INTEREST

Bedford Park

Cefn Cribwr, Nr Kenfig Hill, Bridgend
Tel: (01656) 725155 Follow the A4228 north, continue along the A48 to Pyle and turn right along the B4281. (4.5 Miles)

Centred on a Scheduled Ancient Monument, the former Cefn Cribwr Ironworks established by John Bedford in the 1780's, Bedford Park is an interesting mix of industrial archaeology, rare plants, flowers and country walks and nature trails, all sited within 40 acres of parkland. There are children's play areas and classroom facilities.

Kenfig National Nature Reserve

Ton Kenfig, Bridgend CF33 4PT
Tel: (01656) 743386 Take the A4229 and follow signposts to Ton Kenfig (2.5 Miles)

A fascinating dunescape and a mecca for naturalists, birdwatchers and ramblers. Kenfig contains more than 600 species of flowering plants, including 14 species of orchid. The 28 hectare freshwater lake provides a haven for birds and the Kenfig Reserve Centre has an exhibition, shop and information on places to visit.

Pets Welcome	£ Cash Dispenser	24 24 Hour Facilities	WC Toilets	Alcoholic Drinks
			Food and Drink	Fuel

M4 — JUNCTION 38 & JUNCTION 39

THESE ARE RESTRICTED ACCESS JUNCTIONS

Junction 38:
Vehicles can only exit from the westbound lanes
Vehicles can only enter along the eastbound lanes

Junction 39:
Vehicles can only enter along the westbound lanes.
Vehicles cannot exit at this junction in either direction

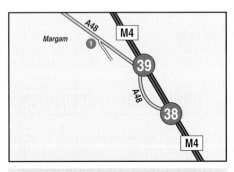

NEAREST EASTBOUND A&E HOSPITAL

The Princess of Wales Hospital, Coity Road, Bridgend CF31 1RQ Tel: (01656) 752752

Proceed to Junction 36 (Signposted from Junction. Distance Approx 10.8 Miles)

NEAREST WESTBOUND A&E HOSPITAL

Neath General Hospital, Penrhiwtyn, Neath SA11 2LQ Tel: (01639) 641161

Proceed to Junction 41, follow the A48 north and turn right along the A474. (Distance Approx 6.8 Miles)

FACILITIES

1   The Twelve Knights

🛏🍴♿📶🍺B   Tel: (01639) 882381
0.9 miles north along the A48, on the left. (Scottish & Newcastle) Open all day. Meals served Sun-Thurs; 12.00-21.30hrs, Fri & Sat; 12.00-22.00hrs.

PLACES OF INTEREST

Margam Abbey & Stones Museum

Margam, Port Talbot SA13 2TA
Margam Abbey Tel: (01639) 871184,
Stones Museum Tel: (029) 20 500200
Follow the A48 east, adjacent to motorway junction
(0.2 Miles)

Margam Abbey Church, founded in 1147, is the only Cistercian Foundation in Wales, whose nave is still intact and used for Christian worship. The Stones Museum, housed in an old school house, contains a collection of Christian memorials from the sub-Roman era right through to the hugely impressive "cart-wheel" crosses of the 10th & 11thC's, the finest example of which is the Cross of Conbelin.

Margam Country Park

Margam, Port Talbot SA13 2TJ
Tel: (01639) 881635 Follow the A48 east, adjacent to motorway junction (Signposted 0.5 Miles)

Set in 850 acres of glorious parklands, Margam Country Park offers something for all the family. The park, centred on Margam Castle, a Tudor/Gothic folly built in 1840, and the magnificent Orangery, constructed in 1789, contains a variety of attractions including splendid walks, Fairytale Land, a children's play area, Margam Maze, Road Train rides, Farm trail, Pets Corner, boating and fishing. Gift Shop. Charlotte's Pantry. Disabled access.

M4   JUNCTION 40

NEAREST EASTBOUND A&E HOSPITAL

The Princess of Wales Hospital, Coity Road, Bridgend CF31 1RQ Tel: (01656) 752752

Proceed to Junction 36 (Signposted from Junction. Distance Approx 13.3 Miles)

NEAREST WESTBOUND A&E HOSPITAL

Neath General Hospital, Penrhiwtyn, Neath SA11 2LQ Tel: (01639) 641161

Proceed to Junction 41, follow the A48 north and turn right along the A474. (Distance Approx 4.3 Miles)

 Large Hotel   Small Hotel   Motel   Guest House/ Bed & Breakfast   Disabled Facilities   Childrens Facilities

M4

PORT TALBOT TOWN CENTRE IS WITHIN ONE MILE OF THE JUNCTION. THERE ARE NO FACILITIES ADJACENT TO THE JUNCTION

PLACES OF INTEREST

Afan Forest Park & Countryside Centre

Cynonville, Port Talbot SA13 3HG
Tel: (01639) 850564 Follow the A4107 east
(Signposted 5.2 Miles)

The gateway to 3250 hectares of forest park, it is ideal for quiet countryside recreation. Facilities are available for walking, cycling, for which cycles are available for hire, orientering, camping and caravanning. Gift Shop. Cafe. Disabled access.

Located within the park ...

South Wales Miners Museum

Tel: (01639) 850564

A small museum depicting the story of the miner and his family in the South Wales valleys. The range of outdoor exhibits includes a blacksmith's shop.

M4 JUNCTION 41

PORT TALBOT TOWN CENTRE IS WITHIN ONE MILE OF THE SOUTH EXIT.

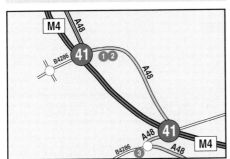

NEAREST A&E HOSPITAL

Neath General Hospital, Penrhiwtyn, Neath
SA11 2LQ Tel: (01639) 641161
Follow the A48 north and turn right along the A474.
(Distance Approx 3.1 Miles)

FACILITIES

1   The Bagle Brook

Tel: (01639) 813017

0.2 miles south along the A48 from the north exit, on the right. (Beefeater) Open all day. Meals served; Mon-Thurs 12.00-14.30hrs & 17.00-22.30hrs, Fri; 12.00-14.30hrs & 17.00-23.00hrs, Sat; 12.00-23.00hrs, Sun; 12.00-22.30hrs

2   Travel Inn

Tel: (01639) 813017

0.2 miles south along the A48 from the north exit, on the right

3   Blanco's Cafe Bar

Tel: (01639) 896378

0.4 miles along the A48 from the south exit, on the right

M4 JUNCTION 42

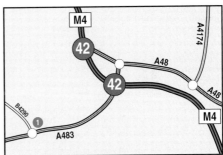

NEAREST A&E HOSPITAL

Neath General Hospital, Penrhiwtyn, Neath
SA11 2LQ Tel: (01639) 641161
Follow the A48 south and turn left along the A474.
(Distance Approx 1.9 Miles)

FACILITIES

1 Shell Swansea Bay

🛢️ ♿ WC 24   Tel: (01792) 326900

1 mile west along the A483, on the right (From eastbound exit), 0.5 miles west along the A483, on the right (From westbound exit) Access, Visa, Overdrive, All Star, Switch, Dial Card, Mastercard, Amex, AA Paytrak, Diners Club, Delta, BP Supercharge, Shell Cards, Smartcard.

PLACES OF INTEREST

Swansea

Swansea Tourist Information Centre, Singleton Street, Swansea SA1 3QG. Tel: (01792) 468321 e-mail; swantrsm@cableol.co.uk. Follow the A483 west (Signposted 5 Miles)

The town was granted its charter in 1210 and the original motte and bailey castle was built in Norman times, only to be replaced by a stone structure in the late 13thC. In its heyday as a port the docks had over six miles of quays, but today these are all but gone and the Maritime Quarter, a modern leisure centre with a 600 berth marina, now occupies the area. A former warehouse on this waterfront has been converted into a Maritime & Industrial Museum (Tel: 01792-650351) tracing the development of Swansea as a seaport and detailing some of the traditional industries. To the south of the town and sweeping west to the unspoilt Victorian village of Mumbles is the impressive Swansea Bay, part of the Gower Peninsula. One of Swansea's most famous sons, Dylan Thomas, is commemorated at the Dylan Thomas Centre (Tel: 01792-463980) on the banks of the River Tawe, whilst his birthplace at 5 Cwmdonkin Drive still exists, although in private ownership as a dwelling.

M4 · JUNCTION 43

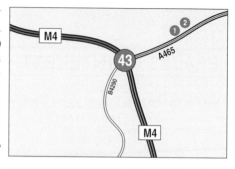

NEAREST A&E HOSPITAL

Neath General Hospital, Penrhiwtyn, Neath SA11 2LQ Tel: (01639) 641161

Follow the A465 north and turn right along the A474 (Distance Approx 3.7 miles)

FACILITIES

1 Sciwen Service Station (Elf)

🛢️ ♿ WC 24   Tel: (01792) 813081

0.4 miles north along the A465, on the left. Keyfuels, Access, Visa, Overdrive, All Star, Switch, Dial Card, Mastercard, Amex, AA Paytrak, Diners Club, Delta, Routex, Elf Cards, Total Fina Cards, UK Fuelcard, Securicor Fuelserv.

2 McDonald's

🍴 ♿   Tel: (01792) 817010

0.4 miles north along the A465, on the left. Open; Sun-Thurs; 07.30-0.00hrs, Fri/Sat & Sat/Sun; 07.30-03.00hrs

PLACES OF INTEREST

Pant-y-Sais Fen

New Road, Jersey Marine SA10 6JR
Tel: (01639) 763207
Follow the A483 west (1.9 Miles)

A nature reserve and site of Special Scientific Interest, Pant-y-Sais Fen is one of the few wetlands remaining in the region and is a must for those who appreciate the flora and fauna only to be found in these areas. The site is adjacent to the Tennant Canal for onward towpath strolls and boardwalks provide environmental-friendly paths through the fen area. Disabled access.

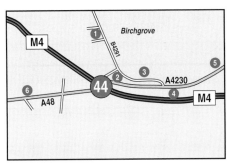

M4 — JUNCTION 44

NEAREST WESTBOUND A&E HOSPITAL

Morriston Hospital, Morriston, Swansea
SA6 6NL Tel: (01792) 702222

Proceed to Junction 45 (Signposted from the junction. Distance Approx 3 miles)

NEAREST EASTBOUND A&E HOSPITAL

Neath General Hospital, Penrhiwtyn, Neath
SA11 2LQ Tel: (01639) 641161

Proceed to Junction 43, follow the A465 north and turn right along the A474 (Distance Approx 3.7 miles)

FACILITIES

1 The Bridge End Inn

Tel: (01792) 321878
0.7 miles north along the B4291, on the left, in Birchgrove. (Scottish & Newcastle) Open all day. Meals served; Tues-Sat; 12.00-14.00hrs & 18.00-20.00hrs, Sun; 12.00-15.00hrs.

2 BF & AM Bevan (BP)

Tel: (01792) 817505
0.2 miles south along the A4230, on the left. Access, Visa, Overdrive, All Star, Switch, Dial Card, Mastercard, Amex, Diners Club, Delta,

Routex, AA Paytrak, Shell Agency, BP Cards. Toilets are available for customer use during daytime only.

3 The Bowen Arms

Tel: (01792) 812321
0.5 miles south along the A4230, on the left. (Entrepreneurs) Open all day. Meals served; Mon-Sat; 12.00-14.30hrs & 17.30-21.30hrs, Sun; 12.00-15.00hrs

4 Shell Baglan Bay

Tel: (01792) 3216910
0.6 miles south along the A4230, on the right. Access, Visa, Overdrive, All Star, Switch, Dial Card, Mastercard, Amex, AA Paytrak, Diners Club, Delta, Routex, BP Supercharge, Shell Cards, Smartcard.

5 The Travellers Well

Tel: (01792) 812002
1 mile south along the A4230, on the left. (Unique Pub Co.) Open all day Fri-Sun; Meals served; Mon-Thurs; 12.00-15.00hrs, Fri-Sun; 12.00-17.00hrs

6 Llansamlet Service Station (Esso)

Tel: (01792) 701330
0.7 miles west along the A48, on the right. Access, Visa, Overdrive, All Star, Switch, Dial Card, Mastercard, Amex, AA Paytrak, Diners Club, Delta, BP Supercharge, Shell Gold, Esso Cards.

PLACES OF INTEREST

Swansea

Follow the A48 west (Signposted 4.5 Miles)

For details please see Junction 42 information

M4 | JUNCTION 45

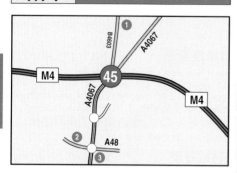

NEAREST A&E HOSPITAL
Morriston Hospital, Morriston, Swansea
SA6 6NL Tel: (01792) 702222
Signposted from the junction (Distance Approx 1 mile)

FACILITIES

1   Millers Arms

Tel: (01792) 842614
0.7 miles north along the B4603, on the right, in Ynystawe. (Whitbread) Open all day on Sat. Meals served; Mon; 11.30-14.00hrs, Tues-Sat; 11.30-14.00hrs & 18.00-21.30hrs, Sun; 12.00-14.30hrs.

2   McDonald's

Tel: (01792) 774295
1 mile south along the A4067, on the right. Open; Sun-Thurs; 07.30-23.00hrs, Fri & Sat; 11.30-0.00hrs

3   Wychtree Service Station (Texaco)

Tel: (01792) 700071
1 mile south along the A4067, on the left. Access, Visa, Overdrive, All Star, Switch, Dial Card, Mastercard, Amex, Diners Club, Delta, Texaco Cards.

PLACES OF INTEREST

Swansea
Follow the A4067 south (Signposted 3.7 Miles)
For details please see Junction 42 information

M4 | JUNCTION 46

THIS IS A RESTRICTED ACCESS JUNCTION

Vehicles can only exit from the westbound lanes
Vehicles can only enter the motorway along the eastbound lanes

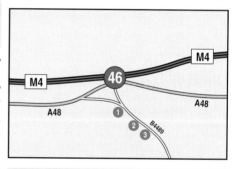

NEAREST A&E HOSPITAL
Morriston Hospital, Morriston, Swansea
SA6 6NL Tel: (01792) 702222
Signposted from the junction (Distance Approx 1.5 miles)

FACILITIES

1   Plough & Harrow

Tel: (01792) 771816
0.2 miles south along the B4489, on the right. (Freehouse) Open all day. Meals served; Mon-Sat; 12.00-14.00hrs & 17.00-20.30hrs, Sun; 12.00-14.00hrs.

2   The Willow Guest House

Tel: (01792) 775948
0.4 miles south along the B4489, on the right

3 Fairyglen Guest House

B   Tel: (01792) 790307

0.6 miles south along the B4489, on the right

PLACES OF INTEREST

Swansea

Follow the B4489 south (Signposted 3.3 Miles)

For details please see Junction 42 information

M4 JUNCTION 47

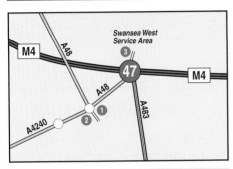

NEAREST A&E HOSPITAL

Morriston Hospital, Morriston, Swansea SA6 6NL Tel: (01792) 702222

Proceed east to Junction 45 (Signposted from the junction. Distance Approx 4.5 miles)

FACILITIES

SERVICE AREA

ACCESSED FROM THIS JUNCTION

Swansea West Services (Granada)

Tel: (01792) 896222   Self Service Restaurant, Burger King, Little Chef, Travelodge & BP Fuel

1 Cross Service Station (Elf)

WC   Tel: (01792) 897447

0.3 miles west along the A48, on the left, in Penllergaer. Access, Visa, Overdrive, All Star, Switch, Dial Card, Mastercard, Amex, AA Paytrak, Diners Club, Delta, BP Supercharge,

Elf Cards, Total Fina Cards. Open; Mon-Sat; 07.00-23.00hrs, Sun; 08.00-22.00hrs.

2 The Old Inn

Tel: (01792) 894097

0.3 miles west along the A48, on the left, in Penllergaer. (Brain's) Restaurant Open; Mon-Fri; 12.00-14.15hrs & 18.00-21.30hrs, Sat; 12.00-14.15hrs & 18.00-22.00hrs, Sun; 12.00-14.30hrs & 19.00-21.15hrs. Bar Meals served; 12.00-14.30hrs & 18.00-21.30hrs daily.

3 McDonald's

Tel: (01792) 898655

Adjacent to north side of roundabout. Open; Sun-Thurs; 07.30-23.00hrs, Sat & Sun; 07.30-0.00hrs.

PLACES OF INTEREST

Swansea

Follow the A483 south (Signposted 5 Miles)

For details please see Junction 42 information

M4 JUNCTION 48

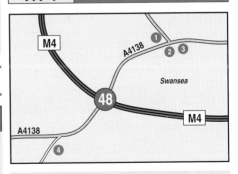

NEAREST A&E HOSPITAL

Morriston Hospital, Morriston, Swansea SA6 6NL Tel: (01792) 702222

Proceed east to Junction 45 (Signposted from the junction. Distance Approx 8.5 miles)

| Pets Welcome | Cash Dispenser | 24 Hour Facilities | Toilets | Alcoholic Drinks | Food and Drink | Fuel |

FACILITIES

1   The Black Horse Inn

🍴 🍽 ♿ 🅿   Tel: (01792) 882239
0.8 miles west along the A4138, on the left.
(Brain's) Open all day. Meals served; Mon;
12.00-14.30hrs & 18.00-20.45hrs, Tues-Sat;
12.00-20.45hrs, Sun; Bookings only.

2   The Gwyn Hotel

🍴 🍽 ♿ 🅿 🛏B   Tel: (01792) 882187
0.9 miles west along the A48, on the right, in
Pontardulais. Open all day. Meals served; Mon-
Sat; 12.00-14.30hrs & 19.00-21.30hrs, Sun;
12.00-15.30hrs

3   Shell Central

🅿 WC   Tel: (01792) 885741
1 mile west along the A48, on the right, in
Pontardulais. Access, Visa, Overdrive, All Star,
Switch, Dial Card, Mastercard, Amex, Diners
Club, Delta, BP Supercharge, Shell Cards,
Smartcard. Open; Mon-Sat; 06.00-22.00hrs,
Sun; 07.00-21.00hrs.

4   Smiths Arms

🍴 🍽 ♿ 🅿   Tel: (01554) 820305
1 mile east along the B4297, on the left, in
Llangennech. (Crown Buckley Taverns) Open
all day. Meals served; Mon-Sat; 12.00-14.00hrs
& 18.00-21.00hrs, Sun; 12.00-18.00hrs

APART FROM PONT ABRAHAM SERVICES THERE ARE NO OTHER FACILITIES AT THIS JUNCTION

SERVICE AREA

ACCESSED FROM THIS JUNCTION

Pont Abraham Services (RoadChef)
Tel: (01792) 884663   Food Fayre Self-Service
Restaurant & BP Fuel

NEAREST A&E HOSPITAL

Morriston Hospital, Morriston, Swansea SA6
6NL Tel: (01792) 702222
Proceed east to Junction 45 (Signposted from the
junction. Distance Approx 11.2 miles)

MOTORWAY ENDS

(TOTAL LENGTH OF MOTORWAY 189.0 MILES OR 190.7 MILES VIA M48)

M4   JUNCTION 49

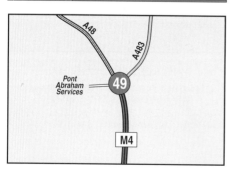

OFF THE MOTORWAY

M5

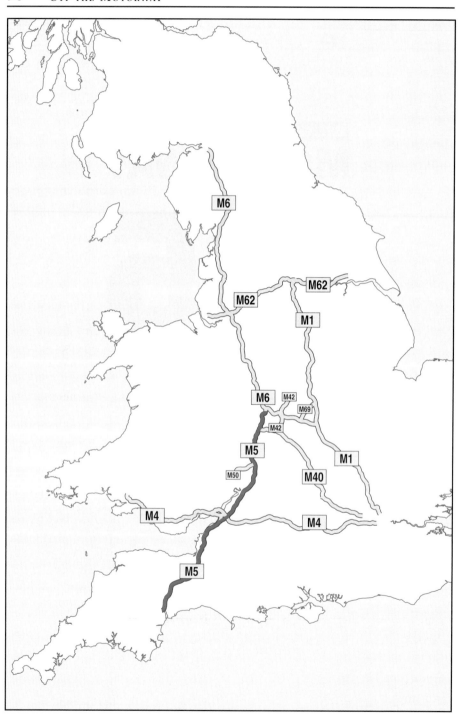

The M5 Motorway

The primary road route into the South West of England, parts of which were surveyed in the 1940s, it extends from the Ray Hall Interchange with the M6 at Junction 8 to the west side of Exeter, linking up with the A30 and A38 onwards into Devon and Cornwall.

From the M6 it passes south between Birmingham to the east and the Black Country, a collection of towns based around Dudley, before striking south west towards Worcestershire. Just south of Frankley, part of the Phoenix (former BMW/Rover) plant at Longbridge can be seen in the distance to the east, at the foot of the Lickey Hills.

The radio masts at Droitwich dominate the skyline as the motorway heads through Worcestershire, passing the town of Droitwich Spa, before reaching the historic County Town. Worcester Cathedral is visible to the west, at the south end of the city, as the carriageways proceed southwards, join up with the M50 and cross the River Avon with the delightful village of Bredon in view on its south bank. Tewkesbury, of 6thC origin and site of the decisive battle of the War of the Roses, is on the west side, and as the motorway continues south the Cotswold Hills are in the distance to the east and the Forest of Dean to the west before it passes between Gloucester and Cheltenham and crosses over the site of RAF Moreton Valence airfield.

The first of two monuments clearly visible from this motorway comes into view on the east side at Nibley Knoll. The Tyndale Monument was constructed in the 17thC as a memorial to William Tyndale who translated the first New Testament into English. Before reaching Bristol, there are panoramic views eastwards across to the Marlborough and Lambourn Downs.

The historic port city of Bristol is passed to the south with the Bristol Channel, straddled by the two Severn Bridges, and the Welsh Mountains beyond in view to the north before the motorway crosses the River Avon via the spectacular Avonmouth Bridge. From the west bank of the river, the motorway clings to the side of Tickenham Hill, with panoramic views across to the north, before dropping down to Clevedon and Weston super Mare and then passes between the Bleadon and Mendip Hills. The 137m high Brent Knoll is clearly visible to the north as the motorway traverses the Somerset Levels and continues south past Bridgwater and thence on to Taunton.

The carriageways continue south west passing the second monument visible from this motorway; The Wellington Monument, a landmark visible for miles around and erected in honour of the Duke of Wellington, and then traverse the Culm Valley before swinging around the south side of Exeter, crossing the River Exe and ending at Junction 31 where the road diverges into the A38 to Plymouth and A30 to Okehampton.

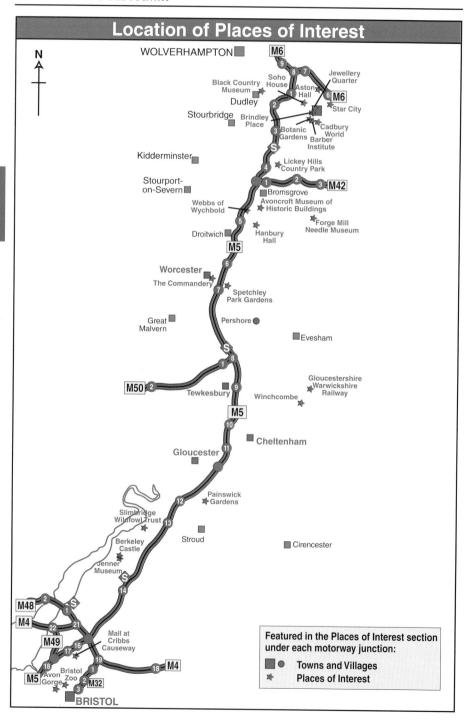

Location of Places of Interest

N

WOLVERHAMPTON

M6

9

8 7

Jewellery Quarter

Soho House

Black Country Museum

Aston Hall

1

M6

6

Dudley

2

Star City

Stourbridge

Brindley Place

Botanic Gardens

Cadbury World

3

Barber Institute

S

Kidderminster

Lickey Hills Country Park

4

Stourport-on-Severn

1

2

3

M42

Bromsgrove

Webbs of Wychbold

Avoncroft Museum of Historic Buildings

5

Forge Mill Needle Museum

Droitwich

Hanbury Hall

M5

6

Worcester

The Commandery

7

Spetchley Park Gardens

Great Malvern

Pershore

Evesham

S

8

1

Gloucestershire Warwickshire Railway

M50

2

Tewkesbury

9

Winchcombe

M5

10

Cheltenham

Gloucester

11

Painswick Gardens

12

Slimbridge Wildfowl Trust

13

Stroud

Cirencester

Berkeley Castle

Jenner Museum

S

14

M48

2

S

M4

1

22

21

M49

Mall at Cribbs Causeway

16

17

19

18

Bristol Zoo

18

M4

M5

Avon Gorge

1

M32

3

BRISTOL

Featured in the Places of Interest section under each motorway junction:

■ ● Towns and Villages

✴ Places of Interest

Location of Places of Interest

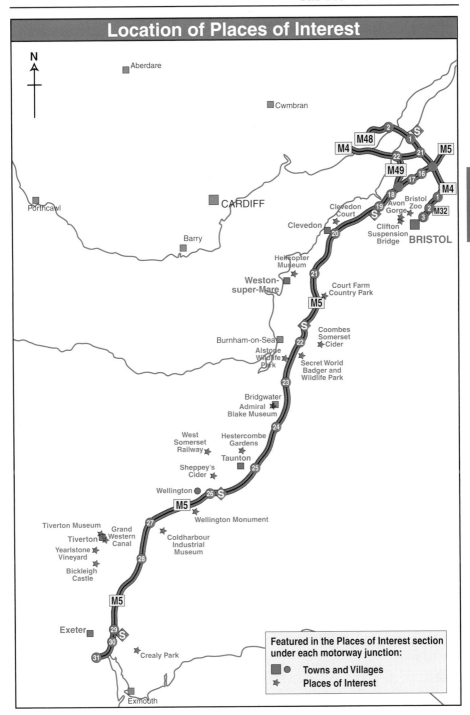

N

Aberdare

Cwmbran

M4 M48

M5

M49

M4

Porthcawl CARDIFF

Clevedon Court

Avon Gorge

Bristol Zoo

M32

Clevedon

Clifton Suspension Bridge

BRISTOL

Barry

Helicopter Museum

Weston-super-Mare

Court Farm Country Park

M5

Coombes Somerset Cider

Burnham-on-Sea

Alstone Wildlife Park

Secret World Badger and Wildlife Park

Bridgwater
Admiral Blake Museum

West Somerset Railway

Hestercombe Gardens

Taunton

Sheppey's Cider

Wellington

M5

Wellington Monument

Tiverton Museum

Grand Western Canal

Tiverton

Coldharbour Industrial Museum

Yearlstone Vineyard

Bickleigh Castle

M5

Exeter

Crealy Park

Exmouth

Featured in the Places of Interest section under each motorway junction:

■ ● Towns and Villages
✦ Places of Interest

M5

M5 JUNCTION 1

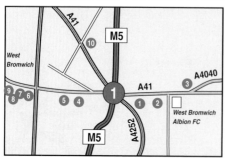

NEAREST A&E HOSPITAL

Sandwell District General Hospital,
Hallam Street, West Bromwich B71 4HJ
Tel: (0121) 553 1831

Take the A41 west towards West Bromwich. Turn right at the roundabout along the A4031 and the hospital is on the right. (Distance Approx 1.8 miles)

FACILITIES

---

1 Hawthornes Petrol Station (BP)

Tel: (0121) 553 7666
0.2 miles east along the A41, on the right. BP Cards, Routex, All Star, Overdrive, Diners Club, Switch, Visa, Mastercard.

2 The Hawthorns Pub

Tel: (0121) 553 0915
0.4 miles east along the A41, on the right. (Bass) Open all day. Lunch; 11.00-14.15hrs, Evening Meals; 17.00-19.00hrs.

3 Albion Filling Station (BP)

Tel: (0121) 551 1767
0.7 miles east along the A41, on the left. BP Cards, Switch, Mastercard, Visa, Dial Card, All Star, Routex, Amex, Diners Club.

4 MI Bhattay & Sons, Esso Service Station

Tel: (0121) 525 4100
0.3 miles west along the Birmingham Road, on the left. Esso Cards, Diners Club, Overdrive, Dial Card, All Star, Amex, Switch, Visa, Mastercard. Open; Mon-Sat; 07.30-22.00hrs, Sun; 09.00-21.00hrs.

5 Total Service Station

Tel: (0121) 553 5513
0.5 miles west along the Birmingham Road, on the left. Visa, Access, Energy Plus, Overdrive, All Star, Dial Card, Switch, Diners Club. Open; Mon-Sat; 07.00-23.00hrs, Sun; 08.00-23.00hrs.

6 Pizza Hut

Tel: (0121) 500 5232
1 mile west along the Birmingham Road, in the Farley Centre, West Bromwich. Open; 12.00-23.00hrs Daily.

7 McDonald's

Tel: (0121) 553 0436
1 mile west along the Birmingham Road, in the High Street, West Bromwich. Open; 07.30-23.00hrs Daily

8 KFC

Tel: (0121) 553 2119
1 mile west along the Birmingham Road, in the High Street, West Bromwich. Open; Sun-Thurs;12.00-0.00hrs, Fri/Sat & Sat/Sun; 11.00-03.00hrs.

9 Shalimar Indian Cuisine

Tel: (0121) 553 1319
1 mile west along the Birmingham Road, in the High Street, West Bromwich. Open 18.00-00.30hrs Daily.

10 The West Bromwich Moat House Hotel

Tel: (0121) 609 9988
0.3 miles north along Europa Avenue, on the right Restaurant Open; Mon-Fri; 12.00-14.00hrs, 18.30-22.00hrs, Sat; 18.30-22.00hrs, Sun; 12.30-14.30hrs, 19.00-21.30hrs.

| Large Hotel | Small Hotel | Motel | Guest House/ Bed & Breakfast | Disabled Facilities | Childrens Facilities |

PLACES OF INTEREST

Birmingham

Birmingham Convention & Visitor Bureau, 130
Colmore Row, Birmingham B3 3AP
Tel: (0121) 693 6300 web:www.birmingham.org.uk
Follow the A41 east (Signposted 4 miles)

Birmingham, Britain's second city, traditionally of industrial origins and the home base of a legion of world renowned brand names, has re-invented itself in the last decade to become a major tourist destination and plays host to millions of conference and exhibition delegates from around the world. Little more than a village at the time of the Domesday Book, its location alongside of the River Rea and adjacent to coal fields made it an ideal launch pad for the Industrial Revolution, and the attraction of entrepreneurs and engineering pioneers led to its rapid growth to become a fine Victorian city. Many of the buildings constructed during this period still remain and some of the earlier structures pre-dating this can still be found. The city centre, a very popular destination for shoppers and day trippers from all over the country, is undertaking a major rebuilding programme with reconstruction of the famous Bull Ring Centre and a brand new facility, The Millenium Point, the first phases of a £1billion plus investment in the expansion and modernization of the city.

Soho House

Soho Avenue, Off Soho Road, Handsworth,
Birmingham B18 5LB Tel: (0121) 554 9122
Follow A41 eastwards towards Birmingham and
turn right into Soho Avenue.(2.3 miles)

The former home of the industrialist pioneer Matthew Boulton who developed the steam engine in partnership with James Watt. Here he met with some of the most important scientists, engineers and thinkers of his time; the Lunar Society.

Jewellery Quarter Discovery Centre

75-79 Vyse Street, Hockley, Birmingham B18 6HA
Tel: (0121) 554 3598 Follow A41 east into the city
centre and the route is signposted (3.3 miles)

The preserved "time-capsule" workshop of the family run firm of Smith & Pepper forms the centre-piece of this highly interesting and fascinating exhibition. Sited in the heart of the Jewellery Quarter, which for over 200 years has helped the city to become known as the "Workshop of the World", it demonstrates the skills and techniques utilized in manufacturing jewellery of the highest standard.

Brindley Place & Broad Street, Birmingham

Follow A41 east into the city centre and the route is
signposted "National Indoor Arena & Convention
Centre" (4.5 miles)

The revitalized and dynamic west end of the city, centred around the International Convention Centre and National Indoor Arena. Enjoy the atmosphere of the waterside bars, night clubs, cafes and restaurants. Canal tours start from Gas Street Basin (Parties Afloat Tel: 0121-236 7057) and the National Sea Life Centre in The Waters Edge (Tel: 0121-633 4700) has spectacular walk-through aquaria. The Ikon Gallery (Tel: 0121-248 0708) holds exhibitions of contemporary art and the Museum & Art Gallery (Tel: 0121-235 2834), nearby, houses one of the world's finest collections of pre-Raphaelite Art. Symphony Hall, within the ICC complex and home of the CBSO, although only recently built is already rated as one of the greatest concert halls in the world.

M5

M5   JUNCTION 2

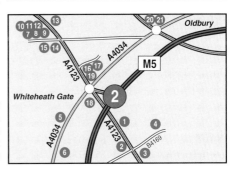

NEAREST A&E HOSPITAL

Sandwell District General Hospital,
Hallam Street, West Bromwich B71 4HJ
Tel: (0121) 553 1831

Proceed to Junction 1 and take the A41 west towards West Bromwich. Turn right at the roundabout along the A4031 and the hospital is on the right. (Distance Approx 4.2 miles)

FACILITIES

1   The New Navigation Pub

0.4 miles south along the A4123, on the left.
(Tetley Domecq) Lunches served.

2   Cin Cin Italian Restaurant

Tel: (0121) 552 1752
0.5 miles south along the A4123, on the right.
Open; Mon-Fri; 12.00-14.00hrs, 18.30-22.30hrs, Sat; 18.30-22.30hrs, Sun 12.00-14.00hrs.

3   The Hen & Chickens Pub

Tel: (0121) 552 1058
0.7 miles south along the A4123, on the left.
Lunches served daily 12.00-14.00hrs.

4   Elf Petrol Station

Tel: (0121) 552 6957
0.8 miles east along the B4169, on the left. Visa, Switch, Access, All Star, Amex, Diners Club, Overdrive, Dial Card.

5   Whiteheath Service Station (Esso)

Tel: (0121) 559 8369
0.6 miles west along the A4034, on the right. Esso Cards, Dial Card, Overdrive, All Star, Visa, Switch, Mastercard, Diners Club, Delta, Amex. Open; Mon-Sat; 07.00-23.00hrs, Sun; 07.00-22.30hrs.

6   Bayford Thrust Service Station

Tel: (0121) 559 1217
0.8 miles west along the A4034, on the left. Switch, Overdrive, Diners Club, Amex. Open; 06.30-23.00hrs Daily.

7   Little Chef

Tel: (0121) 552 2494
0.5 miles north along the A4123, on the left. Open; 07.00-22.00hrs Daily.

8   Forte Travel Lodge

Tel: (0121) 552 2967
0.5 miles north along the A4123, on the left.

9   Osprey Filling Station (BP)

Tel: (0121) 552 3807
0.5 miles north along the A4123, on the left. Mastercard, Routex, BP, Dial Card, Overdrive, Amex, Diners Club.

10   Lakeside Restaurant

Tel: (0121) 552 3031
0.6 miles north along the A4123, on the left. (Travellers Fare) Open; 12.00-21.45hrs Daily

11   Travel Inn

Tel: (0121) 552 3031
0.6 miles north along the A4123, on the left.

12   KFC

Tel: (0121) 544 1819
0.6 miles north along the A4123, on the left. Open: Sun-Thurs; 11.30-00.00hrs, Fri & Sat; 11.30-01.00hrs.

13   Bury Hill Petrol Station (BP)

Tel: (0121) 544 9074
0.9 miles north along the A4123, on the right. Routex, BP, All Star, Overdrive, Dial Card, Amex, Diners Club, Switch, Visa, Mastercard.

14   Newbury Travel Filling Station (UK)

Tel: (0121) 552 3262
0.6 miles along Newbury Lane, on the left. Dial Card, Access, Visa, Switch, Overdrive, Keyfuels. Open: Mon-Thur; 07.30-19.00hrs, Fri; 07.30-20.00hrs, Sat; 08.00-16.00hrs, Sun; 10.00-14.00hrs.

15   Valentino Italian Restaurant

Tel: (0121) 552 4073
0.7 miles along Newbury Lane, on the left. Open for lunch at 12.00hrs and Evening Meals at 19.00hrs Daily.

M5

Large Hotel	Small Hotel	Motel	Guest House/ Bed & Breakfast	Disabled Facilities	Childrens Facilities	

16 The One and Two Halves

🍴 🍽 🚭 ♿   Tel: (0121) 544 9621

0.1 miles north along the A4123, on the right. (Tom Cobleighs) Open all day. Meals served; Mon-Sat; 12.00-22.00hrs, Sun; 12.00-21.30hrs.

17 Holiday Inn

🛏M ♿ 🚗   Tel: (0121) 511 0000

0.1 miles north along the A4123, on the right.

18 McDonald's

🍽 ♿   Tel: (0121) 541 2055

0.1 miles, on the west side of the Roundabout. Open; Sun-Thurs; 07.30-23.00hrs, Fri-Sat; 07.00-23.30hrs

19 Birchley Park Service Station (Elf)

⛽ ♿ WC 24   Tel: (0121) 552 6957

0.1 miles, on the east side of the Roundabout. Visa, Delta, Mastercard, Amex, Diners Club, Dial Card, All Star, Switch, Overdrive, UK Petrol, Maxol, BP Supercharge.

20 McDonald's

🍽 ♿   Tel: (0121) 552 9002

0.6 miles north along the A4034 in Halesowen Street, Oldbury. Open; 07.30-23.00hrs Daily. Drive-Thru Restaurant open until 0.00hrs on Fri & Sat.

21 Savacentre Petrol

⛽   Tel: (0121) 665 2900

0.7 miles north along the A4034 in Halesowen Street, Oldbury. Access, Visa, Switch, All Star, Dial Card, Overdrive. Open; Mon-Sat; 06.00-22.00hrs, Sun; 08.00-18.30hrs.

PLACES OF INTEREST

Birmingham

Follow the A4123 south (Signposted 6.5 miles)

For details please see Junction 1 information.

The Black Country Living Museum

Tipton Road, Dudley DY1 4SQ
Tel: (0121) 557 9643 web: www.bclm.co.uk
Follow A4123 north west towards Dudley and the site is signposted (3 miles)

An open air museum where an old fashioned village has been created by the canal as a living tribute to the skills and enterprise of the people of the Black Country, once the industrial heart of Britain. A full day is required to enjoy all the attractions and tours. Disabled Access to most of site.

Cadbury World

Bournville, Birmingham B30 2LD
Follow Brown & White tourist boards along the A4123 & A4040. (7 miles).

For details please see Junction 4 information.

Brindley Place & Broad Street, Birmingham

Follow A4123 & A456 east into the city centre and the route is signposted "National Indoor Arena & Convention Centre" (6.5 miles)

For details please see Junction 1 information.

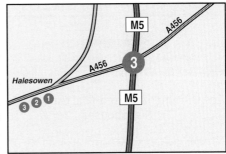

M5   JUNCTION 3

NEAREST A&E HOSPITAL

City Hospital, Dudley Road, Birmingham
B18 7QH Tel: (0121) 554 3801
Follow the A456 east into Birmingham and after about 3 miles turn left at Bearwood along the A4040. Turn right at the end of City Road along the A457 (Dudley Road) and the hospital is on the left. (Distance Approx 5.3 miles)

FACILITIES

1 Little Chef

🍴 ♿ Tel: (0121) 585 5412

0.8 miles west along the A456, on the left. Open; 07.00-22.00hrs Daily

2 Greenfields (BP)

🍴 WC £ Tel: (0121) 585 5105

0.8 miles west along the A456, on the left. BP Plus, Agency, Routex, Amex, Visa, Mastercard, Diners Club, Switch, Overdrive, Dial Card. Open; Mon-Fri; 06.00-23.00hrs, Sat; 07.00-23.00hrs, Sun; 07.00-22.00hrs.

3 The Black Horse Pub

🍴 🍴 ♿ 🔧 Tel: (0121) 550 1465

0.8 miles west along the A456, on the left. (Allied Domecq) Open All Day. Meals served; Mon-Fri; 12.00-15.00hrs and 18.00-22.00hrs (Bar Snacks available between 15.00 and 18.00hrs) Sat & Sun; 12.00-22.00hrs.

PLACES OF INTEREST

Birmingham

Follow the A456 east (Signposted 6 miles)

For details please see Junction 1 information.

The Birmingham Botanical Gardens & Glasshouses

Westbourne Road, Edgbaston,
Birmingham B15 3TR Tel: (0121) 454 1860
web: www.bham-bot-gdns.demon.co.uk
Follow the A456 towards the city centre and turn right onto the B4129 (Norfolk Road). It is signposted Botanical Gardens (5 miles)

Enjoy fifteen acres of magnificent Ornamental gardens including Glasshouses housing exotic plants. Dine in the Pavilion Restaurant where the Prime Minister, the Rt Hon Tony Blair MP and his wife hosted a dinner party on May 16th, 1998 for the G8 leaders and their wives during the Birmingham Summit. Disabled Access.

Brindley Place & Broad Street, Birmingham

Follow A456 east into the city centre and the route is signposted "National Indoor Arena & Convention Centre" (6 miles)

For details please see Junction 1 information.

SERVICE AREA

BETWEEN JUNCTIONS 3 AND 4

Frankley Services (Granada)

Tel: (0121) 550 3131 Little Chef, Burger King, Travelodge and Esso Fuel.

M5   JUNCTION 4

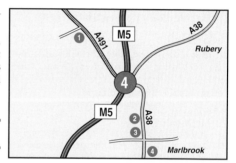

NEAREST A&E HOSPITAL

Selly Oak Hospital, Raddlebarn Road, Selly Oak, Birmingham B29 6JD Tel: (0121) 627 1627

Take the A38 north to Birmingham and into Selly Oak. The hospital is signposted within the city. (Distance Approx 7.4 miles)

NEAREST MINOR INJURY UNIT

The Princess of Wales Community Hospital, Stourbridge Road, Bromsgrove B61 0BB Tel: (01527) 488000

Follow A38 into Bromsgrove. Signposted within the town. Opening Hours; 09.00-17.00, Monday to Friday. (Distance Approx 2.6 miles)

FACILITIES

1 Home Farm B&B

=B Tel: (01527) 874964
0.2 miles west along the A491, on the left.

2 Stakis Country Court Hotel

=H **⋈** **⋈** **&** Tel: (0121) 447 7888
0.8 miles south along the A38, on the right.
Restaurant Open; Mon-Fri; 07.00-10.00hrs,
12.00-14.00hrs, 19.00-21.30hrs, Sat; 07.00-
10.00hrs, 19.00-21.30hrs, Sun; 07.30-10.00hrs,
12.00-14.00hrs, 19.00-21.30hrs.

3 Elf Service Station

⋈ **WC** Tel: (01527) 570178
0.9 miles south along the A38, on the right.
Access, Visa, Overdrive, Amex, All Star, Diners
Club, Dial Card, Switch, Elf, TotalFina. Open;
Mon-Fri; 07.00-23.00hrs, Sun; 08.00-23.00hrs.

4 The Marlbrook

⋈ **⋈** **⋈** **&** Tel: (01527) 878060
1 mile south along the A38, on the left. (Allied
Domecq) Open all day, Meals served 12.00-
21.00hrs Daily

PLACES OF INTEREST

Birmingham

Follow the A38 north (Signposted 10.4 miles)

For details please see Junction 1 information.

Cadbury World

Bournville, Birmingham B30 2LD Booking: (0121)
451 4159 24hr Information: (0121) 451 4180
web: www.cadbury.co.uk/cadworld.htm
Follow Brown & White tourist boards north east
along the A38 and A4040 (7 miles)

For anyone who loves chocolate, a visit to
Cadbury World is a trip to paradise. A fabulous
family experience all about chocolate - sights,
sounds, smells, and of course, tastes. Disabled
Access.

The Barber Institute of Fine Arts

University of Birmingham, Edgbaston, Birmingham
B15 2TS Enquiries: (0121) 414 7333
24hr Information: (0121) 472 0962
web: www.barber.org.uk e-mail: info@barber.org.uk
Follow A38 north east towards city centre.
Immediately after University site turn left along
Edgbaston Park Road and first left into University
Road East (7.6 miles).

One of the finest small picture galleries in the
world with an art reference library of more than
35,000 volumes. The Institute, housed in a
Grade 2 listed building was opened in 1939 and
has acquired a collection of works of
international significance. To those with any
sort of interest in art, this is an unmissable
visual delight. Disabled Access.

Lickey Hills Country Park

Warren Lane, Rednal, Birmingham B45 8ER
Tel: (0121) 447 7106
Follow the A38 east towards Birmingham and after
3 miles turn right along the B4120 towards Rednal.
At Rednal turn right along the B4096. (4.7 miles)

Over 500 acres of woodland, marshes and
heaths with a variety of wildlife. Within the
City of Birmingham the council is responsible
for the maintenance of over a million trees and
many of these can be seen in the panoramic
views across the city from the viewing point
on the top of the hills. The park has a visitor
centre, cafe, children's play area and a trail
suitable for disabled visitors.

M5   JUNCTION 4

**JUNCTION 4A IS A MOTORWAY
INTERCHANGE ONLY AND THERE IS NO
ACCESS TO ANY FACILITIES**

| | Pets Welcome | | Cash Dispenser | | 24 Hour Facilities | | Toilets | | Alcoholic Drinks | | Food and Drink | | Fuel |

M5 JUNCTION 5

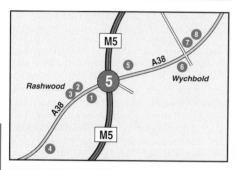

M5

NEAREST A&E HOSPITAL

Worcester Royal Infirmary, Ronkswood Branch,
Newtown Road, Worcester WR5 1HN
Tel: (01905) 763333

Proceed to Junction 6 and take the westbound exit along the A4440. The route is signposted A&E Hospital. (Distance Approx 8.6 miles)

NEAREST MINOR INJURY UNIT

The Princess of Wales Community Hospital, Stourbridge Road, Bromsgrove B61 0BB
Tel: (01527) 488000

Follow the A38 into Bromsgrove. Signposted within the town. Opening Hours; 09.00-17.00, Monday to Friday. (Distance Approx 5 miles)

FACILITIES

1 Robin Hood Inn

Tel: (01527) 861931

0.2 miles west along the A38, on the left. (Free House) Open all day. Meals served between 09.00 and 21.00hrs daily.

2 Little Chef

Tel: (01527) 861594

0.3 miles west along the A38, on the right. Open; 07.00 - 22.00hrs daily.

3 Travelodge

Tel: (01527) 86545

0.3 miles west along the A38, on the right.

4 Chateau Impney

Tel: (01905) 774411

1 mile west along the A38, on the left. Carvery and Restaurant open for Lunch and Evening Meals.

5 Sunnyside Transport Cafe

Tel: (01527) 861374

0.05 miles east along the A38, on the left. Open; Mon-Fri; 07.00-17.00hrs.

6 Wychbold Garage

Tel: (01527) 861861

0.2 miles east along the A38, on the right. Visa, Delta, Mastercard. Attended Service. Open Mon-Fri; 08.00-21.00hrs, Sat; 08.00-20.00hrs, Sun; 10.00-15.00hrs.

7 Poachers Pocket

Tel: (01527) 861413

0.3 miles east along the A38, on the left. (Banks's) Open all day. Meals served Mon-Sat; 12.00-22.00hrs, Sun; 12.00-21.30hrs.

8 Murco Service Station

0.4 miles east along the A38, on the left. All Star, Visa, Mastercard, Switch, Dial Card, Overdrive, Amex. Open; Mon-Sat; 07.00-23.00hrs, Sun; 08.00-22.00hrs.

PLACES OF INTEREST

Webbs of Wychbold

Wychbold, Droitwich Spa, Worcestershire
WR9 0DG Tel: (01527) 86177
e-mail: claire@webbsofwychbold.demon.co.uk
Follow the A38 east. Signposted from Junction 5
(1.2 miles)

One of the leading horticultural centres in the country extending over 55 acres and with excellent disabled access. The extensive facilities include a restaurant and large shop.

Avoncroft Museum of Historic Buildings

Stoke Heath, Worcestershire
Tel: (01527) 831886 Follow the A38 east and continue along the B4091 Signposted from Junction 5 (2.4 miles)

A fascinating collection of threatened buildings which have been carefully re-erected and restored. This open air museum also includes the national Collection of Telephone Kiosks. Picnic Area, Tea Room. Disabled Access.

Hanbury Hall

Droitwich, Worcestershire
Tel: (01527) 821214 4.5 miles west and signposted from Junction 5

An outstanding William and Mary red-brick house famed for its beautiful painted ceilings and staircase. Set in 400 acres of parkland and gardens. Accommodation available in Lodge. Snack Bar and Shop. Limited disabled access.

M5 JUNCTION 6

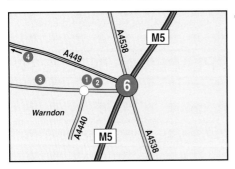

M5

NEAREST A&E HOSPITAL

Worcester Royal Infirmary, Ronkswood Branch, Newtown Road, Worcester WR5 1HN
Tel: (01905) 763333

Take the westbound exit along the A4440 and the route is signposted A&E Hospital. (Distance Approx 2.6 miles)

FACILITIES

1 Travel Inn

Tel: (01905) 451240
0.1 miles west along the A4440, on the right.

2 The Three Pears

Tel: (01905) 451240
0.1 miles west along the A4440, on the right. Open all day. Meals served Sun-Thurs; 12.00-22.30hrs, Fri & Sat; 12.00-23.00hrs.

3 Poachers Pocket

Tel: (01905) 458615
0.8 miles west along the Warndon Road, on the right. (Banks's) Open all day. Meals served 12.00-22.00hrs daily.

4 Blackpole Road Service Station (BP)

Tel: (01905) 456574
1.8 miles west along the A449, in Blackpole, on the left. Access, Visa, Overdrive, All Star, Switch, Dial Card, Mastercard, Amex, Diners Club, Delta, Routex, BP Cards, Shell Agency.

PLACES OF INTEREST

Worcester

Worcester Tourist Information Centre
Tel: (01905) 726311 Follow the A449 westwards (Signposted 4 miles)

The birthplace of Sir Edward Elgar, Worcester has an architectural heritage embracing the splendid Guildhall and magnificent Cathedral. The home of Worcester Porcelain, it is a major crossing point of the River Severn and river trips can be taken from the South Quay (Tel: 01905-422499). The pedestrianized High Street allows for leisurely shopping and for the sports orientated visitors the County Cricket Ground and Racecourse are adjacent to the city centre. The Museum & Art Gallery in Foregate Street (Tel: 01905-25371) features regular programmes of arts and craft exhibitions.

Within the city centre can be found ...

Worcester Cathedral

College Green, Worcester WR1 2LH
Tel: (01905) 28854 or Tel: (01905) 21004

A cathedral city since Saxon days, the magnificent edifice, with its 200ft tower, seen today dates back to Norman times. The crypt, now containing an exhibition of the history

| Pets Welcome | £ Cash Dispenser | 24 24 Hour Facilities | WC Toilets | Alcoholic Drinks | Food and Drink | Fuel |

and archaeology of the early cathedral, was constructed in 1084 and the Chapter House and Cloisters, reminders of its monastic past, were built in the 12thC. King John (d1116) and Prince Arthur (d1502) are both buried here. Cloisters Tea Room. Gift Shop. Part disabled access.

Royal Worcester Visitor Centre

Severn Street, Worcester WR1 2NE (adjacent to Cathedral) Tel: (01905) 21247 or Tel: (01905) 23221.

The Museum of Worcester Porcelain contains rare porcelain displayed in period settings whilst visitors are able to view modern production techniques through the factory tours and take a behind the scenes look at the Design Department. Morning Coffee, Lunch and Afternoon Teas are available in the Warmstry Restaurant and a Factory Shop retails Bargain Seconds and Bestware. Disabled facilities.

M5   JUNCTION 7

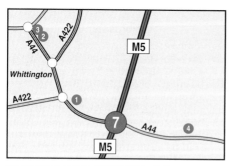

NEAREST A&E HOSPITAL

Worcester Royal Infirmary, Ronkswood Branch, Newtown Road, Worcester WR5 1HN
Tel: (01905) 763333
Take the westbound exit along the A44 and the route is signposted A&E Hospital. (Distance Approx 3 miles)

FACILITIES

1 The Swan

Tel: (01905) 351361
0.3 miles west along the A44, on the right.

(Banks's) Open all day Sat & Sun. Lunch; Mon-Fri; 12.00-14.30hrs, Evening meals; Mon-Fri; 17.30-21.30hrs. Meals served all day Sat; 12.00-21.30hrs and Sun; 12.00-21.00hrs.

2 Whittington Road BP Filling Station

Tel: (01905) 351245
1 mile west along the A44, on the right. Access, Visa, Overdrive, All Star, Switch, Dial Card, Mastercard, Amex, Diners Club, Delta, Routex, BP Cards, Shell Agency. Open; Mon-Fri; 07.00-22.00hrs, Sat; 07.00-21.00hrs, Sun; 08.00 - 21.00hrs

3 The Oak Apple

Tel: (01905) 355121
1 mile west along the A44, on the right. (Banks's) Open all day. Meals served 12.00-21.00hrs daily.

4 Wood View B&B

Tel: (01905) 351893
0.7 miles east along the A44, on the left

PLACES OF INTEREST

Pershore

Pershore Tourist Information Centre, Wanderers World, 19 High Street, Pershore WR10 1AA
Tel: (01386) 554262. Follow the A44 east (Signposted 7 miles)

A genuine old English market town with beautifully preserved Georgian architecture, fine old coaching inns and a magnificent Abbey considered by John Betjeman to be one of the best in Britain. Surrounded by the Vale of Evesham, Pershore nestles on the banks of the beautiful River Severn.

Within the town can be found ...

Pershore Heritage Centre

5 Bridge Street, Pershore WR10 1AJ
Tel: (01386) 552827

An old beamed cottage containing many local displays.

Pershore Abbey

Pershore WR10 1DT Tel: (01386) 561520

Founded by King Oswald in 689AD and although only the choir now remains standing it is an impressive building and treasury of mostly 13thC architecture with a tall pinnacled tower.

Worcester

Follow the A44 north (Signposted 4 miles)

For details please see Junction 6 information.

Within the city centre can be found ...

Worcester Cathedral

College Green, Worcester WR1 2LH

For details please see Junction 6 information.

Royal Worcester Visitor Centre

Severn Street, Worcester WR1 2NE
(adjacent to Cathedral)

For details please see Junction 6 information.

Spetchley Park Gardens

Spetchley Park, Worcester WR5 1RS
Tel: (01905) 345213 or Tel: (01905) 345224
Follow the A44 north, take the A4440 north east and turn east along the A422. (3 miles)

This lovely 30 acre private garden is a Plantsman's Paradise, containing a large collection of trees, shrubs and plants, many of which are rare or unusual. Refreshments available.

The Commandery

Sidbury, Worcester WR1 2HU
Tel: (01905) 361821
Follow the A44 north. (2.4 miles)

A magnificent timber-framed building and Charles II's headquarters during the Battle of Worcester in 1651. Contains period rooms and Civil War displays.

SERVICE AREA

BETWEEN JUNCTIONS 7 AND 8

Strensham Services (Northbound) (RoadChef)

Tel: (01684) 293004 Food Fayre Self-Service Restaurant, Wimpy Bar, RoadChef Lodge & Texaco Fuel

Strensham Services (Southbound) (RoadChef)

Tel: (01684) 293004 Food Fayre Self-Service Restaurant, Wimpy Bar & BP Fuel

M5 JUNCTION 8

JUNCTION 8 IS A MOTORWAY INTERCHANGE ONLY AND THERE IS NO ACCESS TO ANY FACILITIES

M5 JUNCTION 9

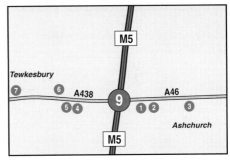

NEAREST A&E HOSPITAL

Cheltenham General Hospital, Sandford Road, Cheltenham GL53 7AN Tel: (01242) 224133

Proceed to Junction 10 and follow the A4019 into Cheltenham. The hospital is signposted from within the town. (Distance Approx 9.1 miles)

NEAREST MINOR INJURY UNIT

Tewkesbury Hospital, Barton Road, Tewkesbury GL20 5QN Tel: (01684) 293303.

Follow A438 into Tewkesbury. Signposted from within the town. Open 24 hours. (Distance Approx 1.5 miles)

FACILITIES

1 BP Express Shopping, Ashchurch Filling Station

🏠 ♿ WC 24 £   Tel: (01684) 293785

0.2 miles east along the A46, on the right. Access, Visa, Overdrive, All Star, Switch, Dial Card, Mastercard, Amex, Diners Club, Delta, Routex, Shell Agency, BP Cards.

| Pets Welcome | £ Cash Dispenser | 24 24 Hour Facilities | WC Toilets | Alcoholic Drinks | Food and Drink | Fuel |

2 Little Chef

🍴 ♿ Tel: (01684) 292037

0.2 miles east along the A46, on the right.
Open; 07.00-22.00hrs Daily

3 Newton Farm B&B

🛏B Tel: (01684) 295903

0.5 miles east along the A46, on the right.

4 Spa Villa B&B

🛏B Tel: (01684) 292487

0.6 miles west along the A438, on the left.

5 Newtown Filling Station (Keyfuels)

⛽ Tel: (01684) 297583

0.6 miles west along the A438, on the left.
Access, Visa, Overdrive, All Star, Switch,
Card, Mastercard, Amex, Diners Club, Delta,
Routex, Keyfuels. Open; Mon-Fri; 07.00-
22.00hrs, Sat; 07.00-21.00hrs, Sun; 08.00 -
21.00hrs

6 The Canterbury

🍴 🍴 🔄 Tel: (01684) 297744

0.7 miles west along the A438, on the right.
(Whitbread) Open all day. Lunch; Mon-Sat;
12.00-14.00hrs, Sun; 12.00-14.30 hrs, Evening
meals; Mon-Thurs; 19.00-21.00hrs, Fri & Sat;
18.00-21.00hrs.

7 Safeway Petrol Station

🍴 ♿ 🚻 Tel: (01684) 273268

1 mile west along the A438, on the right.
Access, Visa, Overdrive, All Star, Switch, Dial
Card, Mastercard, Amex, Diners Club, Delta,
AA Paytrak, UK Fuelcard. Open; Mon-Thurs &
Sat; 07.00-21.00hrs, Fri; 07.00-22.00hrs, Sun;
09.00 -17.00hrs

PLACES OF INTEREST

Tewkesbury

Tewkesbury Tourist Information Centre
Tel: (01684) 295027 Follow the A438 west.
(Signposted 1.5 miles)

Sited at the confluence of the Severn and Avon
Rivers, it was unable to expand outwards and,
as a result, the narrow streets became densely
packed with unusually tall buildings; many of
those built in the 15th and 16th centuries still
remain today. The most notable structure is the
Abbey which stands at the centre of a "Y"
formed by High Street, Church Street and
Barton Street with the area between being filled
with narrow alleyways and courtyards
containing some wonderful old pubs and
mediaeval cottages. The strategic position of
the town ensured that it had a turbulent
military history, with the Battle of Tewkesbury,
which took place on May 4th, 1471 and saw
the Lancastrian army defeated in the
penultimate and most decisive battle in the
Wars of the Roses, taking place to the south of
the town. A Battle Trail which encompases this
area starts near the Abbey. The town also
changed hands several times during the Civil
War.

Tewkesbury Abbey & Museum

Church Street, Tewkesbury GL20 5RZ
Tel: (01684) 850959 Follow the A438 west and
then A38 south (Signposted 1.9 miles)

A parish church of cathedral-like proportions
it was founded in the 8thC and completely
rebuilt at the end of the 11th. It was once the
church of the Benedictine Abbey of Tewkesbury
and was one of the last monasteries to be
dissolved by Henry VIII. In 1540 the abbey was
saved from destruction by the town burghers
who purchased it from the Crown for £453.
The main tower, 132ft high and 46ft square, is
thought to be the largest Norman Church
tower still in existence. There is a shop and
the Refectory serves Tea & Coffee, Lunches and
Afternoon Teas.

Winchcombe

Follow the A46 east, continue along the B4077 and
the route is signposted (9 miles)

The attractive small town of Winchcombe was
once a regional capital of Saxon Mercia and

| | Large Hotel | | Small Hotel | | Motel | | Guest House/ Bed & Breakfast | | Disabled Facilities | | Childrens Facilities |

one of the town's more enduring legends concerns Kenelm, a popular child king who is said to have been martyred here by his jealous sister, Quendrida, in the 8thC. As a means of calming a mob of people who had gathered to voice their anger at this murderous act, legend has it that she recited Psalm 109 backwards, a deed which resulted in her being struck blind in an act of divine retribution. A shrine to St.Kenelm, as he became known, during Mediaeval times was second only to Thomas a Beckett's as one of the foremost places of pilgrimage and Winchcombe prospered to become a walled town with an abbot. However in 1539, the abbey was destroyed by Thomas Seymour of Sudeley following Henry VIII's Dissolution of the Monasteries and all that remains today is a section of a gallery which forms part of The George Inn. In the wake of this event the townspeople were forced to find an alternative source of income, and this they found in the form of a crop which had just been introduced from the New World; Tobacco. Although difficult to grow, the town proceeded to earn a good living from this until an Act of Parliament in 1670, forbidding home grown tobacco in favour of imports from the struggling colony of Virginia, brought it to an abrupt end. With that the town slipped into decline as any sort of commercial centre but many of the buildings erected during its period of prosperity can still be seen today and Winchcombe is well worthy of a visit.

Within the town can be found ...

Winchcombe Folk and Police Museum

Town Hall, High Street, Winchcombe
GL54 5LJ Tel: (01242) 602925

The history of Winchcombe from Neolithic times to the present day is shown in a series of exhibits and there is an international display of police uniforms, caps, badges and equipment through the ages. Shop.

Winchcombe Railway Museum & Garden

23 Gloucester Street, Winchcombe
GL54 5LX Tel: (01242) 602257

One of the largest collections of railway memorabilia and equipment has been assembled here, including a working signal box and a booking office. Many of the artefacts adorn the beautiful half-acre garden which is full of old and rare plants. Shop. Disabled access to most of the site.

Sudeley Castle

Winchcombe GL54 5JD
Tel: (01242) 602308
web: www.stratford.co.uk/sudeley
e-mail: marketing@sudeley.ndirect.co.uk

Once the home of Queen Katherine Parr, this castle was the garrison headquarters of Prince Rupert during the Civil War and was twice besieged, in 1643 and 1644. The ravages of this conflict left it derelict and it was not until it was purchased in 1837 by John and William Dent that the huge restoration task began. The building now houses an impressive collection of furniture and paintings and the nine beautiful gardens that surround it cover 14 acres. St.Mary's Church contains the marble tomb of Queen Katherine Parr, Henry VIII's sixth wife. Gift Shop and Plant Centre. Restaurant. Limited disabled access.

Gloucestershire Warwickshire Railway

Toddington Station, Toddington GL54 5DT
Tel: (01242) 621405 web: www.gwsr.plc.uk
Follow the A46 east and continue along the B4077
to Toddington (8.8 miles)

Running the 13 miles from Toddington to Gotherington, the GWR has splendid views across the Vale of Evesham to the Malvern Hills and beyond to the Welsh Mountains. A variety of motive power is available with vintage steam and diesel locomotives and a diesel railcar utilized throughout the timetable. Originally part of the main line from Birmingham Snow Hill to Cheltenham, the preservation group are currently preparing the line onward to Cheltenham.

M5 JUNCTION 10

THIS IS A RESTRICTED ACCESS JUNCTION

Vehicles can only exit from the southbound lanes and travel east along the A4019

Vehicles can only enter the motorway along the northbound lanes from the A4019 west carriageway

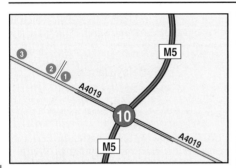

M5

NEAREST A&E HOSPITAL

Cheltenham General Hospital, Sandford Road, Cheltenham GL53 7AN Tel: (01242) 224133

Southbound; Take the A4019 into Cheltenham and turn right along the A46. The hospital is on this road. (Distance Approx 4.4 miles)

Northbound; Proceed north to Junction 9 and return to Junction 10. Take the A4019 into Cheltenham and turn right along the A46. The hospital is on this road. (Distance Approx 14 miles)

NEAREST MINOR INJURY UNIT

Tewkesbury Hospital, Barton Road, Tewkesbury GL20 5QN Tel: (01684) 293303

Proceed to Junction 9 and follow the A438 into Tewkesbury. Signposted from within the town. Open 24 hours. (Distance Approx 6.4 miles)

FACILITIES

1   Stanborough Cottage

📧B   Tel: (01242) 680327
0.6 miles west along the A4019, on the right.

2   The Gloucester Old Spot

🔲 🍴 ♿ 🔧 📧B   Tel: (01242) 680321
0.6 miles west along the A4019, on the right. (Whitbread) Open all day. Lunch, Evening meals.

3   Silver Birches B & B

📧B

0.9 miles west along the A4019, on the right

PLACES OF INTEREST

Cheltenham

Cheltenham Tourist Information Centre,
77 Promenade, Cheltenham GL50 1PP
Tel: (01242) 522878
Follow the A4019 east. (Signposted 4.5 miles)

The chance discovery of a saline spring in 1715 by a local farmer subsequently transformed the character of Cheltenham, a small market town, forever. Some twenty years later Captain Henry Skillicorne saw the potential of this discovery and built an enclosure around the spring along with various buildings and a network of walks and rides which now form the tree-lined Promenade. The town rapidly became a highly fashionable resort and by the 1820s huge amounts of money had been invested in constructing houses and buildings in the Neoclassical Regency style, most of which can be seen today. On a more modern note, Cheltenham is home to the Government tracking station, GCHQ.

Within the town centre can be found ...

Cheltenham Art Gallery & Museum

Clarence Street, Cheltenham GL50 3JT
Tel: (01242) 237431
web: http://www.cheltenham.gov.uk/agm
e-mail: Artgallery@cheltenham.gov.uk

Occupying three floors, the Art Gallery section holds a world-famous collection relating to the Arts and Crafts Movement, including fine furniture and exquisite metalwork, made by Cotswold craftsmen and inspired by William Morris, rare Chinese and English pottery and 300 years of painting by Dutch and English artists. The museum details the story of Edward Wilson, Cheltenham's Arctic explorer and shows the history of Britain's most complete Regency town as well as exhibiting archaeological treasures from the Cotswolds. Full disabled access. Gift Shop. Cafe.

Holst Birthplace Museum

4 Clarence Road, Cheltenham GL52 2AY
Tel: (01242) 524846.

The birthplace of Gustav Holst, composer of The Planets, contains displays detailing the life of the famous musician. This Regency terraced

| H Large Hotel | h Small Hotel | M Motel | B Guest House/ Bed & Breakfast | Disabled Facilities | Childrens Facilities |

house also shows the "upstairs-downstairs" way of life of Victorian and Edwardian times, including a working Victorian kitchen.

M5 | JUNCTION 11

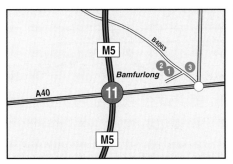

NEAREST A&E HOSPITAL

Cheltenham General Hospital, Sandford Road, Cheltenham GL53 7AN Tel: (01242) 224133
Follow the A40 into Cheltenham. The hospital is signposted from within the town. (Distance Approx 3.5 miles)

OR ALTERNATIVELY

Gloucestershire Royal Hospital, Great Western Road, Gloucester GL1 3PQ Tel: (01452) 528555
Follow the A40 into Gloucester. The hospital is signposted within the town. (Distance Approx 3.7 miles)

FACILITIES

1 Briarfields Motel

Tel: (01242) 235324
0.9 miles east, in Churchdown Road, on the left.

2 White House Hotel

Tel: (01452) 713226
1 mile east, in Churchdown Road, on the left. Restaurant Open; Breakfast; 07.00-09.30hrs daily, Lunch; 12.00-14.00hrs daily, Dinner; Mon-Fri; 19.00-21.30hrs, Sat; 19.00-22.00hrs, Sun; 19.00-21.00hrs.

3 Thistle Cheltenham Hotel

Tel: (01242) 232691
0.9 miles east along the A40, on the left. Restaurant Open; Breakfast; Mon-Fri; 07.00-10.00hrs, Sat; 07.30-10.30hrs, Sun; 07.30-10.30hrs, Lunch; 12.30-13.45hrs daily, Dinner; Mon-Sat; 19.00-21.45hrs, Sun; 19.00-21.15hrs.

PLACES OF INTEREST

Cheltenham
Follow the A40 east. (Signposted 3.8 miles)
For details please see Junction 10 information.

Within the city centre can be found ...

Cheltenham Art Gallery & Museum
Clarence Street, Cheltenham GL50 3JT
For details please see Junction 10 information.

Holst Birthplace Museum
4 Clarence Road, Cheltenham GL52 2AY
For details please see Junction 10 information.

Gloucester
Gloucester Tourist Information Centre, 28 Southgate Street, Gloucester GL1 2DP
Tel: (01452) 421188
e-mail: tourism@glos-city.gov.uk
Follow the A40 west (Signposted 4.4 miles)
Originally a Roman settlement dating from the 1stC and known as Glevum, a Saxon monastery was established here in the 7thC. After the Norman invasion Gloucester Cathedral was established and the building was modified and rebuilt over the years. At 72ft x 38ft the great east window, constructed to celebrate victory at the Battle of Crecy in 1346, is the largest surviving mediaeval window in England. There are still many old buildings surviving within the city and the inland docks which by the end of the 1970s were totally derelict have been thoroughly refurbished and are now a thriving tourist attraction.

Within the city centre can be found ...

Beatrix Potter's House of the Tailor of Gloucester

9 College Court, Gloucester GL1 2NJ
Tel: (01452) 422856

This recently refurbished attraction is housed in the building that Beatrix Potter sketched in 1897 and contains a large range of memorabilia and manuscripts. Shop. Limited Disabled access.

The National Waterways Museum

Llanthony Warehouse, Gloucester Docks,
Gloucester GL1 2EH Tel: (01452) 318054
E-mail paul@nwm.u-net.com.
Follow the A40 west (Signposted "Historic Docks"
4.7 miles)

The museum, which is devoted to the 200 year period when the inland waterways carried the goods of the nation, occupies three floors of the warehouse which, itself, forms part of the museum. The exhibits include working engines and models, live craft demonstrations, archive film presentations and hands-on computer simulations of canal navigations. A number of barges and narrowboats are moored along the two quays and the Queen Boadicea II is utilized for short daily public trips whilst the King Arthur is employed on longer river and canal cruises. A working forge, steam dredger, shire horse stable and canal-related workshops are also encompassed within the site. Gift Shop. Tea Room. Disabled access.

Robert Opie Museum of Advertising & Packaging

Albert Warehouse, Gloucester Docks GL1 2EH
Tel: (01452) 302309. Follow the A40 west
(Signposted "Historic Docks") (4.7 miles)

A collection of some 30,000 items of packaging and advertising accumulated by Robert Opie over the last 20 years forms the nucleus of this fascinating exhibition. It not only offers a real insight into the presentation of consumer packaging since the mid-1800s but, on display, are collections of posters, enamel signs and point-of-sale promotions which together chronicle the changing trends in popular taste. To complete the picture there is also a continuous presentation of vintage television commercials..

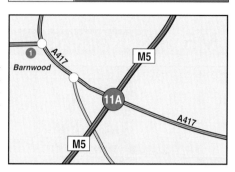

M5   JUNCTION 11A

NEAREST A&E HOSPITAL

Gloucestershire Royal Hospital, Great Western Road, Gloucester GL1 3PQ Tel: (01452) 528555
Follow the A417 into Gloucester. The hospital is signposted within the town. (Distance Approx 2.8 miles)

FACILITIES

1   Forte Post House

Tel: (01452) 613311
1 mile north along the A417, on the left. Traders Bar & Restaurant. Open for meals; Mon-Thurs: 12.00-14.00hrs & 18.00-22.30hrs, Fri: 12.00-14.00hrs & 17.00-22.30hrs, Sat: 17.00-22.30hrs, Sun: 12.00-14.00hrs & 18.00-22.00hrs.

PLACES OF INTEREST

Gloucester

Follow the A417 west (Signposted 3.4 miles)
For details please see Junction 11 information.

Within the city centre can be found ...

The National Waterways Museum

Llanthony Warehouse, Gloucester Docks,
Gloucester GL1 2EH Follow the A417 west
(Signposted "Historic Docks" 4.3 miles)
For details please see Junction 11 information.

Robert Opie Museum of Advertising & Packaging

Albert Warehouse, Gloucester Docks GL1 2EH
Follow the A417 west (Signposted "Historic Docks"
4.3 miles)

For details please see Junction 11 information.

Painswick Rococo Gardens

Painswick House, Painswick,
Gloucestershire GL6 6TH Tel: (01452) 813204
web: www.beta.co.uk/painswick/pwintro.htm
e-mail: Painsgard@aol.com
Follow the A417 east and take the A46 south.
(Signposted along A46 7 miles)

Commissioned by Benjamin Hyett in the 1740's, the gardens survived in that form until they were changed in the early 1800's to a more practical layout to grow fruit and vegetables. Over the years they gradually proved to be too expensive to maintain and eventually were abandoned in the 1950's. Thanks to a painting of the original garden by Thomas Robins in 1748 a reference was available and restoration, with private funding, began in 1984 when its historic importance became apparent as it was found to be the sole, complete, survivor from the brief Rococo period of English garden design (1720-1760). In 1988 control of the garden was handed over to the Painswick Rococo Garden Trust and they continued with the restoration of the 6 acre site. Coffee, Lunches and Teas in the licensed Coach House Restaurant. Gift Shop. Partial Disabled access.

M5 JUNCTION 12

THIS IS A RESTRICTED ACCESS JUNCTION

Vehicles can only exit from the northbound lanes and travel west along the B4008

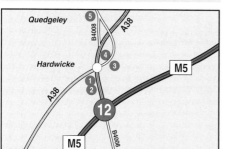

Vehicles can only enter the motorway along the southbound lanes from the B4008 east carriageway

NEAREST A&E HOSPITAL

Gloucestershire Royal Hospital, Great Western Road, Gloucester GL1 3PQ Tel: (01452) 528555

Northbound: Follow the A38 into Gloucester. The hospital is signposted within the town. (Distance Approx 6.3 miles)

Southbound: Proceed south to Junction 13 and return to Junction 12. Follow the A38 into Gloucester. The hospital is signposted within the town. (Distance Approx 12.2 miles)

NEAREST MINOR INJURY UNIT

Stroud General Hospital, Trinity Road, Stroud GL5 2HY Tel: (01453) 562200

Proceed to Junction 13 and follow the A419 to Stroud. Signposted from within the town. Open 24 hours (Doctors in attendance between 09.00-12.30hrs and 14.00-16.30hrs only) (Distance Approx 8.6 miles)

FACILITIES

1 Cross Keys Filling Station (BP)

🛢 ♿ WC 24   Tel: (01452) 721470
0.4 miles north along the B4008, on the left. Access, Visa, Overdrive, All Star, Switch, Dial Card, Mastercard, Amex, Diners Club, Delta, Routex, Shell Agency, BP Card.

2 Little Chef

🍴 ♿   Tel: (01452) 720132
0.4 miles north along the B4008, on the left. Open; 07.00-22.00hrs Daily

3 Hardwicke Garage (UK Petrol)

🛢 WC   Tel: (01452) 720239
0.5 miles north along the A38, on the right. Access, Mastercard. Open; Mon-Fri; 08.00-17.00hrs, Sat; 08.00-13.00hrs, Sun; Closed.

4 The Cross Keys

🍺 🍴 ♿ 🛏   Tel: (01452) 720113
0.5 miles north along the A38, on the right. (Whitbread) Meals served; Sun-Fri; 12.00-

M5

14.00hrs, 19.00-21.00hrs, Sat; 11.00-14.00hrs
& 19.00-21.30hrs

5   The Morning Star

🍴 🍽 🔥   Tel: (01452) 720028
1 mile north along the B4008, in Quedgeley,
on the left. (Whitbread) Open all day. Lunch;
12.00-14.00hrs daily, Evening meals; Mon-Sat;
18.00-20.45hrs.

PLACES OF INTEREST

Local Places of Interest are best accessed from Junctions 11A or 13

M5   JUNCTION 13

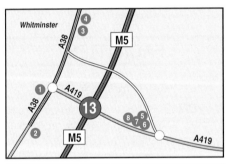

NEAREST A&E HOSPITAL

Gloucestershire Royal Hospital, Great Western
Road, Gloucester GL1 3PQ Tel: (01452) 528555

Proceed to Junction 12 and follow the A38 into
Gloucester. The hospital is signposted within the
town. (Distance Approx 9.2 miles)

NEAREST MINOR INJURY UNIT

Stroud General Hospital, Trinity Road,
Stroud GL5 2HY Tel: (01453) 562200

Follow the A419 to Stroud. Signposted from within
the town. Open 24 hours. (Doctors in attendance
between 09.00-12.30hrs and 14.00-16.30hrs only)
(Distance Approx 5.4 miles)

FACILITIES

1   Fromebridge Mill

🍴 🍽 ♿ 🔥   Tel: (01452) 741796
0.2 miles west along the A419, adjacent to the
roundabout. (Old English Inns) Open all day.
Bar Snacks served Mon-Thurs; 12.00-20.00hrs,
Fri-Sun; 12.00-19.00hrs. Restaurant Open;
12.00-14.30hrs & 18.00-21.30hrs daily.

2   Fromebridge Self Serve (Elf)

⛽ ♿ WC   Tel: (01452) 740753
0.9 miles south along the A38, on the left.
Access, Visa, Overdrive, All Star, Switch, Dial
Card, Mastercard, Amex, Diners Club, Delta,
Elf Cards, Totalfina Cards. Open; Mon-Sat;
07.00-22.00hrs, Sun; 09.00-22.00hrs.

3   The Old Forge

🍴 🍽 🔥   Tel: (01452) 741306
1 mile north along the A38, on the right. (Free
House) Open all day Fri, Sat & Sun. Lunch;
Tues-Sun; 12.00-14.00hrs, Evening meals; Tues-
Sat; 18.00-21.00hrs.

4   The Whitminster

🍴 🍽 🛏B   Tel: (01452) 740234
1 mile north along the A38, on the right. (Free
House) Lunch, Evening Meals.

5   Little Chef

🍽 ♿   Tel: (01453) 828847
0.5 miles east along the A419, on the left. Open;
07.00-22.00hrs

6   Oldbury Service Station (Shell)

⛽ ♿ WC 24   Tel: (01453) 828688
0.5 miles east along the A419, on the left.
Access, Visa, Overdrive, All Star, Switch, Dial
Card, Mastercard, Amex, Diners Club, Delta,
Shell Cards, BP UK Agency, Esso Cards,
Smartcard.

7   Burger King

🍴 ♿   Tel: (01453) 828847
0.5 miles east along the A419, on the left. Open;
07.00-22.00hrs

8   Travelodge Stonehouse

🛏M ♿ 📷   Tel: (01453) 828590
0.5 miles east along the A419, on the left

PLACES OF INTEREST

The Wildfowl & Wetlands Trust
Slimbridge, Gloucestershire GL2 7BT
Tel: (01453) 890065 or Tel: (01453) 890333
Follow the A38 south. (Signposted 5.5 miles)
The Trust was founded by Sir Peter Scott in
1946 and Slimbridge now has the world's
largest collection of exotic wildfowl. It is the
only place in Europe where all six types of
flamingoes can be seen and in winter up to
8000 wild birds fly in to this 800 acre reserve
on the River Severn. Other attractions include
a Tropical House and an indoor Wonder of
Wetlands exhibit and there is a licensed
restaurant as well as picnic areas. Disabled
facilities. No dogs, other than guide dogs, are
allowed into the grounds.

Berkeley Castle
Berkeley, Gloucestershire GL13 9BQ Follow the
A38 south and Berkeley is signposted. (9 miles)
For details please see Junction 14 information

Jenner Museum
The Chantry, Church Lane, Berkeley GL13 9BN
(Adjacent to castle) Follow the A38 south and
Berkeley is signposted. (9 miles)
For details please see Junction 14 information

Painswick Rococo Gardens
Painswick House, Painswick,
Gloucestershire GL6 6TH
Follow the A419 east into Stroud and take the A46
north. (Signposted along A46 8.8 miles)
For details please see Junction 11A information

SERVICE AREA
BETWEEN JUNCTIONS 13 AND 14

Michael Wood Services (Welcome Break)

Tel: (01454) 260631 Granary Restaurant, Burger
King, Welcome Lodge (Northbound side only) BP
Fuel

Footbridge Connection between sites.

M5 | JUNCTION 14

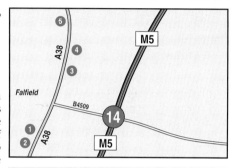

NEAREST A&E HOSPITAL
**Frenchay Hospital, Frenchay Park Road,
Frenchay, Bristol BS16 1LE
Tel: (0117) 970 1212**
Proceed south to Junction 15, take the M4 east to
Junction 19 and then follow the M32 to Junction 1
Take the A4174 east exit and the hospital is
signposted. (Distance Approx 12.3 miles)

FACILITIES

1   The Huntsman House Inn

🔲 🍴 ♿ 🐾   Tel: (01454) 260239
0.5 miles south along the A38, in Falfield, on
the right. (Whitbread) Open all day. Lunch;
12.00-14.30hrs daily, Evening meals; Sun-
Thurs; 18.00-21.00hrs, Fri & Sat; 18.00-
21.30hrs.

2   Mill Lane Filling Station (Elf)

⛽   Tel: (01454) 260286
0.6 miles south along the A38, in Falfield, on
the right. Access, Visa, Overdrive, All Star,
Switch, Dial Card, Mastercard, Amex, Diners

Club, Delta, AA Paytrak, Elf Cards. Toilets available for occasional customer use at staff discretion. Open; Mon-Sat; 07.00-23.00hrs, Sun 08.00-22.00hrs.

3 The Gables Inn

Tel: (01454) 260502

0.5 miles north along the A38, in Falfield, on the right. (Free House) Open all day. Bar: Lunch; 12.00-14.00hrs daily, Evening Meals; Mon-Sat; 18.30-22.00hrs, Sun; 19.00-21.00hrs. Restaurant Open: Lunch; Sundays only; 12.00-14.00hrs, Evening Meals; Mon-Sat; 18.30-22.00hrs, Sun; 19.00-21.00hrs.

4 Green Farm Country House

Tel: (01454) 260319

0.7 miles north along the A38, on the right

5 Stone Garage (UK Fuels)

Tel: (01454) 260551

1 mile north along the A38, in Stone, on the left. Access, Visa, Overdrive, All Star, Switch, Dial Card, Mastercard, Amex, Diners Club, Delta. Open; Mon-Fri; 08.00-18.00hrs, Sat; 09.00-16.00hrs, Sun; Closed.

PLACES OF INTEREST

The Wildfowl & Wetlands Trust

Slimbridge, Gloucestershire GL2 7BT Follow the A38 north and the site is signposted. (10 miles)

For details please see Junction 13 information

Berkeley Castle

Berkeley, Gloucestershire GL13 9BQ
Tel: (01453) 810332 Follow the A38 north and Berkeley is signposted. (5.4 miles)

Built between 1117 and 1153 on the site of a Saxon fort, Berkeley Castle is the oldest castle in England still to be inhabitated. It was here in 1215 that the barons of the West met before setting out to witness the sealing of the Magna Carta by King John at Runnymede, but the incident which gave the castle its greatest notoriety and place in English history was the brutal murder of King Edward II in 1327. The dungeon where this took place can today be viewed along with the 14thC Great Hall, circular keep, state apartments and mediaeval kitchens within the castle as well the Elizabethan terraced garden which surrounds it. These grounds also include a free-flight butterfly house, tea room, gift shop and a large well-stocked deer park.

Jenner Museum

The Chantry, Church Lane, Berkeley GL13 9BN (Adjacent to castle) Tel: (01453) 810631 Follow the A38 north and Berkeley is signposted. (5.4 miles)

The home of Dr Edward Jenner, discoverer of vaccination against Smallpox. The son of a local parson, he was apprenticed to a surgeon in Chipping Sodbury, later moving on to St.George's Hospital in London to become a student under John Hunter, and then returning to Berkeley to practice as a country doctor. It was whilst he was still an apprentice that he noticed that those who had, at some stage, been infected with cowpox did not contract smallpox and he was now able to continue his studies into this phenomenon. His work, over several decades, led to the first vaccination against smallpox and virtually eradicated a disease that in the 1600's is estimated to have killed 60 million people worldwide. The beautiful Georgian house and gardens, including the Temple of Vaccinia, where he vaccinated the poor, remain much as they were in the Doctor's day and a modern display of computer games and CD ROMs show the importance of the science of immunology, which he founded, in modern medicine.

M5 JUNCTION 15

JUNCTION 15 IS A MOTORWAY INTERCHANGE ONLY AND THERE IS NO ACCESS TO ANY FACILITIES

 Large Hotel   Small Hotel   Motel   Guest House/ Bed & Breakfast   Disabled Facilities   Childrens Facilities

M5　JUNCTION 16

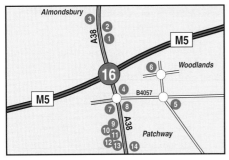

Restaurant Open; Breakfast; Mon-Fri; 07.00-09.30hrs, Sat; 07.30-10.00hrs, Sun; 08.30-10.30hrs, Lunch; Mon-Fri; 12.30-14.00hrs, Sun; 13.00-14.45hrs, Dinner; 19.00-21.45hrs daily.

NEAREST A&E HOSPITAL

**Southmead Hospital, Southmead Road,
Westbury on Trym, Bristol BS10 5NB
Tel: (0117) 950 5050**

Take the A38 south to Filton and the hospital is signposted. (Distance Approx 3.8 miles)

FACILITIES

1　Almondsbury Interchange Hotel & Restaurant

Tel: (01454) 613206

0.3 miles north along the A38, on the right. Restaurant Open; Breakfast; 07.00-09.00hrs daily, Lunch; Mon-Fri; 12.00-13.30hrs, Dinner; Mon-Thurs; 19.00-21.30hrs, Fri & Sat; 19.30-21.30hrs.

2　The Swan Hotel

Tel: (01454) 625671

0.4 miles north along the A38, on the right. (Wizard Inns) Open all day. Meals served; Mon-Sat; 12.00-21.00hrs, Sun; 12.00-16.00hrs

3　Rocklands Restaurant

Tel: (01454) 612208

0.6 miles north along the A38, on the left. Open at weekends. Prior booking only.

4　Stakis Bristol Hotel

Tel: (01454) 201144

0.2 miles south along the A38, on the left.

5　The Bradley Stoke

Tel: (01454) 202193

0.6 miles east along the B4057, on the right. Open all day. Meals served; Mon-Fri; 12.00-15.00hrs & 17.00-21.30hrs, Sat & Sun; 12.00-21.30hrs

6　The Orchard Toby Restaurant

Tel: (01454) 201202

1 mile along Woodlands Lane, on the left. Open all day. Lunch; Mon-Sat; 12.00-14.00hrs, Evening meals; Mon-Sat; 17.30-21.00hrs. Meals served all day Sundays; 12.00-21.00hrs.

7　Aztec Hotel

Tel: (01454) 201090

0.3 miles south along the A38, on the right. Restaurant Open; Breakfast; Mon-Fri; 07.00-09.30hrs, Sat & Sun; 08.30-10.00hrs, Lunch; Mon-Fri; 12.00-14.00hrs, Dinner; Mon-Sat; 19.25-21.45hrs, Sun; 19.15-21.15hrs.

8　Travellers Rest

Tel: (01454) 612238

0.4 miles south along the A38, on the left. (The Hungry Horse) Open all day. Meals served; Mon-Sat; 12.00-22.00hrs, Sun; 12.00-21.30hrs

9　Radnor B&B

Tel: (01454) 618390

0.5 miles south along the A38, on the right.

10　Dove Court B&B

Tel: (01454) 612271

0.6 miles south along the A38, on the right.

11　The Jays Guest House

Tel: (01454) 612771

0.6 miles south along the A38, on the right.

Pets Welcome	Cash Dispenser	24 Hour Facilities	Toilets	Alcoholic Drinks	Food and Drink	Fuel

12 The Willow Hotel & Licensed Restaurant

Tel: (01454) 612276

0.6 miles south along the A38, on the right. Open for meals; Mon-Sat: 19.00-21.30hrs, Sun: Closed

13 Texaco Patchway

Tel: (01454) 453000

0.7 miles south along the A38, on the right. Keyfuels, Switch, Overdrive, Access, Amex, Dial Card, All Star, Visa, Mastercard, Fast Fuel, Texaco Cards.

14 Stokebrook Service Station (Esso)

Tel: (0117) 969 2115

1 mile south along the A38, on the left. Access, Visa, Overdrive, All Star, Switch, Dial Card, Mastercard, Amex, Diners Club, Delta, Shell Cards, BP Cards, Esso Cards.

PLACES OF INTEREST

Bristol

Bristol Tourist Information Centre, St.Nicholas Church, St.Nicholas Street, Bristol BS1 1UE
Tel: (0117) 926 0767
web: http://tourism. bristol.gov.uk
e-mail: bristol@tourism.bristol.gov.uk
Follow the A38 south (Signposted - 7 miles)

Founded during Saxon times, this strategically-important bridging point at the head of the Avon gorge soon became a major port and market centre and, by the early 11thC Bristol had its own mint and was trading with other ports throughout western England, Wales and Ireland. In 1067 the Normans built a massive stone keep, at a site still known as Castle Park, and this structure was all but destroyed at the end of the Civil War. During the middle ages the town expanded enormously to accommodate the huge increases in the export trade in wool and importation of wines from Spain and France. Many of the fine buildings constructed during this period , and subsequently, can be seen today within the city centre.

Within the city centre can be found ...

Bristol City Museum & Art Gallery

Queen's Road, Clifton, Bristol BS8 1RL
Tel: (0117) 922 3571
web: http://www.bristol-city.gov.uk

A treasure house of wonderful exhibits, including an exceptional collection of Chinese glass. Gift Shop. Cafe

SS Great Britain & The Matthew

Great Western Dock, Gas Ferry Road,
Bristol BS1 6TY Tel: (0117) 926 0680

Isambard Kingdom Brunel's SS Great Britain, built in the city in 1843, was the world's first iron-hulled, propeller-driven, ocean-going vessel. After a working life of 43 years it languished in the Falkland Islands, utilized as a storage hulk, until 1970 when it was saved, brought back to its home dock and ultimately restored to its former glory. The Matthew is a replica of John Cabot's 15thC ship which sailed from Bristol on the voyage that discovered Newfoundland in 1497. This replica re-enacted the journey in 1997.

Bristol Cathedral

College Green, Bristol BS1 5TJ
Tel: (0117) 926 4879

Founded in the 12thC as the great church of an Augustine abbey, several original Norman features remain, including the south east transept walls, chapter house, gatehouse and east side of the abbey cloisters. Following the Dissolution of the Monasteries in 1539, Henry VIII elevated the church status to that of cathedral but the structure was not fully completed until the 19thC when a new nave was built in sympathetic style to the existing choir. The building contains some exceptional monuments and tombs.

John Wesley's Chapel

36 The Horsefair, Bristol BS1 3JE
Tel: (0117) 926 4740

Built in 1739 it is the oldest Methodist building in the world. The Chapel, constructed to cope with the huge response to John Wesley's open air preaching to the poor in Bristol, remains wonderfully unspoilt. Visitors are able to see the living rooms, above the Chapel, where John stayed with his brother and some of the first pioneering Methodist preachers.

| Large Hotel | Small Hotel | Motel | Guest House/ Bed & Breakfast | Disabled Facilities | Childrens Facilities | |

Harvey's Wine Cellars

12 Denmark Street, Bristol BS1 5DQ
Tel: (0117) 927 5036

Learn the history of wine in the 13thC cellars and take in the wonderful collection of antique displays. Guided tours and tutored tastings are available. Restaurant. Gift Shop.

Bristol Industrial Museum

Princes Wharf, Wapping Road, Bristol BS1 4RN
Tel: (0117) 925 1470

Houses a fascinating record of the achievements of the city's industrial pioneers, including those with such household names as Harvey, Fry, Wills and McAdam. Visitors can find out about Bristol's history as a port, view the aircraft and aero engines made in the city since 1910, and inspect some of the many famous motor vehicles which have borne the Bristol name since Victorian times. During the summer, working demonstrations of some of the larger exhibits can be seen. These include a giant crane, steam railway, printing workshop and a variety of motor vessels.

M5 JUNCTION 17

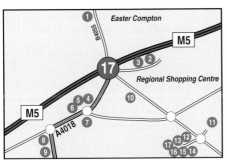

NEAREST A&E HOSPITAL

Southmead Hospital, Southmead Road,
Westbury on Trym, Bristol BS10 5NB
Tel: (01179) 505050

Take the A4018 to Bristol (West) and the hospital is signposted. (Distance Approx 3.1 miles)

FACILITIES

1   The Fox

Tel: (01454) 632220

1 mile north along the B4055, in Easter Compton, on the left. (Unique Pub Co) Meals served; Mon-Sat; 12.00-14.00hrs & 18.30-21.30 hrs, Sun; 12.00-14.00hrs

2   Asda Petrol Station

Tel: (0117) 969 3973

In the Regional Shopping Centre, on the south side of the roundabout, on the left. All major cards except Amex. Toilets in adjacent store. Open; Mon-Fri; 24hrs, Sat; 07.00-22.30hrs, Sun; 09.00-16.30hrs.

3   McDonald's

Tel: (0117) 950 3977

In the Regional Shopping Centre, on the south side of the roundabout, on the left. Open; 07.30-23.00hrs daily

4   Travelodge

Tel: (0117) 950 1530

0.2 miles south along the A4018, on the right.

5   The Lamb & Flag

Tel: (0117) 950 1490

0.2 miles south along the A4018, on the right. (Harvester) Open all day. Meals served; Mon-Fri; 12.00-14.30hrs & 17.00-21.30hrs, Sat; 12.00-21.30hrs, Sun; 12.00-21.00hrs.

6   Cribbs Lodge Hotel

Tel: (0117) 950 0066

0.3 miles south along the A4018, on the right.

7   Harry Ramsdens

Tel: (0117) 959 4100

0.3 miles south along the A4018, on the left. Open; Mon-Thur; 12.00-21.30hrs, Fri & Sat; 12.00-22.00hrs, Sun; 12.00-20.00hrs.

| Pets Welcome | Cash Dispenser | 24 Hour Facilities | Toilets | Alcoholic Drinks | Food and Drink | Fuel |

8 Cribbs Causeway Shell Station

🔲 ♿ WC 24   Tel: (0117) 941 9400
0.6 miles south along the A4018, on the right.
Access, Visa, Overdrive, All Star, Switch, Dial
Card, Mastercard, Amex, Diners Club, Delta,
Shell Cards, BP Agency Card.

9 BP Severnway Petrol Station

🔲 ♿ WC 24   Tel: (0117) 950 0414
1 mile south along the A4018, on the right.
Access, Visa, Overdrive, All Star, Switch, Dial
Card, Mastercard, Amex, Diners Club, Delta,
Routex, AA Paytrak, UK Fuelcard, Shell Agency,
BP Cards.

10 McDonald's

🔲 ♿   Tel: (0117) 950 1523
0.6 miles south along the A4018 on the right.
Restaurant Open; 07.30-23.00hrs daily. "Drive
Thru" remains open until 0.00 hrs, Sun to
Thurs and 02.00hrs Fri/Sat & Sat/Sun.

11 The Mall

🔲 🔲 WC ♿   Tel: (0117) 903 0303
1 mile south along Highway Road, on the left.
(Signposted). There are numerous cafes and
restaurants within the shopping centre. Open;
Mon-Fri; 10.00-20.00hrs, Sat; 09.00-19.00hrs,
Sun; 11.00-17.00hrs.

12 KFC

🔲 ♿   Tel: (0117) 950 5205
1 mile south along Highway Road, in The
Venue Leisure Complex on the right. Open;
10.00-0.00hrs daily

13 Burger King

🔲 ♿   Tel: (0117) 959 0712
1 mile south along Highway Road, in The
Venue Leisure Complex on the right. Open;
10.00-0.00hrs daily

14 Chiquito's (Mexican)

🔲 🔲 ♿ 🔲   Tel: (0117) 959 1459
1 mile south along Highway Road, in The
Venue Leisure Complex on the right. Open;

Mon-Sat; 12.00-23.00hrs, Sun; 12.00-22.30hrs

15 Frankie & Benny's (New York Italian)

🔲 🔲 ♿ 🔲   Tel: (0117) 959 1180
1 mile south along Highway Road, in The
Venue Leisure Complex on the right. Open;
Mon-Sat; 12.00-23.00hrs, Sun; 12.00-22.30hrs

16 Bella Pasta (Italian)

🔲 🔲 ♿ 🔲   Tel: (0117) 959 0982
1 mile south along Highway Road, in The
Venue Leisure Complex on the right. Open;
Mon-Thurs; 12.00-23.30hrs, Fri & Sat; 12.00-
0.00hrs, Sun; 12.00-22.30hrs

17 TGI Fridays (American)

🔲 🔲 ♿ 🔲   Tel: (0117) 959 1987
1 mile south along Highway Road, in The
Venue Leisure Complex on the right. Open;
Sun-Thurs; 12.00-22.30hrs, Fri & Sat; 12.00-
23.30hrs

PLACES OF INTEREST

The Mall at Cribbs Causeway

Bristol BS34 5DG Tel: (0117) 915 5326
Information Line: Tel: (0117) 903 0303
Adjacent to junction and signposted

A new shopping centre with 130 top names,
16 cafes and restaurants and a wealth of
excellent facilities and attractions all under one
roof. Creche. Play Area. Accessibility Unit

Bristol

Follow the A4018 south (Signposted 5.5 miles)

For details please see Junction 16 information.

Within the city centre can be found ...

Bristol City Museum & Art Gallery

Queen's Road, Clifton, Bristol BS8 1RL

For details please see Junction 16 information.

| | Large Hotel | | Small Hotel | | Motel | | Guest House/ Bed & Breakfast | | Disabled Facilities | | Childrens Facilities |

SS Great Britain & The Matthew
Great Western Dock, Gas Ferry Road,
Bristol BS1 6TY
For details please see Junction 16 information.

Bristol Cathedral
College Green, Bristol BS1 5TJ
For details please see Junction 16 information.

John Wesley's Chapel
36 The Horsefair, Bristol BS1 3JE
For details please see Junction 16 information.

Harvey's Wine Cellars
12 Denmark Street, Bristol BS1 5DQ
For details please see Junction 16 information.

Bristol Industrial Museum
Princes Wharf, Wapping Road, Bristol BS1 4RN
For details please see Junction 16 information.

Bristol Zoo Gardens
Clifton, Bristol BS8 3HA Tel: (0117) 973 8951
Follow the A4018 south and A4176 west
(Signposted 5.1 miles)

A great place to enjoy a real life experience and see over 300 species of wildlife in beautiful gardens. Gorilla Island, Twilight World, Bug World and the Seal and Penguin exhibit with underwater viewing provide a fascinating insight into our natural world. Cafe and Restaurant. Garden. Disabled facilities.

Clifton Suspension Bridge Visitor Centre
Bridge House, Sion Place, Bristol BS8 4AP
Tel: (0117) 974 4664 Follow the A4018 south and A4176 west (Signposted 6 miles)

Sited 1.5 miles west of the city centre, the bridge, which is suspended more than 200 feet above the Avon Gorge was designed by the great Victorian engineer, Isambard Kingdom Brunel. This graceful structure was completed in 1864 and offers drivers and pedestrians a magnificent view over the city and surrounding landscape. Before you cross it call in to the Visitor Centre and learn the fascinating story behind the bridge and inspect the superb scale model. Gift Shop.

M5 JUNCTION 18

THERE ARE TWO ACCESS POINTS TO THIS JUNCTION

The distances quoted below have been taken from the southernmost access point

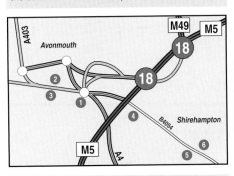

NEAREST A&E HOSPITAL
Southmead Hospital, Southmead Road, Westbury on Trym, Bristol BS10 5NB
Tel: (01179) 505050

Take the A4 into Bristol and the hospital is signposted within the city (Distance Approx 6.8 miles)

FACILITIES

1 Bradford Hotel

🛏h Tel: (0117) 904 7778
0.1 miles, at the corner of Portway and Avonmouth Road.

2 Avonmouth Filling Station (Q8)

🅵 WC Tel: (0117) 982 5921
0.2 miles west along the Avonmouth Road, in Avonmouth, on the right. Access, Visa, Overdrive, All Star, Switch, Dial Card, Mastercard, Amex, Diners Club, Delta, Key Fuels, BP Supercharge, Securicor Fuelserv, UK Fuels, Q8 Cards. Open; Mon-Fri; 06.00-20.00hrs, Sat; 07.00-15.00hrs, Sun; Closed.

3 Miles Arms Hotel

🍴 🛏 ♿ 🛏B Tel: (0117) 982 2317
0.2 miles west along the Avonmouth Road, in

| Pets Welcome | Cash Dispenser | 24 Hour Facilities | Toilets | Alcoholic Drinks | Food and Drink | Fuel |

Avonmouth, on the left. (Bass) Open all day. Meals served; Mon-Fri; 12.00-14.30hrs & 18.00-21.00hrs, Sat & Sun; 12.00-14.00hrs & 19.00-21.00hrs. Bar Snacks served 12.00-21.00hrs daily. Children allowed in pub, with parents and for the consumption of meals, on Sat & Sun prior to 20.00hrs only.

---

4   Hope & Anchor

Tel: (0117) 982 2691
0.2 miles south along the B4054, on the right. (Scottish & Newcastle) Meals served; Mon-Sat; 11.00-21.30hrs, Sun; 12.00-14.00hrs

---

5   Shell Shirehampton

Tel: (0117) 937 9950
1 mile south along the B4054, in Shirehampton, on the right. Access, Visa, Overdrive, All Star, Switch, Dial Card, Mastercard, Amex, Diners Club, Delta, BP Cards, Shell Cards, Esso Card, Smart.

---

6   Shirehampton Lodge Hotel

Tel: (0117) 907 3480
1 mile south along the B4054, in Shirehampton, on the left.

PLACES OF INTEREST

Bristol
Follow the A4 south (Signposted 6 miles)
For details please see Junction 16 information.

Within the city centre can be found ...

Bristol City Museum & Art Gallery
Queen's Road, Clifton, Bristol BS8 1RL
For details please see Junction 16 information.

SS Great Britain & The Matthew
Great Western Dock, Gas Ferry Road, Bristol BS1 6TY
For details please see Junction 16 information.

Bristol Cathedral
College Green, Bristol BS1 5TJ
For details please see Junction 16 information.

John Wesley's Chapel
36 The Horsefair, Bristol BS1 3JE
For details please see Junction 16 information.

Harvey's Wine Cellars
12 Denmark Street, Bristol BS1 5DQ
For details please see Junction 16 information.

Bristol Zoo Gardens
Clifton, Bristol BS8 3HA
Follow the A4 south (Signposted 5.6 miles)
For details please see Junction 17 information.

Bristol Industrial Museum
Princes Wharf, Wapping Road, Bristol BS1 4RN
For details please see Junction 16 information.

Clifton Suspension Bridge Visitor Centre
Bridge House, Sion Place, Bristol BS8 4AP Follow the A4 south (Signposted 5.7 miles)
For details please see Junction 17 information.

M5   JUNCTION 19

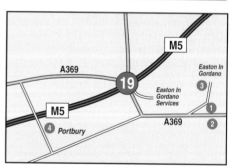

NEAREST A&E HOSPITAL
Southmead Hospital, Southmead Road, Westbury on Trym, Bristol BS10 5NB
Tel: (0117) 950 5050
Proceed to Junction 18 and take the A4 to Bristol. The hospital is signposted within Bristol. (Distance Approx 9.1 miles)

	Large Hotel		Small Hotel		Motel		Guest House/ Bed & Breakfast		Disabled Facilities		Childrens Facilities

FACILITIES

SERVICE AREA

ACCESSED FROM THIS JUNCTION

Gordano Services (Welcome Break) is accessed from this junction Tel: (01275) 373624

Burger King, KFC, Red Hen Restaurant, Granary Restaurant, Welcome Lodge & Shell Fuel.

Please note that at busy periods on Friday evenings and Saturdays during summer months this service area is extremely busy and access may be closed by the police.

---

1   The Rudleigh Inn

Tel: (01275) 372363

0.6 miles south along the A369, on the left. (Courage) Open all day. Food served Mon-Sat;11.00-22.00hrs, Sun; 12.00-21.30hrs.

---

2   Tynings

Tel: (01275) 372608

0.6 miles south along the A369, on the right.

---

3   The Kings Arms

Tel: (01275) 372208

0.5 miles east, in Easton in Gordano, on the left. (Unique Pub Co) Lunch; 12.00-14.00hrs daily, Evening Meals; Mon-Sat; 17.30-21.30hrs

---

4   The Priory

Tel: (01275) 378411

0.8 miles south along the Portbury Road, on the right. (Vintage Inns) Meals served; Mon-Sat; 12.00-22.00hrs, Sun; 12.00-21.00hrs

---

PLACES OF INTEREST

Avon Gorge Nature Reserve

Follow the A369 south to Bristol (3.5 miles)

Sited on the western side of the river it offers some delightful walking through Leigh Woods to the summit of an iron age hill fort.

Clifton Suspension Bridge Visitor Centre

Bridge House, Sion Place, Bristol BS8 4AP
Follow the A369 south to Bristol (4.7 miles)

For details please see Junction 17 information.

M5   JUNCTION 20

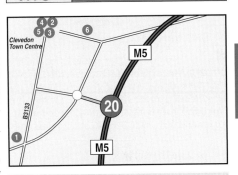

NEAREST A&E HOSPITAL

Weston General Hospital, Grange Road, Uphill, Weston super Mare BS23 4TQ
Tel: (01934) 636363

Proceed to Junction 21 and take the A370 to Weston super Mare. The hospital is signposted from the junction. (Distance Approx 11.4 miles)

NEAREST MINOR INJURY UNIT

Clevedon Hospital, Old Street, Clevedon BS21 6BS Tel: (01275) 872212

Follow the route into Clevedon town centre. (Distance approx 1 mile) Open; 08.00-21.00hrs daily. [No doctor on site].

FACILITIES

CLEVEDON TOWN CENTRE IS WITHIN 1 MILE OF THIS JUNCTION

---

1   Tesco Petrol Station

Tel: (01275) 517400

0.8 miles south along the B3133, on the right. Access, Visa, Overdrive, All Star, Switch, Dial Card, Mastercard, Amex, Delta, AA Paytrak, Shell Cards, BP Cards. Open; 07.00-0.00hrs daily.

M5

2 Grapevine Cafe & Bistro

In the Town Square, in Clevedon town centre.

3 Public Toilets

WC

In the Town Square, in Clevedon town centre.

4 Old Street Garage (Esso)

Tel: (01275) 872596

In Old Street, in Clevedon town centre. Access, Visa, Overdrive, All Star, Switch, Dial Card, Mastercard, Amex, Diners Club, Delta, Shell Gold. Esso Cards, Style Card. Attended service for disabled drivers. Open; 06.30-22.00hrs daily.

5 Ye Olde Bristol Inn

Tel: (01275) 872073

In Chapel Hill, in Clevedon town centre. (Scottish & Newcastle) Open all day. Food served Mon-Sat; 11.00-21.300hrs, Sun; 12.00-21.30hrs.

6 Clevedon Garages (Shell)

Tel: (01275) 873701

0.9 miles along Tickenham Road, in Clevedon, on the right. Access, Visa, Overdrive, All Star, Switch, Dial Card, Mastercard, Amex, Diners Club, Delta, Shell Cards, BP Agency Cards.

PLACES OF INTEREST

Clevedon

Follow the B3133 west (Signposted 0.7 miles)

A genteel seaside town, with a population of 20,000 it has been a stylish holiday resort and residential centre since the latter half of the 18thC. Devoid of most of the popular attractions associated with holiday resorts, it does boast of a recently-restored slim and elegant pier which, amongst other things, is utilized as a landing stage by large pleasure steamers such as the Balmoral and the Waverley, the only surviving seagoing paddle steamers in the world. Clevedon's appeal is in the romantic, Coleridge and Thackeray both lived and worked here, rather than the dramatic and it has managed to retain an atmosphere of tranquil refinement which still has a certain charm.

Clevedon Court NT

Tickenham Road, Clevedon, Somerset BS21 6QU
Tel: (01275) 872257 Follow the B3133 west
(Signposted 1.5 miles)

An outstanding and virtually unaltered, 14thC manor house incorporating a massive 12thC tower and 13thC great hall, it contains many striking Eltonware pots and vases and a fine collection of Nailsea glass. The house looks out over a beautiful 18thC terraced garden. Tea Room.

M5 JUNCTION 21

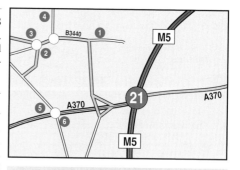

NEAREST A&E HOSPITAL

Weston General Hospital, Grange Road, Uphill, Weston super Mare BS23 4TQ
Tel: (01934) 636363

Take the A370 to Weston super Mare. The hospital is signposted from the junction. (Distance Approx 5.3 miles)

FACILITIES

1 Woolpack Inn

Tel: (01934) 521670

0.5 miles west along the B3440, on the right. (Free House) Lunch; Mon-Fri; 12.00-14.00hrs, Sat & Sun; 12.00-14.30hrs, Evening meals; Mon-Sat; 18.00-22.00hrs, Sun; 19.00-21.00hrs.

 H Large Hotel   h Small Hotel   M Motel   B Guest House/ Bed & Breakfast   Disabled Facilities   Childrens Facilities

2 The Summer House

🍴 🍽 ♿ 🚬 Tel: (01934) 520011

0.9 miles west along the B3440, on the left. (Banks's) Open all day. Restaurant open; Mon-Wed; 12.00-14.30hrs, Thurs-Sat; 12.00-22.00hrs, Sun; 12.00-21.00hrs. Bar meals served 12.00-17.30hrs daily.

3 Weston Motoring Centre (BP)

⛽ WC 24 Tel: (01934) 511414

0.9 miles west along the B3440, on the right. Access, Visa, Overdrive, All Star, Switch, Dial Card, Mastercard, Amex, Diners Club, Delta, Routex, BP Cards, Shell Agency, Statoil, AA Paytrak.

4 Sainsbury's Petrol Station

⛽ £ Tel: (01934) 516088

1 mile west along Queen's Way, on the left. Visa, Access, Mastercard, Switch, Eurocard, Delta, All Star, Overdrive, Amex, Dialcard, AA Paytrak. Disabled Toilets in adjacent Store. Open; Mon-Wed; 06.00-21.30hrs, Thurs & Fri; 06.00-22.30hrs, Sat; 06.00-20.30hrs, Sun; 08.00 -18.00hrs.

5 Safeway Petrol Station

⛽ ♿ WC £ Tel: (01934) 515067

0.9 miles west along the A370, on the right. Access, Visa, Overdrive, All Star, Switch, Dial Card, Mastercard, Amex, Delta, BP Supercharge. Open; Mon-Sat; 06.00-22.30hrs, Sun; 08.00-22.00hrs. (24hr Credit Card operated pumps available when shop is closed)

6 The Bucket & Spade

🍴 🍽 🚬 ♿ Tel: (01934) 521235

0.9 miles along the A370, on the left. (Punch Taverns) Meals served; Mon-Sat; 12.00-22.00hrs, Sun; 12.00-21.30hrs

PLACES OF INTEREST

Weston super Mare

Weston super Mare Tourist Information Centre, Beach Lawns, Weston super Mare
Tel: (01934) 888800

From existing as a small fishing hamlet in 1811, within 100 years it had grown to become Somerset's second largest town with a population of well over 50,000. The Grand Pier, the focal point of a seafront packed with assorted attractions, as well as the splendid Winter Gardens and Pavilion overlooking the long safe and sandy beach combine to make this a very popular holiday resort.

Court Farm Country Park

Wolvershill Road, Banwell, Weston super Mare BS24 6DL Tel: (01934) 822383
Follow the A370 west and turn south west towards Barnwell (Signposted 1.9 miles)

A super day out for all the family on this working farm with over 20,000 ft^2 of undercover entertainment, great in all weathers. Enjoy the hands on experience of bottle feeding of the young animals and there are free tractor rides, cider tasting and an Indoor Adventure Playground.

The Helicopter Museum

The Airport, Locking Moor Road, Weston super Mare BS22 8PL Tel: (01934) 635227
web: dialspace.dial.pipex.com/town/terrace/aaa76
Follow the A370 towards Weston super Mare (Signposted with "propeller" 3 miles)

The home of the world's largest collection of helicopters and autogyros and the only museum in Britain dedicated to the rotary wing aircraft. Over 60 exhibits are on view and there are also displays on their history and development, a flight simulator and a conservation hangar where the machines are restored. Gift Shop. Cafe. Disabled facilities.

M5

SERVICE AREA

BETWEEN JUNCTIONS 21 AND 22

Sedgemoor Services (Southbound) (RoadChef) Tel: (01934) 750888   Food Fayre Self-Service Restaurant, Wimpy Bar & Esso Fuel

Sedgemoor Services (Northbound) (Welcome Break) Tel: (01934) 750730   Granary Self-Serve Restaurant, Burger King, Welcome Lodge & Shell Fuel

M5   JUNCTION 22

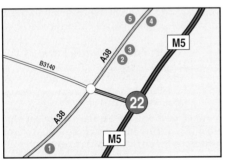

NEAREST A&E HOSPITAL

Weston General Hospital, Grange Road, Uphill, Weston super Mare BS23 4TQ Tel: (01934) 636363

Take the A38 north to Weston super Mare. The hospital is signposted within the town. (Distance Approx 7.2 miles)

NEAREST MINOR INJURY UNIT

Bridgwater Hospital, Salmon Parade, Bridgwater TA6 5AH Tel: (01278) 451501

Proceed to Junction 23 and take the A38 south to Bridgwater and the hospital is signposted within the town. Open: 09.00-22.00hrs daily (Doctor not in attendance 17.00-22.00hrs) (Distance Approx 8.2 miles)

FACILITIES

1   The Bristol Bridge Inn

Tel: (01278) 787269

1 mile south along the A38, on the left. (Free House) Open all day. Bar Snacks served; Mon-Sat; 11.00-23.00hrs, Sun; 12.00-22.30hrs.

2   The Fox & Goose

Tel: (01278) 760223

0.4 miles north along the A38, on the right. (Free House) Meals served; Mon-Fri; 19.00-21.00hrs, Sat & Sun; 12.00-14.00hrs & 19.00-21.00hrs.

3   Mr Useful Brent Knoll Garages

Tel: (01278) 760563

0.4 miles north along the A38, on the right. Access, Visa, Overdrive, All Star, Switch, Dial Card, Mastercard, Delta. Open; Mon-Sat; 07.00-20.00hrs, Sun; 08.00-20.00hrs

4   The Goat House Cafe & Restaurant

Tel: (01278) 760995

0.8 miles north along the A38, on the right. Open; Mon-Fri; 09.00-16.00hrs, Sat & Sun; 09.00-17.00hrs

5   Battleborough Grange Hotel

Tel: (01278) 760208

0.8 miles north along the A38, on the left. Restaurant Open; Lunch; 12.00-14.00hrs, Dinner 19.00-21.30hrs daily

PLACES OF INTEREST

Alstone Wildlife Park

Alstone Road, Highbridge TA9 3DT Tel: (01278) 782405   Follow the A38 south through Highbridge (Signposted on A38 2.8 miles)

A small non-commercial family run park with a huge variety of interesting and amusing fur and feathered friends including a herd of Red Deer, Llamas, Emu, Ponies, Pigs, Owls, Waterfowl and Theodore the friendly camel. Picnic Area. Refreshments. Disabled facilities.

Coombes Somerset Cider

Japonica Farm, Mark TA9 4QD Tel: (01278) 641265   Follow the A38 north towards Cheddar and after 200 yards turn right to Mark (Signposted "Mark, Cider Farm & Tea Rooms" 2.7 miles)

Founded in 1919, this family-owned firm still

	Large Hotel		Small Hotel		Motel		Guest House/ Bed & Breakfast		Disabled Facilities		Childrens Facilities

makes cider in the traditional way, using local cider apple varieties and maturing the cider in oak vats. Visitors can see this process from start to finish and sample the products. There is also an interesting video of the cider maker's year, a museum and a tea room. Disabled facilities.

M5 JUNCTION 23

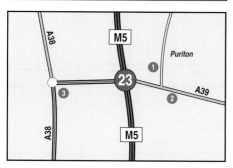

NEAREST NORTHBOUND A&E HOSPITAL

Weston General Hospital, Grange Road, Uphill, Weston super Mare BS23 4TQ
Tel: (01934) 636363

Proceed to Junction 22 and take the A38 north to Weston super Mare. The hospital is signposted within the town. (Distance Approx 13 miles)

NEAREST SOUTHBOUND A&E HOSPITAL

Musgrove Park Hospital, Taunton TA1 5DA
Tel: (01823) 333444

Proceed to Junction 25 and take the A358 west to Taunton. The hospital is signposted within the town. (Distance Approx 14.7 miles)

NEAREST MINOR INJURY UNIT

Bridgwater Hospital, Salmon Parade, Bridgwater TA6 5AH Tel: (01278) 451501

Take the A38 south to Bridgwater and the hospital is signposted within the town. Open: 09.00-22.00hrs daily (Doctor not in attendance 17.00-22.00hrs) (Distance Approx 3.1 miles)

FACILITIES

1   The Puriton Inn

🍴 🍺 ♿ 🐾 ✂   Tel: (01278) 683464
0.2 miles along the Puriton Road, on the left. (Whitbread) Open all day. Bar Snacks and meals

served; 11.30-14.00hrs daily, Mon-Sat; 18.00-21.00hrs, Sun; 19.00-21.00hrs.

2   Rockfield House

📧 B   Tel: (01278) 683561
0.2 miles east along the A39, on the right.

3   The Admiral's Table

🍴 🍺 ♿ 🐾 📧 h   Tel: (01278) 685671
0.6 miles south along the A38, on the left. (Eldridge, Pope & Co) Meals served 07.00-22.00hrs daily.

PLACES OF INTEREST

Admiral Blake Museum

Blake Street, Bridgwater TA6 3NB
Follow the A38 south to Bridgwater and it is signposted within the town. (3.4 miles)

For details please see Junction 24 information

Secret World Badger & Wildlife Rescue Centre

East Huntspill, Somerset TA9 3PZ
Tel: (01278) 783250 Follow the A39 east towards Street and turn north along the B3134 to East Huntspill (5.3 miles)

Offers a wide range of attractions and activities related to the wildlife of the area, including a badger observation sett, bee hives and nature trail. Featured on HTV "Cross Country" and BBC "Animal People" programmes during 1999. Tea Rooms. Play Areas. Shop. Disabled access.

Alstone Wildlife Park

Alstone Road, Highbridge TA9 3DT
Follow the A38 north towards Highbridge (Signposted on A38 4.4 miles)

For details please see Junction 22 information.

M5   JUNCTION 24

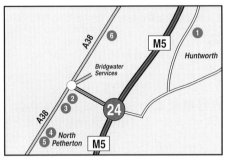

NEAREST A&E HOSPITAL

Musgrove Park Hospital, Taunton TA1 5DA
Tel: (01823) 333444

Proceed to Junction 25 and take the A358 west to Taunton. The hospital is signposted within the town. (Distance Approx 9.9 miles)

NEAREST MINOR INJURY UNIT

Bridgwater Hospital, Salmon Parade
Bridgwater TA6 5AH Tel: (01278) 451501

Take the A38 north to Bridgwater and the hospital is signposted within the town. Open: 09.00-22.00hrs daily (Doctor not in attendance 17.00-22.00hrs) (Distance Approx 2.2 miles)

FACILITIES

SERVICE AREA

ACCESSED FROM THIS JUNCTION

Bridgwater Services (First) Tel: (01278) 456800
BP Fuel, Burger King, Lodge and Restaurant.

1   The Boat & Anchor Inn

Tel: (01278) 662473
1 mile along the Huntworth Road, on the right. (Free House) Meals served; 12.00-14.00hrs & 19.00-21.00hrs daily.

2   Compass Tavern

Tel: (01278) 662283
0.3 miles south along the A38, on the left. (Wayside Inns) Open all day. Meals served all day; Mon-Sat; 12.00-22.00hrs, Sun; 12.00-21.00hrs.

3   Quantock View Guest House

Tel: (01278) 663309
0.3 miles south along the A38, on the left.

4   Grahams Transport Stop Cafe

Tel: (01278) 663052
0.5 miles south along the A38, on the left. Open; Mon; 07.00-20.30hrs, Tues-Thurs; 06.30-20.30hrs, Fri; 06.30-17.00hrs, Sat; 07.00-11.00hrs.

5   Woods Filling Station (UK Petrol)

Tel: (01278) 663076
0.5 miles south along the A38, on the left. Access, Visa, Overdrive, All Star, Switch, Dial Card, Mastercard, Delta, Keyfuels, UK Fuelcard, Totalfina Cards, Securicor. Open; Mon-Sat; 07.00- 21.00, Sun; 09.00-18.00hrs.

6   Taunton Road Service Station (BP)

Tel: (01278) 427486
0.7 miles north along the A38, on the right. Access, Visa, Overdrive, All Star, Switch, Dial Card, Mastercard, Amex, Diners Club, Routex, Shell Agency.

PLACES OF INTEREST

Admiral Blake Museum

Blake Street, Bridgwater TA6 3NB Tel: (01278) 456127 Follow the A38 north to Bridgwater and it is signposted within the town. (2.1 miles)

The birthplace of Admiral Blake is now utilized as the Town Museum and, as well as featuring events in his life, houses a variety of exhibits related to the local history. Blake was an important officer in Cromwell's army and twice defended Taunton against overwhelming Royalist odds. In his fifties he was given command of the British navy and went on to win important battles against the Dutch and Spanish and re-established Britain's naval supremacy in Europe. One of the exhibits is a three-dimensional model of one of his most famous victories, the Battle of Santa Cruz. There is also a similar diorama of the Battle of Sedgemoor and exhibitions of artefacts dating from the Neolithic period to the Second World War and a section dealing with shipping.

M5 JUNCTION 25

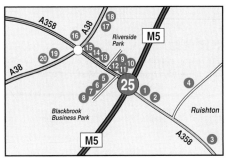

NEAREST A&E HOSPITAL

Musgrove Park Hospital, Taunton TA1 5DA
Tel: (01823) 333444

Take the A358 west to Taunton and the hospital is signposted within the town. (Distance Approx 3.4 miles)

FACILITIES

1   The Black Brook Tavern

Tel: (01823) 443121
0.1 miles east along the A358, on the left. (Scottish & Newcastle) Open all day. Meals served in Tavern; Mon-Sat; 12.00-22.00hrs, Sun; 12.00-15.00hrs & 18.00-21.30hrs. Carvery Restaurant; Open Mon-Sat; 12.00-14.00hrs, 18.00-21.30hrs, Sun; 12.00-21.30hrs.

2   Lodge Inn

Tel: (01823) 443121
0.1 miles east along the A358, on the left.

3   Countryways Accommodation

Tel: (01823) 442326
0.9 miles east along the A358, on the left.

4   Ruishton Inn

Tel: (01823) 442285
0.5 miles east, in Ruishton, on the left. (Free House) Meals served; 12.00-13.45hrs daily, Mon-Sat; 19.00-21.45hrs

5   Forte Post House, Taunton

Tel: (01823) 332222
0.1 miles west along the A358, in Blackbrook Business Park, on the left. Junction Restaurant; Breakfast; Mon-Fri; 06.30-09.30hrs, Sat & Sun; 07.30-10.00hrs, Lunch; Sun-Fri; 12.30-14.30hrs, Dinner; Mon-Sat; 18.30-22.00hrs, Sun; 19.00-22.00hrs

6   Murco Petrol Station

Tel: (01823) 351507
0.1 miles west along the A358, in Blackbrook Business Park, on the left. Access, Visa, Overdrive, All Star, Switch, Dial Card, Mastercard, Amex, Diners Club, Delta, BP Supercharge, Murco Cards.

7   The Taunton Harvester

Tel: (01823) 442221
0.1 miles west along the A358, in Blackbrook Business Park, on the left. (Bass) Meals served; Sun-Fri; 12.00-22.30hrs, Sat; 12.00-23.00hrs (Only Bar snacks are available, Mon-Fri; 15.00-17.00hrs)

8   Holiday Inn Express, Taunton

Tel: (01823) 624000
0.1 miles west along the A358, in Blackbrook Business Park, on the left.

9   Sainsburys Petrol Station

Tel: (01823) 443163
0.3 miles west along the A358, in Hankridge Farm, on the right. Visa, Access, Mastercard, Eurocard, Switch, Delta, Overdrive, All Star, Dial Card, Amex. Disabled Toilets in adjacent Store. Open; 06.00-0.00hrs daily

M5

| Pets Welcome | Cash Dispenser | 24 Hour Facilities | Toilets | Alcoholic Drinks | Food and Drink | Fuel |

M5

10 Pizza Hut

🍴 ♿ Tel: (01823) 444747

0.3 miles west along the A358, in Hankridge Farm, on the right. Open; Sun-Thurs; 11.30-23.00hrs, Fri & Sat; 11.30-0.00hrs.

11 The Hankridge Arms

🍷 🍴 ♿ ᴬ Tel: (01823) 444405

0.3 miles west along the A358, in Hankridge Farm, on the right. (Hall & Woodhouse) Open all day. Meals served in Bar & Restaurant; 12.00-14.30hrs daily, Sun-Thurs; 18.30-21.30hrs, Fri & Sat; 18.30-22.00hrs.

12 Travelodge Taunton

🛏M ♿ Tel: (01823) 444702

0.3 miles west along the A358, in Hankridge Farm, on the right.

13 McDonald's

🍴 ♿ Tel: (01823) 443765

0.3 miles west along the A358, in Riverside Park, on the right. Open; Sun-Thurs; 07.30-23.30hrs, Fri & Sat; 07.30-00.00hrs. ("Drive Thru" service only after 23.00hrs daily)

14 Fatty Arbuckles American Diner

🍷 🍴 ♿ ᴬ Tel: (01823) 354343

0.3 miles west along the A358, in Riverside Park, on the right. Open; Mon-Sat; 12.00-23.30hrs, Sun; 12.00-23.00hrs.

15 Tramonte's Italian Restaurant

🍷 🍴 ♿ ᴬ Tel: 0800 389 2101

0.3 miles west along the A358, in Riverside Park, on the right. Open; Tues-Fri; 17.00-23.00hrs, Sat & Sun; 12.00-15.00hrs & 18.00-23.00hrs.

16 Creech Castle

🍷 🍴 ♿ ᴬ Tel: (01823) 333512

0.5 miles west along the A358, on the right. (Free House) Meals served in Pub; 11.00-15.00hrs & 18.30-21.30hrs daily. Restaurant Open; 19.00-21.30hrs daily.

17 Central Service Station (Esso)

⛽ WC Tel: (01823) 333587

0.8 miles north along the A38, on the right. Access, Visa, Overdrive, All Star, Switch, Dial Card, Mastercard, Amex, Diners Club, Delta, Shell Gold, Esso Cards. Open; Mon-Sat; 07.00-20.00hrs, Sun; 08.30-18.00hrs

18 The Bathpool Inn

🍷 🍴 ♿ ᴬ Tel: (01823) 272545

0.9 miles north along the A38, on the right. (Whitbread) Open all day. Meals served all day from 11.30-22.00hrs.

19 White Lodge Travel Inn

🛏M ♿ Tel: (01823) 321112

0.6 miles along Corfe Road, on the right.

20 White Lodge Restaurant

🍷 🍴 ♿ ᴬ Tel: (01823) 321112

0.6 miles along Corfe Road, on the right. (Whitbread) Open all day. Meals served; Mon-Fri; 12.00-14.30hrs & 17.00-21.45hrs, Sat & Sun; 12.00-23.00hrs.

PLACES OF INTEREST

Taunton

Taunton Tourist Information Centre, The Library, Paul Street, Taunton TA1 3PF
Tel: (01823) 336344. Follow the A358 west.
(Signposted 1.9 miles)

The county town of Somerset, Taunton, was founded in the 8thC as a military camp by the Saxon King Ine and, by Norman times it had grown to have its own Augustine monastery, minster and castle. This castle was the focus of two important sieges during the Civil War and in 1685 the Bloody Assizes were held here in which the infamous Judge Jeffreys sentenced 150 followers of the Duke of Monmouth to death. The much-altered castle today houses the Somerset County Museum (Tel: 01823-320201) and part of the old monastic gatehouse in the Priory grounds, nearby, has been restored and is now in use as The Somerset County Cricket Museum (Tel: 01823-272946). Taunton was a thriving wool and textile centre and much of the wealth generated was expended on many of the fine buildings still to be seen in the town centre.

West Somerset Railway

Bishops Lydeard Station, Bishops Lydeard TA4 3BX
Tel: (01643) 704996
web http://www.West-Somerset-Railway.co.uk
e-mail: info@West-Somerset-Railway.co.uk
Follow the A358 west and follow the WSR signposts. (7.1 miles)

The West Somerset Railway recreates the era of a Great Western Railway country branch line, connecting Bishops Lydeard with the holiday resort of Minehead. Enjoy a steam hauled train ride through some 20 miles of glorious Somerset scenery as it weaves its way past the Quantock Hills and along the Bristol Channel. There is a visitor centre, giftshop and refreshment room at the Bishops Lydeard Station and a large giftshop and refreshment room at Minehead. Disabled access to trains at all stations except Doniford Halt and disabled toilets are provided at Minehead and Bishops Lydeard.

Hestercombe Gardens

Hestercombe, Cheddon Fitzpaine, Taunton TA2 8LG
Tel: (01823) 413923 Follow the A358 west, A38 north and follow signposts to Cheddon Fitzpaine (3.3 miles)

Fifty acres of formal gardens and parkland, including the famous Edwardian Gardens designed by Sir Edwin Lutyens and Gertrude Jekyll, and the Georgian Landscape Garden with lakes, temples and delightful woodland walks. Tea Room. Gift Shop. Limited Disabled access.

SERVICE AREA

BETWEEN JUNCTIONS 25 AND 26

Taunton Deane Services (RoadChef)

Tel: (01823) 271111 Food Fayre Self-Service Restaurant, Wimpy Bar, RoadChef Lodge & Shell Fuel

M5 — JUNCTION 26

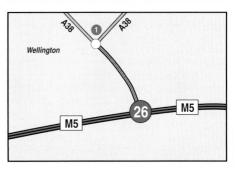

M5

NEAREST A&E HOSPITAL

**Musgrove Park Hospital, Taunton TA1 5DA
Tel: (01823) 333444**

Follow the A38 route to Taunton. The hospital is signposted within the town. (Distance Approx 4.7 miles)

FACILITIES

1   Piccadilly Filling Station (Keyfuels)

🏧 WC   Tel: (01823) 662148
1 mile north along the A38, on the left. Diesel Direct, All Star, Switch, Diners Club, Overdrive, Dial Card, Amex. Open; Mon-Sat; 07.00-21.00hrs, Sun; 08.00 -21.00hrs

PLACES OF INTEREST

Wellington

Follow the A38 south (Signposted 2.2 miles)

An old market town with broad streets containing fine Georgian buildings including the neo-classical town hall. Once an important producer of woven cloth and serge, the prosperity of the town owed much to Quaker entrepreneurs and the Fox banking family. The much-altered perpendicular style church contains the ostentatious tomb of Sir John Popham, the judge who presided at the trial of Guy Fawkes. Present-day Wellington is a pleasant and prosperous shopping centre.

| Pets Welcome | Cash Dispenser | 24 Hour Facilities | WC Toilets | Alcoholic Drinks | Food and Drink | Fuel |

Wellington Monument

Follow the A38 south and it is signposted from Wellington. (4.4 miles)

An 170ft obelisk constructed in honour of the Duke of Wellington on the estate bought for him by the nation following his victory at the Battle of Waterloo. Although the foundation stone was laid in 1817, lack of funds led to a radical simplification of the design and the edifice was not finally completed until two years after his death, in 1854. Visitors who would like to make the 235 step climb to the top should telephone Mr Pocock at Monument Farm on Tel: (01823) 680303 prior to arrival to arrange for provision of the key. It is strongly advised to take a torch as there are no windows in the monument!

Sheppy's Cider Family Centre

Three Bridges, Bradford on Tone, Taunton TA4 1ER Tel: (01823) 461233 Follow the A38 north to Bradford on Tone (2.8 miles)

A 370 acre farm with 42 acres of cider orchards producing a wide variety of apples. The attractions include nature walks, cider sampling and tours of the cellars and press room. There is a museum with cidermaking video, cider shop, licensed tea room and picnic and childrens play areas. Disabled toilets.

West Somerset Railway

Bishops Lydeard Station, Bishops Lydeard TA4 3BX Follow the A38 north and follow the WSR signposts. (9.6 miles)

For details please see Junction 25 information

M5   JUNCTION 27

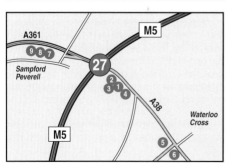

NEAREST SOUTHBOUND A&E HOSPITAL

Royal Devon & Exeter Hospital, Barrack Road, Wonford, Exeter EX2 5DW. Tel: (01392) 411611

Proceed to Junction 30 and take the A379 west to Exeter. The hospital is signposted within the town. (Distance Approx 17 miles)

NEAREST NORTHBOUND A&E HOSPITAL

Musgrove Park Hospital, Taunton TA1 5DA Tel: (01823) 333444

Proceed to Junction 26 and take the A38 route to Taunton. The hospital is signposted within the town. (Distance Approx 13.5 miles)

NEAREST MINOR INJURY UNIT

Tiverton District Hospital, William St. Tiverton EX16 6DJ Tel: (01884) 253251

Take the A36 west into Tiverton and the hospital is signposted within the town. Open 24 hours (Distance Approx 6 miles)

FACILITIES

---

1   Little Chef

🍴 ♿ Tel: (01884) 821205

0.1 miles east along the A38, on the right. Open; 07.00-22.00hrs Daily.

---

2   Burger King

🍴 ♿ Tel: (01884) 821205

0.1 miles east along the A38, on the right. Open 10.00-22.00hrs Daily.

---

3   Forte Travelodge

🏨M ♿ 🛏 Tel: (01884) 821087

0.1 miles east along the A38, on the right.

---

4   Tiverton Petrol Station (Shell)

⛽ ♿ WC 24 Tel: (01884) 822010

0.1 miles east along the A38, on the right. Access, Visa, Overdrive, All Star, Switch, Dial Card, Mastercard, Amex, Diners Club, Shell Cards, Smart Card.

---

5   The Waterloo Cross

🛏 🍴 ♿ 🛏 🏨B 🛏 Tel: (01884) 840328

0.4 miles east along the A38, at Waterloo Cross,

| 🏨H | Large Hotel | 🏨h | Small Hotel | 🏨M | Motel | 🏨B | Guest House/ Bed & Breakfast | ♿ | Disabled Facilities | Childrens Facilities |

on the right. (Eldridge, Pope & Co) Lunch; 12.00-14.00hrs Daily, Evening Meals; Mon-Sat; 18.00-21.00hrs. Sun; 19.00-21.00hrs.

6   The Old Well Petrol Station & Coffee Shop (Anglo)

[P] **[WC]** **[ll]** **[&]** Tel: (01884) 840873
0.5 miles east along the A38, at Waterloo Cross, on the right. Visa, Overdrive, Access, All Star, Dial Card, Switch, Mastercard. Open; 08.00-18.00hrs daily. (Coffeee Shop Open; 09.30-17.00hrs daily)

7   Parkway House Hotel

[H] **[tl]** **[ll]** **[~]** Tel: (01884) 82025
0.9 miles south of the A373, in Sampford Peverell, on the right. Cezanne's Restaurant & Bar; Food served 12.00-21.30hrs daily. Disabled access to restaurant.

8   Kellands Garage (Texaco)

[P] **[WC]** Tel: (01884) 820264
1 mile south of the A373, in Sampford Peverell, on the right. Overdrive, All Star, Call Card, Amex, Visa, Texaco Fuel Card, Diesel Direct, Switch, Dial Card, Mastercard, Diners Club. Open; Mon-Sat; 07.00-22.00hrs, Sun; 08.00-22.00hrs

9   Public Toilets

[WC]
1 mile south of the A373, in Sampford Peverell, on the right.

PLACES OF INTEREST

Tiverton

Tiverton Tourist Information Centre, Phoenix Lane, Tiverton Tel: (01884) 255827 Follow the A361 west (Signposted 6 miles)

A town of pre-Saxon origin, it stands high above the junction of the Rivers Exe and Lowman, leading to its original name of Twyford-ton (two ford town). Firmly established by the 16thC as a thriving centre for the woollen industry it was almost

destroyed by fire on no less than three occasions 1598, 1612 and 1731. Tiverton Castle, originally constructed in the 12thC and taken by siege by the Roundheads in 1645 is just one of many interesting buildings to be found within the town centre.

Within the town centre can be found ...

Tiverton Museum

St.Andrew Street, Tiverton EX16 6PH
Tel: (01884) 256295

Covering nearly half an acre and illustrating the history of the district from Roman times to the present day, it is one of the largest social history museums in the South West. Particularly noted for its collection of agricultural equipment, other galleries include civic displays, domestic artefacts, railway items and toys and dolls. Gift Shop. Disabled Access.

Grand Western Canal

Canal Hill, Tiverton EX16 4HX Follow the A361 to Tiverton and it is signposted within the town. (6.1 miles)

It was originally conceived as a barge canal linking Taunton and Topsham and thereby making a navigable link between the Bristol and English Channels. In the end only a short section, opened in 1814, was built to this standard with a smaller "tub boat" canal being constructed between Lowdwells and Taunton and officially opening in 1838. This tub boat section closed in 1864 and the original barge canal, disused since 1924, was officially closed in 1962. It is this stretch, restored and re-opened in the 1980's, that forms the centrepiece of the Grand Western Canal and Country Park. The Grand Western Horseboat Company (Tel: 01884-253345) runs horse drawn barge trips, to which there is disabled access, and hires out rowing boats and self-drive boats by the day. There is a Gift Shop, Restaurant Barge and picnic area as well as walks within the country park. Adjacent to the car park is Lime Kiln Cottage, a listed 16thC thatched building, where the Canal Tea Rooms and Gardens (Tel: 01884-252291), specializing in home-made cakes and traditional Devon cream teas, may be found.

M5

Coldharbour Industrial Museum

Uffculme, Cullompton EX15 3EE
Tel: (01884) 840960 Follow the A38 east and
follow the signs to Willand. Signposted along this
route. (2 miles)

A working wool museum demonstrating how old Victorian spinning, carding, and machines produce knitting wool and the Devon tartan. The 1910 steam engine is regularly run and the recently restored 1867 beam engine is on display as is the New World Tapestry which tells the story of the colonization of the Americas in cartoon style. The Mill is surrounded by a water garden and delightful walks and there is a mill shop and restaurant.

M5 | JUNCTION 28

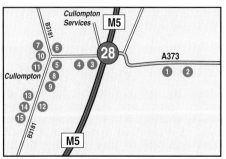

NEAREST A&E HOSPITAL

Royal Devon & Exeter Hospital, Barrack Road,
Wonford, Exeter EX2 5DW. Tel: (01392) 411611

Proceed to Junction 30 and take the A379 west to
Exeter. The hospital is signposted within the town.
(Distance Approx 13.3 miles)

NEAREST MINOR INJURY UNIT

Tiverton District Hospital, William Street,
Tiverton EX16 6BJ Tel: (01884) 253251

Take the road into Cullompton and follow the B3181
north to Willand. Take the B3391 west to Tiverton
and the hospital is signposted within the town. Open
24 hours. (Distance Approx 8.6 miles)

FACILITIES

SERVICE AREA

ACCESSED FROM THIS JUNCTION

Cullompton Services (Margram)
Tel: (01884) 38054

BP Fuel, Little Chef and McDonald's.

---

1 Oakdale B&B

🅱 ♿ 🏠   Tel: (01884) 33428

0.6 miles east along the A373, on the right.

---

2 Aller Barton Farm

🅱   Tel: (01884) 32275

0.7 miles east along the A373, on the right.

---

3 The Weary Traveller

🍴 🛏 🚶   Tel: (01884) 32317

0.1 miles west along the Cullompton Road, on the left. (Whitbread) Lunch, Evening meals.

---

4 CCS Ford, Jet Petrol Station

⛽ WC   Tel: (01884) 33795

0.2 miles west along the Cullompton Road, on the left. Overdrive, Dial Card, All Star, Visa, Access, Jet Card, Amex, Mastercard, Switch, Diners Club. Open; Mon-Fri; 06.00-22.30hrs, Sat; 06.00-22.00hrs, Sun; 07.30-22.00hrs.

---

5 Public Toilets

WC

0.4 miles west along the Cullompton Road, in Cullompton, on the left.

---

6 Court House B&B

🅱   Tel: (01884) 32510

0.4 miles west along the Cullompton Road, on the right.

7 Tandoori Spice

■ ▮ & ☒   Tel: (01884) 32060
0.4 miles north along the B3181, in Cullompton, on the left. Open; Wed-Mon; 12.00-14.00hrs & 18.00-23.30hrs

8 The King's Head

▮ ▮   Tel: (01884) 32418
0.5 miles south along the B3181, in Cullompton, on the left. (Courage) Open all day. Sunday Lunches only; 12.00-15.00hrs.

9 The Market House Inn

■ ▮ & ☒   Tel: (01884) 32339
0.5 miles south along the B3181, in Cullompton, on the left. Open all day. Lunches served daily.

10 The Gallery Bistro

■ ▮ &   Tel: (01884) 33336
0.5 miles south along the B3181, in Cullompton, on the right. Open; Cafe; Mon-Sat; 09.30-16.30hrs, Sun (for fixed Sunday lunch); 12.00-15.00hrs. Bistro; Thurs-Sat; 19.00-0.00hrs.

11 The Manor House Hotel

■ ▮ & ☒ ☒B   Tel: (01884) 32281
0.5 miles south along the B3181, in Cullompton, on the right. (Free House) Open all day. Bar snacks only.

12 White Hart Inn

■ ▮ ☒ & ☒B   Tel: (01884) 33260
0.6 miles south along the B3181, in Cullompton, on the left. (Courage) Meals served; Mon-Sat; 12.00-14.30hrs & 18.00-21.00hrs Daily

13 Exeter Road Garage (Esso)

▮ & WC   Tel: (01884) 32726
0.8 miles south along the B3181, in Cullompton, on the right. Essocard, Shell Card, All Star, Overdrive, Diners Club, Dial Card, Amex, Switch, Mastercard, Visa. Open; 07.00-21.00hrs Daily

14 The Bell Inn

■ ▮ ☒   Tel: (01884) 35672
0.8 miles south along the B3181, in Cullompton, on the right. (Heavitree) Open all day. Lunch, Evening meals.

15 The Inglenook Wine Bar & Restaurant

■ ▮ & ☒   Tel: (01884) 33685
1 mile south along the B3181, in Cullompton, on the right. Meals served Mon-Thurs 18.30-21.30, Fri & Sat 18.30-22.00, Sun 12.00-21.00 [Sunday Roast 12.00-18.00]

PLACES OF INTEREST

Bickleigh Castle

Bickleigh, Tiverton, Devon EX16 8RP
Tel: (01884) 855363 Follow the west exit into Cullompton and follow the signs to Bickleigh (7.5 miles)

A Royalist stronghold for over 900 years the castle includes an 11thC Chapel, guard room, great hall and a "Tudor" bedroom. The displays include Cromwellian arms and armour and a museum of 19thC rural life. Bickleigh Castle stands in a picturesque moated garden and there is a garden centre and delicious Devonshire cream teas with fresh baked scones and cakes are available.

Yearlstone Vineyard

Bickleigh, Tiverton, Devon EX16 8Rl
Tel: (01884) 855700 Follow the west exit into Cullompton and follow the signs to Bickleigh (7.5 miles)

Free tasting sessions are available here at Devon's oldest vineyard, magnificently sited in the Exe Valley. Pick up a tour guide and stroll among the rows of vines, see how West Country wines are produced and then enjoy comparing vintages and varieties.

M5

Pets Welcome	Cash Dispenser	24 Hour Facilities	Toilets	Alcoholic Drinks	Food and Drink

M5 | JUNCTION 29

THIS IS A RESTRICTED ACCESS JUNCTION

Vehicles can only enter the motorway (northbound) from the eastbound carriageway of the A30

Vehicles can only enter the motorway (southbound) from the westbound carriageway of the A30

Northbound and southbound vehicles can only exit along the eastbound carriageway of the A30

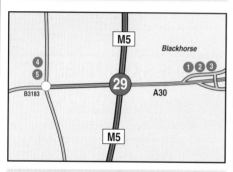

NEAREST A&E HOSPITAL

Royal Devon & Exeter Hospital, Barrack Road, Wonford, Exeter EX2 5DW. Tel: (01392) 411611

Southbound: Proceed to Junction 30 and take the A379 west to Exeter. The hospital is signposted from the junction. (Distance Approx 3.4 miles)

Northbound: Take the A30 east, turn around at the end of the dual carriageway section and follow the A30 into Exeter. The hospital is signposted within the town. (Distance Approx 2.5 miles)

FACILITIES

1 Hill Croft B&B

🛏️B   Tel: (01392) 367881
0.7 miles east along London Road, on the left.

2 Cheval Noir

🍴 🍽️ 🚻 🔧   Tel: (01392) 366649
0.8 miles east along London Road, on the left.
Open; 12.00-14.00hrs & 18.00-21.30hrs daily.

3 The Firs

🛏️B   Tel: (01392) 361821
0.9 miles east along London Road, on the left.

4 Holiday Inn Express

🛏️H 🚻 🛏️   Tel: (01392) 261000
0.4 miles west along the A30, on the right.

5 The Barn Owl

🍴 🍽️ 🚻 🔧   Tel: (01392) 449011
0.4 miles west along the A30, on the right.
(Bass) Meals served; Mon-Sat; 12.00-22.00hrs,
Sun; 12.00-21.30hrs.

M5 | JUNCTION 30

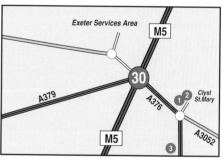

NEAREST A&E HOSPITAL

Royal Devon & Exeter Hospital, Barrack Road, Wonford, Exeter EX2 5DW Tel: (01392) 411611

Take the A379 west to Exeter and the hospital is signposted from the junction. (Distance Approx 2.4 miles)

FACILITIES

SERVICE AREA

ACCESSED FROM THIS JUNCTION

Exeter Services (Granada)
Tel: (01392) 274044 Esso Fuel, Harry Ramsden's, Burger King and Travelodge

1 The Halfmoon Inn

🔲 🔲 🔲B   Tel: (01392) 873515

0.5 miles south along the A376, in Clyst St Mary. (Heavitree) Lunch 12.00-14.00 and Evening Meals 19.00-21.30hrs Daily

2 The Maltsters

🔲 🔲 🔲B   Tel: (01392) 873445

0.5 miles south along the A376, in Clyst St Mary. (Whitbread) Meals served all day (May close for one hour during mid-afternoon). Disabled access to pub.

3 Redlands BP Service Station

🔲 £   Tel: (01392) 873040

1 mile south along the A376, on the right. Access, Visa, Overdrive, All Star, Switch, Dial Card, Mastercard, Amex, Diners Club, Delta, Routex, BP Cards, Shell Agency. Open; Mon-Sat; 07.00-22.00hrs, Sun; 09.00-19.00hrs. There is a 24 hour credit card operated pump.

PLACES OF INTEREST

Exeter

Exeter Tourist Information Centre
Tel: (01392) 265700 web: www.exeter.gov.uk
e-mail: econ.dvmt@exeter.gov.uk
Follow the signboards (3.1 miles)

An historic city with a majestic Norman cathedral, many fine old buildings and a wealth of excellent museums. It was established by the Celtish tribe of Dumnonii some 200 years before the Romans arrived and established it as the city of Isca, a walled and fortified stronghold. In the dark ages the city became a major ecclesiastical centre with King Cenwealh founding an abbey in 670AD on the site of the present cathedral. The city was ransacked by the Vikings in the 9thC and they occupied it twice before they were defeated by King Alfred. William the Conqueror took the city in 1086 and the Normans then constructed Rougemont Castle and commenced work on St.Peter's Cathedral, a task that was not completed until 1206. A port of some importance, the town prospered for many years and this is reflected

in the quality and variety of buildings within the city.

Crealy Park

Clyst St.Mary, Exeter EX5 1DR
Tel: (01395) 233200 web: www.crealy.co.uk
Follow the A3052 east (Signposted along this route 2.6 miles)

A huge outdoor adventure ground featuring many attractions for children and adults and the Farm Nursery and World of Pets allow for hands-on experience with small animals. Restaurant, Cafes, Kiosks, Gifts & Toy Shop.

M5 JUNCTION 31

M5

MOTORWAY ENDS

(TOTAL LENGTH OF MOTORWAY 162.0 MILES)

| Pets Welcome | £ Cash Dispenser | 24 24 Hour Facilities | WC Toilets | Alcoholic Drinks | Food and Drink | Fuel |

M5

 Large
Hotel
 Small
Hotel
 Motel
Guest House/
Bed & Breakfast
 Disabled
Facilities
 Childrens
Facilities

OFF THE MOTORWAY

M6

M6

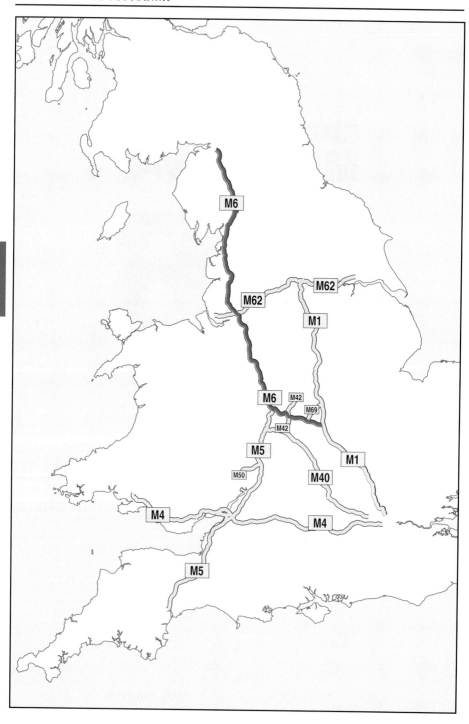

The M6 Motorway

Commencing at its junction with the M1 near Rugby, the motorway heads due east, passing Coventry to the south before connecting with the M42 and traversing through the northern suburbs of Birmingham, accessed via the famed Junction 6 an interchange better known as Spaghetti Junction, a term coined by a local journalist upon seeing the first aerial view of the proposed scheme.

The motorway connects with the M5 at Ray Hall Interchange before passing the railway yards at Bescot and turning north, bisecting Wolverhampton and Walsall as it heads through Staffordshire. The M6 was originally designed to carry 80,000 vehicles/day but, as any traveller who has struggled to make progress along the section between Junctions 4 and 10 would testify, it is hopelessly over-capacity with some 120,000 vehicles using it daily and making this the busiest stretch of motorway in Europe.

Stafford is by-passed along the west side with the 11thC Stafford Castle visible on a hill and the electrified London Euston to Glasgow, West Coast Main line is crossed for the first of numerous times as both routes make their way towards Carlisle. Further north, Stoke and the Potteries, the manufacturing base for some of the world's finest tableware, are passed on the east side before the motorway heads north across Cheshire and crosses the Manchester Ship Canal and River Mersey by the largest construction project on the route; The Thelwall Viaduct. Originally just one bridge, a second was opened in 1995, panoramic views are available as it is crossed with the cooling towers of Fiddlers Ferry Power Station and Runcorn Bridge visible in the distance beyond Warrington to the west and Altrincham and the western outskirts of Manchester visible on the east side.

The route continues northwards through West Lancashire, by-passing Wigan, Preston and Lancaster before the scenery starts to change dramatically with the Cumbrian Mountains visible on the west side and the Yorkshire Dales National Park, incorporating the Pennine Mountains to the east. The town of Kendal can be glimpsed between the hills to the west as the carriageways weave their way through the fells and a stone viaduct that carried the Tebay to Hellifield railway line over the River Lune can be seen on the east side before the motorway enters the Lune Valley, At the northern end of the valley, the village of Tebay is passed on the east side before the motorway ascends to the 1033ft high Shap Summit.

Continuing north there are panoramic views across the east in which Maulds Meaburn Moor, the Lune Forest and Dufton Fell can be seen with the Pennines beyond in the distance and the motorway carriageways diverge revealing the unusual spectacle of sheep grazing on the central reservation of a motorway, albeit wide enough to enclose a field. Penrith is passed on the east side and, as the motorway proceeds north various fells, with the Pennines beyond, continue to be visible to the east before the city of Carlisle, originally a Roman station, is by-passed to the west, the M6 ending at Junction 44 and diverging into the A74 to Glasgow and the A7 to Edinburgh via the Borders.

Location of Places of Interest

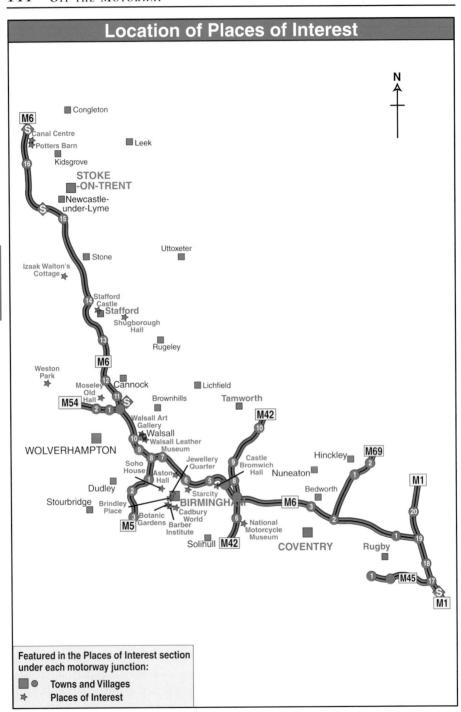

Featured in the Places of Interest section
under each motorway junction:

■ ● Towns and Villages
★ Places of Interest

Location of Places of Interest

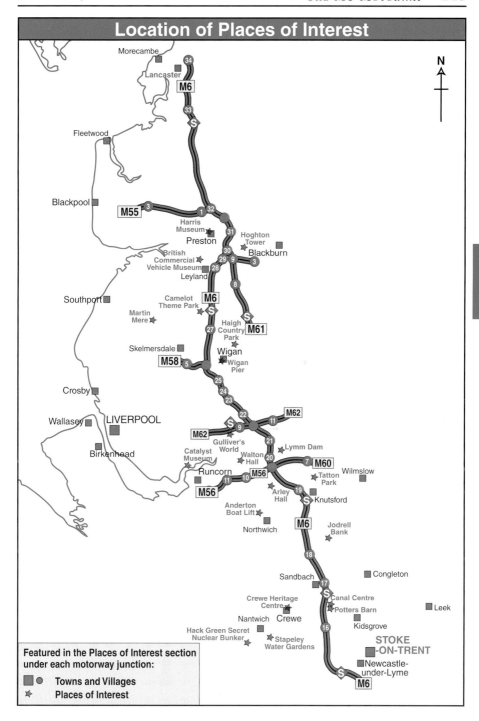

N

Morecambe

Lancaster

M6

34

33

S

Fleetwood

Blackpool

M55

3

32

1

Harris
Museum

Preston

31

Hoghton
Tower

British
Commercial
Vehicle Museum

30

29

9

Blackburn

3

28

Leyland

8

Southport

Camelot
Theme Park

M6

S

Martin
Mere

Haigh
Country
Park

S

M61

27

Skelmersdale

Wigan

M58

5

Wigan
Pier

25

Crosby

24

23

Wallasey

LIVERPOOL

22

9

M62

11

M62

S

Gulliver's
World

21

Lymm Dam

Birkenhead

Catalyst
Museum

Walton
Hall

20

7

M60

Runcorn

M56

Tatton
Park

Wilmslow

11

10

M56

19

Arley
Hall

Knutsford

Anderton
Boat Lift

M6

Jodrell
Bank

Northwich

18

Sandbach

Congleton

17

Crewe Heritage
Centre

S

Canal Centre

Potters Barn

Leek

Nantwich

Crewe

16

Kidsgrove

Hack Green Secret
Nuclear Bunker

Stapeley
Water Gardens

STOKE
-ON-TRENT

Newcastle-
under-Lyme

S

M6

Featured in the Places of Interest section
under each motorway junction:

■ ● Towns and Villages
✴ Places of Interest

Location of Places of Interest

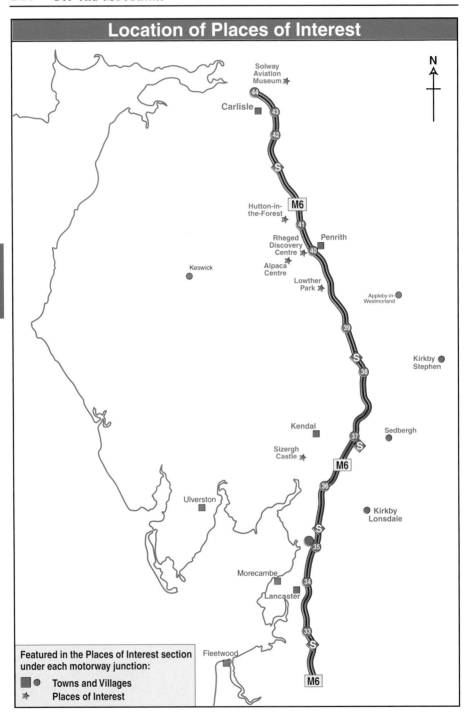

Featured in the Places of Interest section under each motorway junction:

■ ● Towns and Villages
✯ Places of Interest

M6 | JUNCTION 1

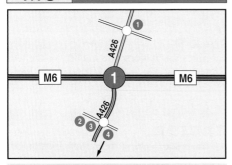

NEAREST A&E HOSPITAL

Hospital of St.Cross, Barby Road, Rugby CV22 5PX Tel: (01788) 572831

Take the A426 into Rugby and turn left along the A428. Turn right into Barby Road and the hospital is on the left. (Distance Approx 3.5 miles)

FACILITIES

THERE ARE NO FACILITIES WITHIN ONE MILE OF THIS JUNCTION

---

1 Star Gibbets Cross (Texaco)

📶 ♿ WC £   Tel: (01788) 861500
1.3 miles north along the A426, on the right. Access, Visa, Overdrive, All Star, Switch, Dial Card, Mastercard, Amex, AA Paytrak, Diners Club, Delta, BP Supercharge, Fast Fuel, Texaco Cards. Open; Mon-Fri; 06.30-22.30hrs, Sat & Sun; 07.00-22.00hrs

---

2 Holiday Inn Express, Rugby

🏨 ♿   Tel: (01788) 550333
1.3 miles south along the A426, on the right

---

3 Bell & Barge Harvester

🍽 🍴 ♿ 🐾   Tel: (01788) 569466
1.3 miles south along the A426, on the right (Bass Taverns) Open all day. Meals served; 12.00-22.00hrs daily

---

4 Rugby Shell

📶 WC 24   Tel: (01788) 862920
2 miles south along the A426 on the right, in

Leicester Road. Access, Visa, Overdrive, All Star, Switch, Dial Card, Mastercard, Amex, Diners Club, Delta, BP Agency, Shell Cards.

PLACES OF INTEREST

James Gilbert Rugby Football Museum

5 St.Matthews Street, Rugby CV21 3BY
Follow the A426 into Rugby and turn east along the A428 (Signposted 3.0 Miles)

For details please see M1 Junction 18 information.

Rugby School Museum

10 Little Church Street, Rugby CV21 3AW

For details please see M1 Junction 18 information.

M6 | JUNCTION 2

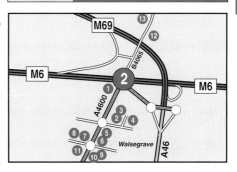

NEAREST A&E HOSPITAL

Coventry & Warwickshire Hospital, Stoney Stanton Road, Coventry CV1 4FH Tel: (024) 7622 4055

Follow the A4600 south and turn right along the A4053 (Signposted in city) (Distance Approx 4.3 Miles)

FACILITIES

---

1 Hilton National Hotel

🏨 🍴 🍴 ♿ 🚭   Tel: (024) 7660 3000
Adjacent to south side of roundabout. Restaurant open; Breakfast; Mon-Sat; 07.00-

| | Pets Welcome | | Cash Dispenser | | 24 Hour Facilities | WC | Toilets | | Alcoholic Drinks | | Food and Drink | | Fuel |

10.00hrs, Sun; 07.30-10.30hrs, Lunch; Mon-Fri; 12.00-14.00hrs, Dinner; 19.00-22.00hrs daily

2 Big W Cafe

⊞   Tel: (024) 7660 4519

0.3 miles south along the A4600, on the left in Cross Point Business Park. Open; Mon-Sat; 08.00-22.00hrs, Sun; 11.00-17.00hrs

3 Burger King

⊞   Tel: (024) 7660 3661

0.3 miles south along the A4600, on the left in Cross Point Business Park. Open; Mon-Fri; 11.00-22.00hrs, Sat & Sun; 11.00-22.30hrs

4 Pizza Hut

⊞   Tel: (024) 7660 3040

0.3 miles south along the A4600, on the left in Cross Point Business Park. Open; Sun-Fri; 12.00-23.00hrs, Sat; 11.30-23.30hrs

5 Posthouse Coventry

⊞   Tel: 0870 400 9021

0.4 miles south along the A4600, on the left. Restaurant open; Breakfast; Mon-Fri; 06.30-09.30hrs, Sat; 07.00-10.00hrs, Sun; 07.30-10.30hrs, Lunch; Mon-Fri; 12.30-14.30hrs, Sat; Closed, Sun; 12.30-15.00hrs, Dinner; Mon-Sat; 18.30-22.30hrs, Sun; 19.00-22.00hrs.

6 Asda Petrol Station

⊡   Tel: (024) 7661 3426

0.5 miles south along the A4600, on the left. Access, Visa, Overdrive, All Star, Switch, Dial Card, Mastercard, Amex, Diners Club, Delta, AA Paytrak, Asda Cards. Open; Mon-Sat; 05.00-0.00hrs, Sun; 06.00-22.00hrs. Toilets available in adjacent store during shop opening hours.

7 McDonald's

⊞   Tel: (024) 7661 2492

0.6 miles south along the A4600, on the left. Open; 07.30-23.00hrs [Drive Thru open until 0.00hrs Sat & Sun]

8 Hotel Campanile

⊞   Tel: (024) 7662 2311

0.6 miles south along the A4600, on the left.

Bistro & Restaurant open; Lunch; 12.00-14.00hrs, Dinner 19.00-22.00hrs Daily.

9 Mount Pleasant

⊞   Tel: (024) 7661 2406

0.7 miles south along the A4600, on the left. (Greene King) Open all day. Food served Mon-Sat;12.00-21.45hrs, Sun;12.00-21.30hrs.

10 Woodway BP Petrol Station

⊡   WC 24 £   Tel: (024) 7660 2928

0.7 miles south along the A4600, on the left. Access, Visa, Overdrive, All Star, Switch, Dial Card, Mastercard, Amex, Diners Club, Delta, Routex, AA Paytrak, UK Fuelcard, Shell Agency, BP Cards, Mobil Cards, IDS, Keyfuels, AS24

11 Star Walsgrave (Texaco)

⊡   WC 24   Tel: (024) 7684 1990

0.7 miles south along the A4600, on the right. Access, Visa, Overdrive, All Star, Switch, Dial Card, Mastercard, Amex, Diners Club, Delta, BP Supercharge, Texaco Cards.

12 Rose & Castle

⊞   Tel: (024) 7661 2822

0.8 miles north along the B4065, in Ansty, on the right. (Freehouse) Open all day on Sun. Meals served; Mon-Sat; 12.00-15.00hrs & 18.00-23.00hrs, Sun; 12.00-22.30hrs

13 Ansty Hall Hotel

⊞   Tel: (024) 7661 2222

1 mile north along the B4065, on the left. The Shilton Restaurant; Open; 12.00-14.30hrs & 19.00-21.00hrs daily.

PLACES OF INTEREST

Coventry

Coventry Tourist Information Centre, Bayley Lane, Coventry CV1 5RN. Tel: (024) 7683 2303/4
Follow the A4600 south (Signposted 4.5 Miles)

Standing on the River Sherbourne, Coventry is one of the oldest cities in England with a history that begins with a 6thC convent, destroyed in 1016 and a monastery constructed by Earl Leofric and his wife Godgifu (Godiva) for the Benedictines in 1043. For over a thousand years Coventry has been a manufacturing centre, initially with wool and cloth, ribbons and watches and then bicycles, motor cycles and cars. Six hundred years ago it was rated fourth amongst England's cities in power and importance and, despite its large scale destruction during World War II, many of the early buildings still survive. The notable exception, of course, is Coventry Cathedral which was all but razed to the ground on November 14th, 1940 but gave rise to a dynamic new building which became the focal point for the post-war rebuilding of the city. The legend of Lady Godiva, and her naked ride in protest at the high taxes imposed by her husband, is inextricably linked with Coventry and there is a statue of her in Broadgate.

Within the city centre ...

Coventry Cathedral & Visitors Centre

7 Priory Row, Coventry CV1 5ES
Tel: (024) 7622 7597
website; www.coventrycathedral.org
e-mail; information@coventrycathedral.org

The impressive, post-war, St Michael's Cathedral, designed by Sir Basil Spence in 1954, stands alongside the ruins of the 14thC edifice destroyed in the blitz during 1940 and contains a superb tapestry by Graham Sutherland. The Visitors Centre features an historic exhibition with an audio visual display "The Spirit of Coventry" in the undercroft of the cathedral. The international department continues the work of reconciliation and the cathedral community prays daily for peace. Disabled access.

Museum of British Road Transport

St Agnes Lane, Hales Street, Coventry CV1 1PN
Tel: (024) 7683 2425 website; www.mbrt.co.uk
e-mail; museum@mbrt.co.uk

The evolution of transport from the earliest of cycles through to the latest high-tech motor industry developments are shown in this display of British cars, cycles and motorcycles, the largest in the world. Other attractions include the story of the record breaking Thrust 2 and the Blitz Experience, a re-enactment of the bombing raids on Coventry during 1940. Gift Shop. Disabled access.

Herbert Art Gallery & Museum

Jordan Well, Coventry CV1 5QP
Tel: (024) 7683 2381

The "Godiva City" exhibition, a display of 1000 years of Coventry's history, is the centrepiece of the Museum whilst Graham Sutherland's working drawings for the Cathedral tapestry, paintings by LS Lowry, John Collier and David Cox and sculpture by Henry Moore and Jacob Epstein are featured in the Art Gallery. Tea Room. Gift Shop. Disabled access.

St Mary's Guildhall

Bayley Lane, Coventry CV1 5RN
Tel: (024) 7683 3041

An important mediaeval building that has been pivotal in the history of the city since the 14thC. Superb craftsmanship is evident in the stone, glass, timber and thread work and six royal portraits hang in the hall that at one time had played host to Elizabeth I, incarcerated Mary Queen of Scots and been used as an arsenal during the Civil War.

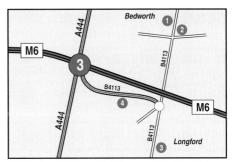

M6   JUNCTION 3

NEAREST A&E HOSPITAL

**Coventry & Warwickshire Hospital,
Stoney Stanton Road, Coventry CV1 4FH
Tel: (024) 7622 4055**

Take the A444 south exit to Coventry and follow this route to the city centre. Turn left along the St.Nicholas Ringway, take the first exit and follow the B4109 north into Stoney Stanton Road. The hospital is on the left. (Distance Approx 3.9 miles)

 Pets Welcome   Cash Dispenser   24 Hour Facilities   WC Toilets   Alcoholic Drinks   Food and Drink   Fuel

FACILITIES

1   Star Exhall (Texaco)

🚻 WC   Tel: (024) 7649 4300
1 mile north along the B4113, on the left.
Access, Visa, Overdrive, All Star, Switch, Dial
Card, Mastercard, Amex, Diners Club, Delta,
Texaco Cards. Open; 06.45-22.30hrs daily

2   The Lord Raglan

Tel: (024) 7636 0260
0.9 miles north along the B4113, on the right.
(Vanguard) Open all day. Meals served; Mon-
Fri; 12.00-14.00hrs

3   The Longford Engine

Tel: (024) 7636 5556
0.9 miles south along the B4113, in Longford,
on the left. (Tetley's) Open all day. Breakfast/
Lunch; 09.00-15.00hrs daily, Evening meals;
17.00-20.00hrs daily

4   Novotel Hotel

Tel: (024) 7636 5000
0.5 miles east along the B4113, on the right.
Restaurant Open; Breakfast; Mon-Fri; 06.00-
09.30hrs, Sat & Sun; 06.00-10.30hrs, Lunch;
Mon-Sat; 12.00-14.00hrs, Sun; Closed, Dinner;
18.00-0.00hrs. Bar Snacks served 12.00-0.00hrs
daily.

PLACES OF INTEREST

Coventry

Follow the A444 south (Signposted 4.7 Miles)
For details please see Junction 2 information.

Within the city centre ...

Coventry Cathedral & Visitors Centre
7 Priory Row, Coventry CV1 5ES
For details please see Junction 2 information.

Museum of British Road Transport
St Agnes Lane, Hales Street, Coventry CV1 1PN
For details please see Junction 2 information.

Herbert Art Gallery & Museum
Jordan Well, Coventry CV1 5QP
For details please see Junction 2 information.

St Mary's Guildhall
Bayley Lane, Coventry CV1 5RN
For details please see Junction 2 information.

SERVICE AREA

BETWEEN JUNCTIONS 3 AND 4

Corley Services (Westbound) (Welcome Break)
Tel: (01676) 540111 Granary Restaurant, Burger
King & Shell Fuel

Corley Services (Eastbound) (Welcome Break)
Tel: (01676) 540111 KFC, La Brioche Doree,
Garfunkel's, Granary Restaurant, Burger King &
Shell Fuel

M6   JUNCTION 4

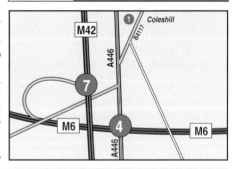

NEAREST A&E HOSPITAL

Heartlands Hospital Bordesley Green East,
Birmingham B9 5SS Tel: (0121) 766 6611

Take the A446 north exit and proceed to the first
roundabout. Turn left along the B4114 and continue
into the B4128. Folow this route for about 4 miles
and the hospital is on the left hand side. (Distance
Approx 6.6 miles)

FACILITIES

1   The George & Dragon

Tel: (01675) 466586
0.8 miles north along the B4117, in Coleshill,

on the left. (Bass) Meals served; Lunch; 12.00-14.00hrs daily, Evening Meals; Mon-Fri; 18.00-22.00hrs, Sat; 17.30-22.00hrs, Sun; 19.00-22.00hrs.

PLACES OF INTEREST

National Motorcycle Museum

Coventry Road, Bickenhill, Solihull B92 0EJ
Tel: (0121) 704 2784.
Follow the M42 south to Junction 6. Leave the motorway and the entrance is adjacent to the roundabout. (2.5 Miles)

A unique collection of over 700 beautifully restored British motorcycles dating from 1898 to the 1990's and outlining the history of a once world-dominating industry.

M6 | JUNCTION 4A

JUNCTION 4A IS A MOTORWAY INTERCHANGE WITH THE M42 ONLY AND THERE IS NO ACCESS TO ANY FACILITIES

M6 | JUNCTION 5

THIS IS A RESTRICTED ACCESS JUNCTION

Vehicles can only exit from the westbound lanes. Vehicles can only enter the motorway along the eastbound lanes.

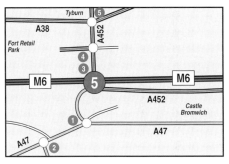

NEAREST EASTBOUND A&E HOSPITAL

Heartlands Hospital Bordesley Green East, Birmingham B9 5SS Tel: (0121) 766 6611
Continue to Junction 4, take the A446 north exit and proceed to the first roundabout. Turn left along the B4114 and continue into the B4128. Follow this route for about 4 miles and the hospital is on the left hand side. (Distance Approx 11.4 miles)

NEAREST WESTBOUND A&E HOSPITAL

Heartlands Hospital Bordesley Green East, Birmingham B9 5SS Tel: (0121) 766 6611
Take the A47 west towards Birmingham and turn left along the A4040. Follow this route for about 1 mile and turn right along Bordesley Green East (B4128). The hospital is on the left along this road. (Distance Approx 3.6 miles)

FACILITIES

1 BP Clock Garage

🅿 24   Tel: (0121) 749 1833
0.6 miles south along the A452, on the right. Access, Visa, Overdrive, All Star, Switch, Dial Card, Mastercard, Amex, Diners Club, Delta, Routex, AA Paytrak, Shell Agency, BP Cards, Mobil Cards.

2 The Hunters Moon

🍴 🍽 ♿ 🐾   Tel: (0121) 748 8951
1 mile west along the A47, on the left. (Bass) Meals served; Mon-Sat; 12.00-15.00hrs & 18.00-22.00hrs [Meals not served on Sun]

3 The Fort Jester

🍴 🍽 ♿ 🐾   Tel: (0121) 747 2908
0.1 miles north along the A452, on the left. (Punch Taverns) Open all day. Meals served; Mon-Sat; 12.00-22.00hrs, Sun; 12.00-21.30hrs

4 Holiday Inn Express

🛏H ♿ 🐾   Tel: (0121) 747 6633
0.1 miles north along the A452, on the left

5 The Tyburn House

🍴 🍽 ♿ 🐾   Tel: (0121) 747 2128
1 mile north along the A452, on the right. (Bass) Open all day. Meals served; Mon-Sat;

12.00-14.30hrs & 17.30-21.30hrs, Sun; 12.00-21.00hrs.

PLACES OF INTEREST

Castle Bromwich Hall Gardens Trust

Chester Road, Castle Bromwich, Birmingham B36 9BT Tel: (0121) 749 4100 Follow the A47 east and bear left along the B4118 (Birmingham Road). (Signposted 1.2 Miles)

An oasis of tranquility within 10 acre walled gardens lovingly recreated and restored to their 18thC splendour. Features include a maze, summer house, greenhouse and Holly Walk. Refreshments.

M6 JUNCTION 6

GRAVELLY HILL INTERCHANGE

This interchange is better known as Spaghetti Junction, a term coined by a local journalist upon seeing the first aerial view of the proposed scheme, and is the largest motorway interchange in Europe. Such is its renown and notoriety that the name features as a term in "20th Century Words" published by Oxford University Press. It was opened on May 24th, 1972 to complete the M6, includes seven miles of road and was built utilizing 175,000 yds^3 of concrete, 12,500 tonf of steelwork and 3,000 pillars. At its highest point the road is more than 80ft above ground level.

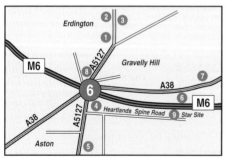

NEAREST A&E HOSPITAL

City Hospital, Dudley Road, Birmingham B18 7QH Tel: (0121) 554 3801

Take the A38(M) into Birmingham and at the second exit turn right along the Newtown Middleway (A4540). Follow this route to its junction with the A457 and turn right. The hospital is on the right

hand side of this route (Dudley Road). (Distance Approx 4 miles)

ALTERNATIVE A&E DEPARTMENT FOR CHILDREN UP TO THE AGE OF 16

The Birmingham Children's Hospital, Steelhouse Lane, Birmingham B4 6NH Tel: (0121) 333 9999

Take the A38(M) into Birmingham and it is signposted at the end of this short motorway. (Distance Approx 2.2 miles)

FACILITIES

1   The Rossmore Hotel

Tel: (0121) 377 7788

0.7 miles north along the A5127, on the left

2   BP Sixways

Tel: (0121) 373 1973

1 mile north along the A5127, on the left. Access, Visa, Overdrive, All Star, Switch, Dial Card, Mastercard, Amex, Diners Club, Delta, Routex, AA Paytrak, UK Fuelcard, Shell Agency, BP Cards, Smartcard, Securicor Fuelserv, IDS, Keyfuels.

3   Kennings (UK Fuels)

Tel: (0121) 377 6413

1 mile north along the A5127, on the right. Access, Visa, Overdrive, All Star, Switch, Mastercard, Amex, Diners Club, Delta, UK Fuelcard, Shell Cards, BP Cards. Open; Mon-Fri; 08.00-18.00hrs, Sat; 08.00-12.30 hrs, Sun; 09.00-12.00hrs.

4   BJ Banning (Elf)

Tel: (0121) 327 2741

0.2 miles south along the A5127, on the left. Access, Visa, Overdrive, All Star, Switch, Dial Card, Mastercard, Diners Club, Delta, AA Paytrak, Elf Cards, Total Fina Cards. Open; Mon-Fri; 07.30-18.30hrs, Sat; 07.30-13.30hrs, Sun; Closed.

5   Star Expressway Service Station (Texaco)

Tel: (0121) 328 1186

0.6 miles south along the A5127, on the left. Access, Visa, Overdrive, All Star, Switch, Dial Card, Mastercard, Amex, Diners Club, Delta, AA Paytrak, Texaco Card, UK Fuelcard.

6 Gravelly Park Service Station (Total Fina)

 Tel: (0121) 327 0026

0.5 miles east along the A38, on the right. Access, Visa, Overdrive, All Star, Switch, Dial Card, Mastercard, Amex, Diners Club, Delta, Routex, Keyfuels, Solo, Electron, Total Fina Cards. Open; Mon-Fri; 07.00-21.00hrs, Sat; 08.00-17.00hrs, Sun; 10.00-16.00hrs. Staff Toilets only, but available for customer use at discretion of staff.

7 Midland Link Service Station (Esso)

 Tel: (0121) 350 4493

0.7 miles east along the A38, on the left. Access, Visa, Overdrive, All Star, Switch, Dial Card, Mastercard, Amex, Diners Club, Delta, AA Paytrak, Shell Gold, BP Supercharge, Esso Cards.

8 The Armada

 Tel: (0121) 327 9900

Adjacent to north side of roundabout. (Bass) Open all day. Meals served; Mon-Sat; 11.00-14.30hrs & 17.30-19.30hrs, Sun; 12.00-15.00hrs

9 StarCity

0.8 miles south along the Heartlands Spine Road, on the left. There are numerous cafes and restaurants within the leisure complex. (Due to open in July 2000)

PLACES OF INTEREST

StarCity

Heartlands Spine Road, Birmingham B7 5TR
Follow the A5127 south towards Birmingham and turn left along the Heartlands Spine Road.
(Signposted 0.8 Miles)

The first of a new generation of state-of-the-art US-style urban entertainment centres, designed by the Jerde Partnership and containing the largest cinema in Europe, the Warner Village, a 30 screen megaplex seating 6,200 people. Vibrant leading edge facilities incorporating themed pubs, bars and restaurants, retail shops, casino, tenpin bowling, pool and snooker, electronic games and a leisure club can all be found on two levels and under one roof. The 395,000ft^2 complex, laid out in a street and walkways format surrounding a central atrium and events plaza, is due to open in July 2000.

Aston Hall

Trinity Road, Aston, Birmingham B6 6JD
Tel: (0121) 327 0062 Take the A6127 south towards Birmingham and follow the signboards to Aston Villa FC. (1.8 Miles)

The region's finest Jacobean country house, built between 1618 and 1635 by Sir Thomas Holte. Containing elaborate plasterwork ceilings and friezes, a magnificent carved oak staircase and a spectacular 136ft Long Gallery. The period rooms contain fine furniture, paintings, textiles and metalwork.

Birmingham

Follow the A38(M) south (Signposted 2.3 Miles)

For details please see Junction 1 (M5) information.

Brindley Place & Broad Street, Birmingham

Follow A38(M) south into the city centre and the route is signposted "National Indoor Arena & Convention Centre" (3 Miles)

For details please see Junction 1 (M5) information.

Soho House

Soho Avenue, Off Soho Road, Handsworth, Birmingham B18 5LB.
Follow A38(M) south towards Birmingham and turn west along the A4540. Turn north along the A41 and left into Soho Avenue.(3.7 Miles)

For details please see Junction 1 (M5) information.

Jewellery Quarter Discovery Centre.

75-79 Vyse Street, Hockley, Birmingham B18 6HA.
Follow A38(M) south towards Birmingham and turn west along the A4540. Turn south along the A41 and the route is signposted. (3.6 Miles)

| Pets Welcome | £ Cash Dispenser | 24 24 Hour Facilities | WC Toilets | Alcoholic Drinks | Food and Drink | Fuel |

For details please see Junction 1 (M5) information.

Cadbury World

Bournville, Birmingham B30 2LD
Follow the A38(M) south and the brown & white tourist boards along the A38 & A4040. (7 miles).

For details please see Junction 4 (M5) information.

M6 JUNCTION 7

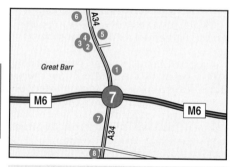

NEAREST WESTBOUND A&E HOSPITAL

Walsall Manor Hospital, Moat Road,
Walsall WS2 9PS Tel: (01922) 721172

Proceed to Junction 9 and take the A4148 north. The hospital is on the left hand side of Pleck Road on this route. (Distance Approx 4.6 miles)

NEAREST EASTBOUND A&E HOSPITAL

Sandwell District General Hospital
Hallam Street, West Bromwich B71 4HJ
Tel: (0121) 553 1831

Take the A34 towards Birmingham and after 0.5 miles turn right along the A4041 to West Bromwich. Turn left along All Saints Way (A4031), first left into Church Vale, continue into Hallam Street and the hospital is on the left. (Distance Approx 3.3 miles)

FACILITIES

1 Posthouse Birmingham

Tel: 0870 400 9009
0.4 miles north along the A34, on the right. Traders Bar & Restaurant; Open all day. Meals served all day from 07.00-22.30hrs.

2 The Beacon

Tel: (0121) 357 2567
0.6 miles north along the A34, on the left. (Harvester) Open all day. Meals served all day from 12.00-22.30hrs.

3 Holiday Inn Express

Tel: (0121) 358 4044
0.6 miles north along the A34, on the left

4 Great Barr Service Station (Elf)

Tel: (0121) 358 5616
0.6 miles north along the A34, on the left. Access, Visa, Overdrive, All Star, Switch, Dial Card, Mastercard, Amex, Diners Club, Delta, Elf Cards, Total Fina Cards. Open; 07.00-23.00hrs daily.

5 Beacon Service Station (Total Fina)

Tel: (0121) 358 2628
0.6 miles north along the A34, on the right. Access, Visa, Overdrive, All Star, Switch, Dial Card, Mastercard, Amex, Diners Club, Delta, BP Supercharge, Elf Cards, Total Fina Cards.

6 The Bell Inn

Tel: (0121) 357 7461
1 mile north along the A34, on the left. (Bass) Open all day. Meals served; Mon-Fri; 12.00-14.30hrs & 18.00-21.00hrs, Sat; 12.00-21.00hrs, Sun; 12.00-17.00hrs.

7 Shell Great Barr

Tel: (0121) 358 4622
0.2 miles south along the A34, on the right (Actual distance 1.8 miles) Access, Visa, Overdrive, All Star, Switch, Dial Card, Mastercard, Amex, Diners Club, Delta, BP Supercharge, Esso Eurocard, Smartcard. Open; 07.00-23.00hrs daily.

8 Scott Arms

Tel: (0121) 526 2548
0.6 miles south along the A34, on the right (Avery Taverns) Meals served; Mon-Sat; 12.00-14.00hrs

	Large Hotel		Small Hotel		Motel	
			Guest House/ Bed & Breakfast		Disabled Facilities	Childrens Facilities

M6 | JUNCTION 8

**RAY HALL INTERCHANGE.
THIS JUNCTION IS A MOTORWAY
INTERCHANGE ONLY WITH THE M5 AND
THERE IS NO ACCESS TO ANY
FACILITIES**

M6 | JUNCTION 9

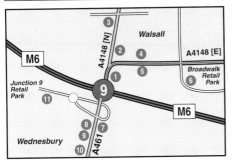

NEAREST WESTBOUND A&E HOSPITAL

**Walsall Manor Hospital, Moat Road,
Walsall WS2 9PS Tel: (01922) 721172**

Proceed to Junction 9 and take the A4148 north.
The hospital is on the left hand side of Pleck Road
on this route. (Distance Approx 2.5 miles)

NEAREST EASTBOUND A&E HOSPITAL

**Sandwell District General Hospital
Hallam Street, West Bromwich B71 4HJ
Tel: (0121) 553 1831**

Proceed to Junction 7, take the A34 towards
Birmingham and after 0.5 miles turn right along
the A4041 to West Bromwich. Turn left along All
Saints Way (A4031), first left into Church Vale,
continue into Hallam Street and the hospital is on
the left. (Distance Approx 4.8 miles)

NEAREST SOUTHBOUND A&E HOSPITAL (VIA M5)

**Sandwell District General Hospital
Hallam Street, West Bromwich B71 4HJ
Tel: (0121) 553 1831**

Proceed to Junction 1 and take the A41 west
towards West Bromwich. Turn right at the
roundabout along the A4031 and the hospital is on
the right. (Distance Approx 4.8 miles)

NEAREST A&E HOSPITAL

**Walsall Manor Hospital, Moat Road, Walsall WS2
9PS Tel: (01922) 721172**

Take the A4148 north and the hospital is on the left
hand side of Pleck Road on this route. (Distance
Approx 1.4 miles)

FACILITIES

1 Bescot House Hotel & Restaurant

Tel: (01922) 622447

0.1 miles north along the A461, on the right.
Restaurant open; Mon-Sat; 19.00-21.30hrs.

2 Abberley Hotel

Tel: (01922) 627413

0.3 miles north along the A4148[N], on the
right

3 The Forge & Fettle Tap House

Tel: (01922) 622499

1 mile north along the A4148[N], on the left.
(Banks's) Open all day. Meals served; 11.00-
22.00hrs daily.

4 Morrissons Store

Tel: (01922) 616177

0.5 miles east along the A4148[E], on the left.
Access, Visa, Overdrive, All Star, Switch, Dial
Card, Mastercard, Delta, AA Paytrak, BP
Supercharge, Morrissons Account Card. Toilets
in adjacent store. Open; Mon-Wed; 07.00-
21.00hrs, Thu-Fri; 07.00-22.30hrs, Sat; 07.00-
20.30hrs, Sun; 09.00-19.00hrs.

5 Grange Garage (Q8)

Tel: (01922) 626734

0.5 miles east along the A4148[E], on the right.
Access, Visa, Overdrive, All Star, Switch, Dial
Card, Mastercard. Open; Mon-Sat; 07.30-
20.00hrs, Sun; 09.00-20.00hrs.

6 McDonald's

Tel: (01922) 635747

0.7 miles east along the A4148 on the right, in
Broadwalk Retail Park. Open; 07.30-23.00hrs

| Pets Welcome | £ Cash Dispenser | 24 24 Hour Facilities | WC Toilets | Alcoholic Drinks | Food and Drink | Fuel |

daily. Fri-Sun; "Drive-Thru" remains open until 0.00hrs.

7   Stones Restaurant

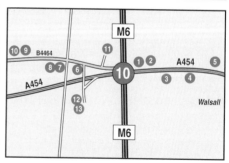

Tel: (0121) 502 2218
0.4 miles south along the A461, on the left.
(Free House) Lunch; Mon-Fri; 12.00-14.00hrs,
Sat; Closed, Sun; 12.00-15.00hrs, Evening
Meals; Mon-Sat; 18.00-21.30hrs

8   The Cottage Inn

Tel: (0121) 556 2253
0.5 miles south along the A461, on the right.
(Punch Inns) Open all day. Bar snacks served
all day.

9   The Horse & Jockey

Tel: (0121) 556 0464
0.5 miles south along the A461, on the right.
(Punch Taverns) A la Carte Restaurant; Open;
Sat; 19.00-22.00hrs, Sun; 12.00-15.00hrs. Steak
Bar 2-for1 Restaurant; Open; Mon-Sat; 12.00-
15.00hrs & 18.00-22.00hrs, Sun; 12.00-
22.00hrs.

10 Safeway BP Wednesbury

Tel: (0121) 505 2205
0.9 miles south along the A461, on the right.
Access, Visa, Overdrive, All Star, Switch, Dial
Card, Mastercard, Amex, Diners Club, Delta.
Open; 07.00-23.00hrs daily

11 Burger King

Tel: (0121) 556 2100
0.4 miles south, on the right, in Junction 9
Retail Park (Access gained 0.1 miles south along
the A461, on the left). Open; 09.00-21.00hrs
daily

PLACES OF INTEREST

Walsall Leather Museum

Littleton Street West, Walsall WS2 8EQ
Tel: (01922) 721153 Follow the A4148 north into
Walsall (1.9 Miles)

From the earliest of days, leather has been a
vital material in the daily life of Britain and
this is reflected in the wide range of samples of
the craft exhibited here. Jugs, bottles, bridles,
saddles, luggage, clothing, musical instruments
and forge bellows are just some of the examples
on display. Walsall is renowned as the British
leathergoods capital, over a hundred
companies in the area are still involved in the
manufacture of leather goods, and this
museum captures the atmosphere of the
original workshops. Disabled access. Gift Shop.
Coffee Shop.

Also in the town centre ...

The New Art Gallery Walsall

Gallery Square, Walsall WS2 8LG
Tel: (01922) 654400
website; www.artatwalsall.org.uk
e-mail; info@artatwalsall.org.uk

Designed by Caruso St John Architects the
landmark building was completed in 1999 and
opened in 2000. The whole of the third floor
is dedicated to the display of contemporary and
historic art in beautiful naturally lit galleries
of international specification. The first and
second floors, a series of intimate,
interconnected rooms house the Garman Ryan
Collection and, at ground floor level, a
Discovery Gallery with a 3-storey childrens
house and art gallery create an active
involvement in the creative processes of art.
Disabled access. Cafe. Gift Shop. Restaurant.

M6   JUNCTION 10

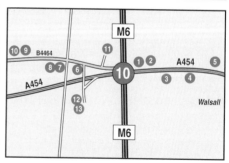

NEAREST A&E HOSPITAL

Walsall Manor Hospital, Moat Road, Walsall WS2
9PS Tel: (01922) 721172

Take the A454 east towards Walsall and turn right
along Pleck Road (A4148) The hospital is on the
right. (Distance Approx 0.8 miles)

| Large Hotel | Small Hotel | Motel | Guest House/ Bed & Breakfast | Disabled Facilities | Childrens Facilities |

FACILITIES

1   Hills Transport Cafe

🍴 Tel: (01922) 722593
0.2 miles east along the A454, on the left. Open; Mon-Fri; 07.30-16.00hrs, Sat; 07.30-10.30hrs

2   The Parkbrook

🍺🍴♿🌳 Tel: (01922) 622970
0.3 miles east along the A454, on the left. (Punch Taverns) Open all day. Meals served; Mon-Sat; 12.00-14.00hrs & 18.00-20.00hrs, Sun; 12.00-14.30hrs.

3   Primley Service Station (Esso)

⛽ WC Tel: (01922) 634532
0.4 miles east along the A454, on the right. Access, Visa, Overdrive, All Star, Switch, Dial Card, Mastercard, Amex, Diners Club, Delta, AA Paytrak, Shell Gold, BP Supercharge, Esso Cards. Open; Mon-Fri; 06.00-23.00hrs, Sat; 06.00-22.00hrs, Sun; 07.00-22.00hrs

4   Save Petrol Station

⛽ WC Tel: (01922) 638054
0.6 miles east along the A454, on the right. Overdrive, Savecard, Mastercard, Switch, Visa, All Star. Open; 07.00-23.00hrs daily

5   The Orange Tree Pub and B&B

🍺🍴♿🌳🛏B Tel: (01922) 625119
0.9 miles east along the A454, on the left. (Inn Partnership) Open all day. Meals served; 12.00-20.30hrs daily

6   Save Petrol Station

⛽ Tel: (01922) 641351
0.4 miles west along the B4464, in Bentley, on the left. Access, Visa, Switch, Mastercard, Savecard. Open; 07.00-23.00hrs daily.

7   The Greedy Pig Cafe

🍴 Tel: (01922) 722998
0.5 miles west along the B4464, in Bentley, on the left. Open; Mon-Fri; 08.00-14.00hrs, Sat; 08.00-12.00hrs

8   The Golden Bengal

🍴♿ Tel: (01922) 746786
0.5 miles west along the B4464, in Bentley, on the left. Open; 17.30-0.00hrs daily

9   County Bridge Service Station (Esso)

⛽ WC Tel: (01902) 605734
1 mile west along the B4464, on the right. Access, Visa, Overdrive, All Star, Switch, Dial Card, Mastercard, Amex, Diners Club, Delta, UK Fuelcard, Shell Gold, BP Supercharge, Esso Cards. Open; 06.00-23.00hrs daily.

10 The Red Lion

🍺🍴♿🌳 Tel: (01902) 365921
1 mile west along the B4464, on the right. (Bass) Open all day. Meals served; Mon-Sat; 12.00-14.30 hrs & 18.00-20.30hrs, Sun; 12.00-15.00hrs

11 Quality Friendly Hotel

🛏H🍺🍴♿ Tel: (01922) 724444
On the west side of the roundabout (Signposted "Hotel"). Restaurant Open; Breakfast; Mon-Fri; 07.00-09.30hrs, Sat & Sun; 08.00-10.00hrs, Lunch; Sun-Fri; 12.00-14.00hrs, Sat; Closed, Dinner; 19.00-22.00hrs daily.

12 Deep Pan Pizza

🍺🍴♿🌳 Tel: (0121) 568 8053
0.6 miles south along Bentley Mill Way, on the right. Open; Sun-Thurs; 12.00-23.00hrs, Fri & Sat; 12.00-0.00hrs.

13 Fatty Arbuckles

🍺🍴♿🌳 Tel: (0121) 568 6910
0.6 miles south along Bentley Mill Way, on the right. Open; Mon-Sat; 12.00-23.30hrs, Sun; 12.00-23.00hrs.

PLACES OF INTEREST

Walsall Leather Museum

Littleton Street West, Walsall WS2 8EQ Follow the A454 east into Walsall (1.3 Miles)

For details please see Junction 10 information

| Pets Welcome | £ Cash Dispenser | 24 24 Hour Facilities | WC Toilets | Alcoholic Drinks | Food and Drink | Fuel |

Also in town centre ...

The New Art Gallery Walsall

Gallery Square, Walsall WS2 8LG

For details please see Junction 10 information

M6 | JUNCTION 10A

JUNCTION 10A IS A MOTORWAY
INTERCHANGE ONLY WITH THE M54
AND THERE IS NO ACCESS TO ANY
FACILITIES

SERVICE AREA

BETWEEN JUNCTIONS 10A AND 11

Hilton Park Services (Northbound) (Granada)

Tel: (01922) 412237 Fresh Express Self Service Restaurant, Burger King, Little Chef & BP Fuel (& Shell Diesel)

Hilton Park Services (Southbound) (Granada)

Tel: (01922) 412237 Fresh Express Self Service Restaurant, Harry Ramsden's, Burger King, Travelodge & BP Fuel (& Shell Diesel)

Footbridge connection between sites.

M6 | JUNCTION 11

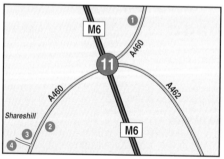

NEAREST NORTHBOUND A&E HOSPITAL

Staffordshire General Hospital, Weston Road, Stafford ST16 3SA Tel: (01785) 257731

Proceed to Junction 13. Take the A449 exit north into Stafford, continue along the A49 and then the A34. At the roundabout turn right along the A518 east towards Weston and the hospital is on the left. (Distance Approx 11.6 miles)

NEAREST SOUTHBOUND A&E HOSPITAL

Walsall Manor Hospital, Moat Road, Walsall WS2 9PS Tel: (01922) 721172

Proceed to Junction 10, take the A454 east towards Walsall and turn right along Pleck Road (A4148). The hospital is on the right. (Distance Approx 6.5 miles)

FACILITIES

1   The Wheatsheaf

Tel: (01922) 412304

0.2 miles north along the A460, on the left. (Banks's) Open all day. Meals served 12.00-21.00hrs daily.

2   M6 Diesel

Tel: (01922) 412995

0.7 miles west along the A460, on the left. Securicor Fuelserv, Keyfuels, IDS, UK Fuels, Overdrive, AS24, Morgan Fuels, Access, Visa, Amex, Diners Club, Mastercard, Switch. Open; Continuously between 17.00 hrs on Sunday to 22.00hrs Friday. Sat; 07.00-15.00hrs. (Fuel Card operated pumps when kiosk closed)

3   Shareshill Service Station (Jet)

Tel: (01922) 419338

0.8 miles west along the A460, on the right. Access, Visa, Overdrive, All Star, Switch, Dial Card, Mastercard, Amex, Diners Club, BP Supercharge, Jet Cards. Open; Mon-Fri; 07.00-21.00hrs, Sat-Sun; 08.00-21.00hrs.

4   The Elms

Tel: (01922) 412063

1 mile west, in Shareshill, on the left. (Bass) Open all day Fri-Sun. Meals served; Mon-Sat; 12.00-14.00hrs, Sun; 12.00-14.30hrs.

PLACES OF INTEREST

Moseley Old Hall (NT)

Moseley Old Hall Lane, Fordhouses, Wolverhampton WV10 7HY Tel: (01902) 782808 Follow the A460 west, proceed past Junction 1

| H | Large Hotel | h | Small Hotel | M | Motel | B | Guest House/ Bed & Breakfast | | Disabled Facilities | | Childrens Facilities |

(M54) and turn right along Moseley Road. (2.9 Miles)

An Elizabethan house where the Whitgreave family sheltered King Charles II after the Battle of Worcester in 1651. The richly panelled walls cover ingenious hiding-holes that were originally designed to accommodate Catholic priests and it was in one of these that he was concealed from Cromwell's troops. An exhibition recounts this event and fine furniture, documents, portraits and other relics of the Whitgreaves are on display. The garden is recreated in 17thC style with a formal knot and planted with varieties of herbs and plants from three centuries ago. Shop. Tea Room. Disabled access.

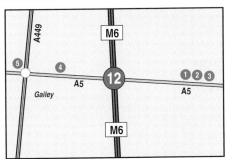

NEAREST NORTHBOUND A&E HOSPITAL

Staffordshire General Hospital, Weston Road, Stafford ST16 3SA Tel: (01785) 257731

Proceed to Junction 13. Take the A449 exit north into Stafford, continue along the A49 and then the A34. At the roundabout turn right along the A518 east towards Weston and the hospital is on the left. (Distance Approx 8.9 miles)

NEAREST SOUTHBOUND A&E HOSPITAL

Walsall Manor Hospital, Moat Road, Walsall WS2 9PS Tel: (01922) 721172

Proceed to Junction 10, take the A454 east towards Walsall and turn right along Pleck Road (A4148). The hospital is on the right. (Distance Approx 9.2 miles)

FACILITIES

1   IDS International Diesel Service (Q8)

📞 24   Tel: (01543) 503435

1 mile east along the A5, on the left. 24 Hour card operated pumps. IDS Card and account holders only

2   The Hollies Transport Cafe

🍴 WC 24   Tel: (01543) 503435

1 mile east along the A5, on the left

3   Oak Farm Hotel & Restaurant

🛏🍴🍴♿   Tel: (01543) 462045

1 mile east along the A5, on the left. Open for Evening Meals only; Mon-Sat; 19.00-21.15. Sun; Closed.

4   Gailey Service Station (Texaco)

📞 WC   Tel: (01902) 791172

0.8 miles west along the A5, on the right. Access, Visa, Overdrive, All Star, Switch, Dial Card, Mastercard, Amex, Diners Club, Delta, AA Paytrak, BP Supercharge, Texaco Cards. Open; Mon-Fri; 06.00-23.00hrs, Sat; 07.00-22.00hrs, Sun; 08.00-23.00hrs.

5   The Spread Eagle

🍺🍴♿🐾   Tel: (01902) 790212

1 mile west along the A5, on the right. (Milestone) Bar Opening hours; Summer [May to September]; Open all day, Bar Meals served from 12.00 to 22.00hrs daily, Winter [October-April]; Bar Meals 12.00-15.00hrs & 18.00-22.00hrs daily. Restaurant Opening hours; Lunch; Mon-Sat; 12.00-14.30hrs, Evening Meals; Mon-Thur; 17.30-22.00hrs, Fri-Sat; 17.30-22.30hrs. Meals served all day on Sundays from 12.00-21.00hrs.

PLACES OF INTEREST

Weston Park

Weston under Lizard, Near Shifnal,
Shropshire TF11 8LE Tel: (01952) 850207
Follow the A5 east (Signposted along route. 8.2
Miles)

Built in 1671 in the restoration style and the home of the Earls of Bradford, the house contains paintings by Holbein, Van Dyck, Bassano, Reynolds, Gainsborough and Lely, Tapestries by Gobelins and Aubusson and fine 17thC silver. The parkland that surrounds the house has matured over several hundred years into a masterpiece of unspoilt landscape and contains fallow deer and rare breeds of sheep. There are some wonderful architectural features within the park too, including the Roman Bridge and Temple of Diana, both designed and built by James Paine for Sir Henry Bridgeman in c1760. For the children, there is a Woodland Adventure Playground, Pets Corner and Deer Park as well as a 1.5 mile long miniature railway. Weston Park was chosen to host a meeting of the world leaders during the Birmingham G8 Summit of 1998. Old Stables Tea Room. Gift Shop.

M6   JUNCTION 13

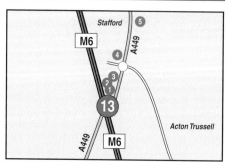

NEAREST A&E HOSPITAL

Staffordshire General Hospital, Weston Road,
Stafford ST16 3SA Tel: (01785) 257731

Take the A449 exit north into Stafford, continue along the A49 and then the A34. At the roundabout turn right along the A518 east towards Weston and the hospital is on the left. (Distance Approx 3.7 miles)

FACILITIES

1   Holiday Inn Express

[=H] [disabled] [childrens]   Tel: (01785) 212244
0.1 miles north along the A449, on the left

2   Fatty Arbuckles

[M] [fork] [disabled] [childrens]   Tel: (01785) 212221
0.1 miles north along the A449, on the left
Open; Mon-Sat; 11.00-23.30hrs, Sun; 11.00-23.00hrs.

3   Catch Corner

[M] [fork] [disabled] [childrens]   Tel: (01785) 245867
0.1 miles north along the A449, on the left.
(Tom Cobleigh) Open all day. Meals served 12.00-21.30hrs daily.

4   Acton Gate Service Station (Elf)

[i] [WC]   Tel: (01785) 254428
0.5 miles north along the A449, on the left.
Access, Visa, Overdrive, All Star, Switch, Dial Card, Mastercard, Amex, Diners Club, Delta, BP Supercharge, Elf Cards, Total Fina Cards. Open; 07.00-21.00hrs Daily.

5   The Garth Hotel

[=H] [M] [fork] [disabled] [childrens]   Tel: (01785) 256124
1 mile north along the A449, on the right.
(Tavern) Bar meals served 12.00-21.00hrs daily. Restaurant; Lunch; Sun-Fri; 12.00-14.00hrs, Sat; Closed. Evening Meals; Mon-Sat; 19.00-21.45hrs, Sun; 19.00-21.00hrs.

PLACES OF INTEREST

Stafford

Stafford Tourist Information Centre,
Ancient High House, Greengate Street,
Stafford ST16 2HS Tel: (01785) 240204
Follow the A449 north (Signposted 3.6 Miles)

Sited on the banks of the River Sow, the county town of Staffordshire is of Saxon origin but very few traces of this period remain. There are, however, some excellent period buildings within the town centre including the Ancient

High House, an Elizabethan building, William Salt Library, Chetwynd House, built in 1750 and now the town's Post Office, and The Noell Almshouses. A castle was erected here during Norman times.

Within the town centre ...

Ancient High House

Greengate Street, Stafford ST16 2JA
Tel: (01785) 240204

The largest timber framed house in England, it was built in 1595 by the Dorrington family and is now a heritage and exhibition centre housing the Yeomanry Museum and Tourist Information Centre. Shop.

William Salt Library

Eastgate Street, Stafford ST16 2LT
Tel: (01785) 252276

An 18thC town house now containing the county archive and history section.

Stafford Castle & Visitor Centre

Newport Road, Stafford ST16 1DJ
Tel: (01785) 257698
Follow the A449 into Stafford and take the A518 southwest towards Newport. (4.3 Miles)

Originally a motte and bailey castle of the Norman period, the stone keep was destroyed during the Civil War. The purpose built Visitor Centre displays artefacts from archaeological digs and contains a fascinating audio-visual presentation, scale models and hands-on equipment. Shop.

Shugborough Hall

Shugborough, Milford, Nr Stafford ST17 0XB
Tel: (01889) 881388
website; www.staffordshire.gov.uk
Follow the A446 north (Signposted 6.3 Miles)

The 17thC seat of the Earls of Lichfield, the magnificent 900 acre estate includes Shrugborough Park Farm, a Georgian farmstead built in 1805 for Thomas, Viscount Anson, and now home to rare breeds and demonstrations of traditional farming methods. The mansion itself is a splendid piece of architecture, altered several times over its 300 years, but always retaining the distinct grandeur. The vast rooms, with their ornate plasterwork and cornicing, contain an impressive collection of paintings, ceramics, silverware and a wealth of elegant

French furniture. The Staffordshire County Museum has been established in the former stable and kitchen wing of the house whilst in the beautiful parkland surrounding the mansion, can be found an outstanding collection of neoclassical monuments and the Lady Walk leads along the banks of the River Sow to the delightful terraced lawns and rose garden. Restaurant. Gift Shop.

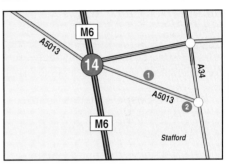

M6 JUNCTION 14

NEAREST A&E HOSPITAL

Staffordshire General Hospital, Weston Road, Stafford ST16 3SA Tel: (01785) 257731

Take the A5013 into Stafford, proceed along the A34 and follow the A518 east towards Weston. The hospital is on the left. (Distance Approx 2.7 miles)

FACILITIES

1 Tillington Hall Hotel

Tel: (01785) 253531
0.5 miles south along the A5013, on the left. Restaurant Open; Lunch; Sun-Fri; 12.30-13.45hrs, Sat; Closed. Evening Meals; Mon-Sat; 19.00-21.45hrs, Sun; 19.00-21.00hrs.

2 Brookhouse Service Station (Elf)

Tel: (01785) 244329
1 mile south along the A5013, in Stafford, on the right. Access, Visa, Overdrive, All Star, Switch, Dial Card, Mastercard, Amex, Diners Club, Delta, Routex, Elf Cards, Total Fina Cards. Open; Mon-Sat; 07.00-23.00hrs, Sun; 08.00-22.00hrs.

| Pets Welcome | £ Cash Dispenser | 24 24 Hour Facilities | WC Toilets | Alcoholic Drinks | Food and Drink | Fuel |

PLACES OF INTEREST

Stafford

Follow the A5013 south (Signposted 2.4 Miles)
For details please see Junction 13 information

Within the town centre ...

Ancient High House

Greengate Street, Stafford ST16 2JA
For details please see Junction 13 information

William Salt Library

Eastgate Street, Stafford ST16 2LT
For details please see Junction 13 information

Stafford Castle & Visitor Centre

Newport Road, Stafford ST16 1DJ
Follow the A5013 south into Stafford and take the
A518 southwest towards Newport. (3.6 Miles)
For details please see Junction 13 information

Shugborough Hall

Shugborough, Milford, Nr Stafford ST17 0XB
Follow the A5013 south into Stafford and take the
A51 east (Signposted 7.1 Miles)
For details please see Junction 13 information

Izaak Walton's Cottage

Worston Lane, Shallowford, Nr Great Bridgeford,
Staffordshire ST15 0PA Tel: (01785) 760278
Follow the A5013 west through Great Bridgeford
and turn right (Signposted 3.1 Miles)

A pretty 17thC half timbered cottage, once
owned by the famous biographer and author
of "The Compleat Angler" and bequeathed to
Staffordshire Borough Council by him. Today
it is a registered museum on angling with
period room displays and there is an authentic
17thC herb garden, picnic area and orchard.
Souvenir Shop.

SERVICE AREA

BETWEEN JUNCTIONS 14 AND 15

Stafford Services (Northbound) (Granada)
Tel: (01785) 811188   Self Service Restaurant,
Burger King, Little Chef, Travelodge & BP Fuel

Stafford Services (Southbound) (RoadChef)
Tel: (01785) 826300   Food Fayre Self-Service
Restaurant, Cafe Continental Coffee Shop, Wimpy
Bar, RoadChef Lodge & Esso Fuel

M6   JUNCTION 15

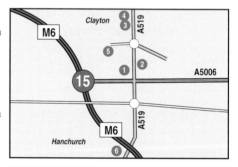

NEAREST A&E HOSPITAL

**North Staffordshire Royal Infirmary, Princes
Road, Hartshill, Stoke ST4 7JN Tel: (01782)
715444**

Take the A5006 exit east and proceed along the
A500 to Stoke. Follow the A52 signposts towards
Newcastle-under-Lyme and continue into Hartshill
Road. The hospital is in Princes Road on the left.
(Distance Approx 4 miles)

FACILITIES

1   Posthouse Stoke on Trent

Tel: 0870 400 9077
0.2 miles north along the A519, on the left.
Traders Bar & Restaurant; Open all day. Bar
meals served 07.00-22.30hrs daily. Restaurant
opening hours; Lunch; 10.30-14.30hrs daily,
Evening Meals; Mon-Fri; 18.30-22.15hrs,
Sundays; 18.00-22.00hrs.

2 BP Swift Service Station

🅿 WC Tel: (01782) 713308
0.4 miles north along the A519, on the right. Access, Visa, Overdrive, All Star, Switch, Dial Card, Mastercard, Amex, Diners Club, Delta, Routex, Shell Agency, BP Cards. Open; 07.00-22.00hrs daily.

3 Clayton Lodge Hotel

🛏H 🍽 🍴 ⛲ Tel: (01782) 613093
1 mile north along the A519, on the left. (Jarvis) Restaurant Open; Lunch; Mon-Thu; 12.00-13.45hrs, Sun; 12.00-14.45hrs, Evening Meals; Mon-Sat; 19.00-21.00hrs, Sun; 18.30-20.30hrs. Bar Snacks available on Fri & Sat.

4 Clayton Service Station (Esso)

🅿 ♿ WC Tel: (01782) 613189
1 mile north along the A519, on the left. Access, Visa, Overdrive, All Star, Switch, Dial Card, Mastercard, Amex, Diners Club, Delta, Shell Gold, BP Supercharge, Esso Cards. Open; Mon-Fri; 06.00-23.00hrs, Sat-Sun; 07.00-22.00hrs.

5 Westbury Tavern

🍽 🍴 ♿ 🐾 Tel: (01782) 638766
0.8 miles north along the Westbury Road, on the left. (Punch Taverns) Open all day Fri-Sun; Meals served; Mon-Sat; 12.00-14.00hrs & 17.00-20.00hrs, Sun; 12.00-14.30hrs.

6 Hanchurch Manor Hotel

🛏B Tel: (01782) 643030
0.8 miles south along the A519, on the right.

PLACES OF INTEREST

Stoke on Trent

Stoke on Trent Tourist Information Centre,
Quadrant Road, Hanley, Stoke on Trent ST1 1RZ
Tel: (01782) 236000
website; www.stoke.gov.uk/tourism
e-mail; stoke.tic@virgin.net
Follow the A500 east (Signposted 3.4 Miles)

It was the presence of the essential raw materials for the manufacture and decoration of ceramics, in particular marl clay, coal and water that led to the concentration of pottery manufacture in this area. Production started in the 17thC but it was the entrepreneurial skills of Josiah Wedgwood and Thomas Minton that created a form of flow line production, bringing individual potters together into large factories, and caused a massive rise in output in the 18thC. Alongside of these, hundreds of small establishments were thriving and producing a whole range of more utilitarian chinaware creating, what was described in the late 19thC when production was at its height, the most unhealthy area in the country! The city of Stoke on Trent was only established as recently as 1910 when Fenton joined the five towns immortalized by Arnold Bennett; Tunstall, Burslem, Hanley, Longton and Stoke. For those interested in Victorian and Industrial architecture this is a wonderful place to visit. Factory shops, many selling high quality seconds, abound in this area and visitors are advised to obtain the leaflet "Visit the Potteries for a China Experience" either from within the area or by contacting the TIC, 'phone number above.

There are also many museums and visitor centres devoted to pottery and ceramics and these are all signposted within the area;

The Potteries Museum and Art Gallery
Bethesda Street, Hanley ST1 3DE
Tel: (01782) 232323

Gladstone Pottery Museum
Uttoxeter Road, Longton ST3 1PQ
Tel: (01782) 319232

Etruria Industrial Museum
Lower Bedford Street, Etruria ST4 7AF
Tel: (01782) 233144

The Ceramica Experience
Burslem Old Town Hall, Burslem ST6 4AR
Tel: (01782) 832001

Wedgwood Visitor Centre
Barlaston ST12 9ES Tel: (01782) 204218

Royal Doulton Visitor Centre
Nile Street, Burslem ST6 2AJ Tel: (01782) 292434

Spode Museum & Visitor Centre
Spode Works, Church Street,
Stoke on Trent ST4 1BX Tel: (01782) 744011

| Pets Welcome | £ Cash Dispenser | 24 24 Hour Facilities | WC Toilets | Alcoholic Drinks | Food and Drink | Fuel |

The Dudson Centre

Hop Street, Hanley ST1 5DD Tel: (01782) 683000

SERVICE AREA

BETWEEN JUNCTIONS 15 AND 16

Keele Services (Northbound) (Welcome Break)

Tel: (01782) 626221 Granary Restaurant & Shell Fuel

Keele Services (Southbound) (Welcome Break)

Tel: (01782) 626221 Granary Restaurant, La Brioche Doree & Shell Fuel

KFC and Burger King restaurants on Footbridge connecting both sites.

M6   JUNCTION 16

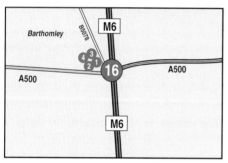

NEAREST A&E HOSPITAL

North Staffordshire Royal Infirmary, Princes Road, Hartshill, Stoke ST4 7JN Tel: (01782) 715444

Follow the A500 east into Stoke and at the first roundabout turn right along the A34. After about 1.3 miles turn left at the roundabout along the A52, Proceed along Hartshill Road and the hospital is on the right in Princes Road. (Distance Approx 8.6 miles)

FACILITIES

1   Little Chef

Tel: (01270) 883115

0.1 miles north along the B5078, on the left. Open; 07.00-22.00hrs Daily.

2   Travelodge Crewe

Tel: (01270) 883157

0.1 miles north along the B5078, on the left

3   Burger King

Tel: (01270) 883115

0.1 miles north along the B5078, on the left Open; Mon-Fri; 11.00-21.00hrs, Sat & Sun; 10.00-22.00hrs

4   Barthomley Service Station (Elf)

Tel: (01270) 883212

0.1 miles north along the B5078, on the left Access, Visa, Overdrive, All Star, Switch, Dial Card, Mastercard, Amex, Diners Club, Delta, UK Fuelcard, Elf Cards, Total Fina Cards, Keyfuels, Securicor Fuelserv. Open; 07.00-23.00hrs Daily.

PLACES OF INTEREST

The Railway Age

Crewe Heritage Centre, Vernon Way, Crewe CW1 2DB Tel: (01270) 212130 Follow the A500 west, and A5020 northwest to Crewe (Signposted Crewe Station 5.6 Miles)

Built on the site of the former huge Crewe North Engine Sheds, the exhibition shows the railway history of the town. There are three full sized signal boxes on view, including Crewe North Junction which overlooks the station and the locomotives on site include the former L&NWR "Cornwall" and the Gas Turbine Locomotive No.18000 which was built as a prototype in 1950. The Advanced Passenger Train, an abortive attempt to perfect the tilting train as the next generation of express passenger rolling stock for the 1980's, is also in the yard. There are a considerable number of models and hands-on displays. Cafe. Souvenir Shop. Limited Disabled access.

Hack Green Secret Nuclear Bunker

PO Box 127, Nantwich CW5 8AQ Information Line Tel: (01270) 629219 e-mail; coldwar@dial.pipex.com Follow the A500 west into Nantwich and A530 south. (Signposted from Nantwich, 11.2 Miles)

One of the more unusual tourist destinations!

RAF Hack Green was a WWII operational radar station and an extensive underground concrete command centre was built here in the 1950's, becoming a secret Regional Government Headquarters in the 1980's. Had nuclear war broken out, this is where 130 civil servants and military commanders would have ruled what was left of north west England. The maze of passages and rooms contains fascinating displays of all the paraphernalia required for communications and administration and an exhibition of authentic equipment shows the awesome power of nuclear weapons. NAAFI style canteen. Some disabled access.

Stapeley Water Gardens

London Road, Stapeley, Nantwich, Cheshire CW5 7LH Tel: (01270) 623868
e-mail; StapeleyWG@BTInternet.com
Follow the A500 west (Signposted 7.9 Miles)

Set within 64 acres it is now the world's largest water garden centre. There is also an Angling Centre and the Palms Tropical Oasis incorporating the "World of Frogs" exhibition. Restaurant. Cafe. Gift Shop. Disabled access.

SERVICE AREA

BETWEEN JUNCTIONS 16 AND 17

Sandbach Services (Northbound) (RoadChef)
Tel: (01270) 767134 Food Fayre Self-Service Restaurant, Cafe Continental Coffee Shop, Wimpy Bar & Esso Fuel
Sandbach Services (Southbound) (RoadChef)
Tel: (01270) 767134 Food Fayre Self-Service Restaurant, Wimpy Bar & Esso Fuel
Footbridge connection between sites.

M6   JUNCTION 17

NEAREST NORTHBOUND A&E HOSPITAL

South Manchester University Hospital,
Southmoor Road, Wythenshawe, Manchester
M23 9LT Tel: (0161) 998 7070

Proceed to Junction 19 and take the A556 north. Continue along the A56 into Altrincham and turn right along the A560. At the crossroads with the A5144 turn right and after about 0.1 miles turn left into Clay Lane. Turn left at the end and continue along Dobbinetts Lane. The hospital is along this road. (Distance Approx 21 miles)

NEAREST SOUTHBOUND A&E HOSPITAL

Leighton Hospital, Middlewich Road, Leighton, Nr Crewe CW1 4QJ Tel: (01270) 255141

Take the A5022 south towards Crewe and the hospital is signposted along this route. (Distance Approx 9 miles)

FACILITIES

SANDBACH TOWN CENTRE IS WITHIN ONE MILE OF THIS JUNCTION

1   Saxon Cross Hotel

Tel: (01270) 763281
0.2 miles north along the A5022, on the left. Restaurant Open; Mon-Sat; 19.00-21.30hrs

2   Chimney House Hotel & Restaurant

Tel: (01270) 764141
0.3 miles east along the A534, on the right. Restaurant Open; Breakfast; 07.00-09.30hrs daily, Lunch; Sun-Fri; 12.00-14.00hrs, Sat; Closed. Dinner; Mon-Sat; 18.30-21.45hrs, Sun; 19.00-21.30hrs.

3   Star Saxon Cross Service Station (Texaco)

Tel: (01270) 758980
0.1 miles west along the A534, on the left. Access, Visa, Overdrive, All Star, Switch, Dial Card, Mastercard, Amex, Diners Club, Delta, Texaco Cards. NB Not open on Christmas Day.

4   Cafe Symphony Restaurant & Wine Bar

Tel: (01270) 763664
0.7 miles west along Congleton Road, on the right. Open; Mon-Fri; 12.00-14.00hrs & 17.30-

| Pets Welcome | £ Cash Dispenser | 24 24 Hour Facilities | WC Toilets | Alcoholic Drinks | Food and Drink | Fuel |

22.00hrs, Sat; 17.30-22.00hrs, Sun; 11.00-14.00hrs & 17.30-22.00hrs

5 Star Sandbach Service Station (Texaco)

WC **£**   Tel: (01270) 758990

0.8 miles west along Congleton Road, on the left. Access, Visa, Overdrive, All Star, Switch, Dial Card, Mastercard, Amex, Diners Club, Delta, UK Fuelcard, Texaco Cards. Open; 07.00-21.00hrs Daily.

6 The Military Arms

                    Tel: (01270) 765442

0.9 miles west along the Sandbach Road, in Sandbach, on the right. (Inn Partnership) Open all day. Lunch; Wed-Fri; 12.00-14.30hrs only.

7 Fortune Palace Cantonese Restaurant

                    Tel: (01270) 763255

0.9 miles west along Congleton Road, in Sandbach, on the right in Green Street. Open; 17.00-0.00hrs Daily

8 Sallys Cafe

                    Tel: (01270) 761985

0.9 miles west along Congleton Road, in Sandbach, on the right in Green Street. Open; Mon-Sat; 09.15-15.30hrs.

9 Sandbach Service Station Co Ltd

WC   Tel: (01270) 763395

1 mile west along Congleton Road, in Sandbach, in Bradwall Road. Access, Visa, Switch, Mastercard, Amex, Diners Club, Delta, Electron, Solo. Open; Mon-Fri; 07.30-17.30hrs, Sat; 07.30-13.00hrs, Sun; Closed.

PLACES OF INTEREST

The Canal Centre

Hassall Green, Nr Sandbach CW11 4YB
Tel: (01270) 762266 Follow the A534 west and turn left along the A533. (Signposted "Canal Centre & Potters Barn" along A533, 3.4 Miles)

An ideal place to take time out to relax, at this 200 year old canal centre on the Trent & Mersey Canal. Watch boats passing through the locks whilst enjoying lockside refreshments in the tea room, browse in the gift shop which includes handmade and painted canal and craft ware or dine in the licensed restaurant. Limited disabled access.

Connected by a short walk along the canal towpath ...

The Potters Barn

Roughwood Lane, Hassall Green, Nr Sandbach
CW11 4XX Tel: (01270) 884080

A traditional pottery set amidst beautiful Cheshire countryside. There are workshops, a showroom gallery and a picnic area. Limited disabled access..

M6   JUNCTION 18

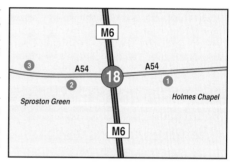

NEAREST NORTHBOUND A&E HOSPITAL

South Manchester University Hospital, Southmoor Road, Wythenshawe, Manchester M23 9LT Tel: (0161) 998 7070

Proceed to Junction 19 and take the A556 north. Continue along the A56 into Altrincham and turn right along the A560. At the crossroads with the A5144 turn right and after about 0.1 miles turn left into Clay Lane. Turn left at the end and continue along Dobbinetts Lane. The hospital is along this road. (Distance Approx 17 miles)

NEAREST SOUTHBOUND A&E HOSPITAL

Leighton Hospital, Middlewich Road, Leighton, Nr Crewe CW1 4QJ Tel: (01270) 255141

Take the A54 exit west and after about 2.7 miles turn left along the A530. The hospital is along this road. (Distance Approx 8.6 miles)

| Large Hotel | Small Hotel | Motel | Guest House/ Bed & Breakfast | Disabled Facilities | Childrens Facilities |

FACILITIES

1 Star Croco Service Station (Texaco)

🏧 ♿ WC £   Tel: (01477) 536910
0.6 miles east along the A54, on the right.
Access, Visa, Overdrive, All Star, Switch, Dial
Card, Mastercard, Amex, Diners Club, Delta,
UK Fuelcard, Texaco Cards. Open; Mon-Fri;
07.00-21.00hrs, Sat & Sun; 08.00-21.00hrs.

2 The Fox & Hounds

🍴 🍴 🔌 ♿   Tel: (01606) 832303
0.6 miles west along the A54, on the left. (Inn
Partnership) Home cooked food served daily;
12.00-14.30hrs and 18.30-21.30hrs.

3 Sproston Service Station (Save)

🏧   Tel: (01606) 832387
0.8 miles west along the A54, on the right.
Access, Visa, Overdrive, All Star, Switch, Dial
Card, Mastercard, Delta, Save Card. Open;
06.00-23.00hrs daily.

PLACES OF INTEREST

Jodrell Bank Science Centre, Planetarium & Arboretum

Lower Withington, Nr Macclesfield SK11 9DL
Tel: (01477) 571339 Follow the A54 east through
Holmes Chapel and take the A535 north
(Signposted 4.6 Miles)

The Science Centre, alongside of the Lovell
Radio Telescope, contains many hands-on
science exhibits relating to space, science and
the environment and the Planetarium gives a
vivid visual experience of the universe. A 35
acre Arboretum contains many beautiful trees
and wildlife and there is an Environmental
Discovery Centre. Picnic Area. Play Area. Shop.
Cafe. Disabled access.

SERVICE AREA

BETWEEN JUNCTIONS 18 AND 19

Knutsford Services (Granada)
Tel: (01565) 634167 Burger King & BP Fuel. Self-
Service Restaurant and Little Chef on Footbridge
connecting both sites

M6   JUNCTION 19

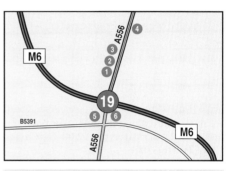

NEAREST NORTHBOUND A&E HOSPITAL

**South Manchester University Hospital,
Southmoor Road, Wythenshawe, Manchester
M23 9LT Tel: (0161) 998 7070**
Take the A556 exit north, continue along the A56
into Altrincham and turn right along the A560. At
the crossroads with the A5144 turn right and after
about 0.1 miles turn left into Clay Lane. Turn left at
the end and continue along Dobbinetts Lane. The
hospital is along this road. (Distance Approx 9.1
miles)

NEAREST SOUTHBOUND A&E HOSPITAL

**Leighton Hospital, Middlewich Road, Leighton,
Nr Crewe CW1 4QJ Tel: (01270) 255141**
Proceed to Junction 18 and take the A54 west exit.
Turn left along the A530 and the hospital is on this
road. (Distance Approx 16.8 miles)

FACILITIES

1 Little Chef

🍴 ♿   Tel: (01565) 755049
0.1 miles north along the A556, on the left.
Open; 07.00-22.00hrs Daily.

 Pets Welcome   £ Cash Dispenser   24 24 Hour Facilities   WC Toilets   Alcoholic Drinks   Food and Drink   Fuel

M6

2   Travelodge Knutsford

 Tel: (01565) 652187

0.1 miles north along the A556, on the left.

3   BP Tabley Mere

 Tel: (01565) 755127

0.1 miles north along the A556, on the left. Access, Visa, Overdrive, All Star, Switch, Dial Card, Mastercard, Amex, Diners Club, Delta, Routex, AA Paytrak, UK Fuelcard, Shell Agency, BP Cards. Open; 06.00-22.00hrs daily

4   The Old Vicarage Private Hotel

 Tel: (01565) 652221

0.2 miles north along the A556, on the right

5   The Windmill

 Tel: (01565) 632670

0.1 miles south along the A556, on the right. (Robinson's) Meals served; 12.00-14.00hrs & 19.00-21.00hrs daily.

6   Tabley Hill Service Station (Repsol)

 Tel: (01565) 650958

0.1 miles south along the A556, on the left. Access, Visa, Overdrive, All Star, Switch, Dial Card, Mastercard, Amex, Diners Club, Delta, Keyfuels, Repsol Cards. Toilets available for customer use during shop opening hours only.

PLACES OF INTEREST

Anderton Boat Lift

Anderton, Northwich CW9 6FA Visitor Centre
Tel: (01606) 77699 Follow the A556 west into Northwich, take the A533 north and it is signposted along this route (7.6 miles)

At Anderton, the River Weaver and Trent & Mersey Canal, completed in 1777, are adjacent but there is a height difference of 50ft. A transfer centre was established here, but it required the costly and time consuming process of the unloading and trans-shipping of goods and it was obvious that some sort of physical link was required between the two, but the solution turned out to be unique in this country. Edward Leader Williams, the Weaver Navigations Trust Engineer, suggested the use of a "boat carrying lift" and in consultation with Edwin Clarke, a prominent civil engineer, he produced a design that was a magnificent example of the Victorian's mastery of cast iron and hydraulics. The first boat lift was completed in 1875 and replaced by the existing one in 1908. Commercial traffic ceased in the mid-1960's and it closed, life expired, in 1982. Today, it is in the hands of the Anderton Boat Lift Trust & The Friends of the Anderton Boat Lift, an independent association dedicated to its restoration and management.

Tatton Park

Knutsford, Cheshire WA16 6QN
Tel: (01625) 534400 Follow the A556 north and turn right along the A50 (Signposted 2.8 Miles)

The home of the Egerton family since the late 16thC, the present Neo-classical mansion was begun in the late 18thC and built around an earlier house from Charles II's time. The 54 acres of ornamental and woodland gardens were laid out by Humphry Repton and featured the 1 mile long Tatton Mere, winding through them and the 400 yards long Broad Walk of tall trees leading to an 1811 replica of a Greek monument on the edge of the 1000 acres of deerpark. There is a 1930's working farm, children's playground and speciality shops selling local produce. Gift Shop.

Arley Hall & Gardens

Great Budworth, Nr Northwich CW9 6NA
Tel: (01565) 777353 Follow the B5391 west (Signposted 6.9 Miles)

Famous for its 12 acres of award winning gardens, Arley Hall is a fine example of the early Victorian Jacobean style. Lunches and light refreshments are provided in the Tudor Barn. Gift Shop. Plant Nursery.

| **M6** | **JUNCTION 20** |

THIS IS ALSO PART OF A MOTORWAY INTERCHANGE WITH THE M56 JUNCTION 9

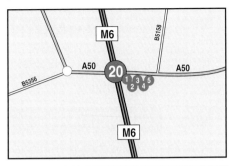

M6
B5158
A50
A50
B5356
20 ① ③ ⑤ ② ④
M6

NEAREST A&E HOSPITAL

Warrington Hospital, Lovely Lane, Warrington WA5 1QG Tel: (01925) 635911

Take the A50 north into Warrington and continue through the town centre along the A570 towards Prescot. At the roundabout junction of the A570, B5210 and Lovely Lane, turn right into Lovely Lane and the hospital is on the right. (Distance Approx 5.4 miles)

FACILITIES

1 Poplar 2000 Services Restaurant

🍴 ♿ 24   Tel: (01925) 757777
0.1 miles east along the A50, on the right.

2 McDonald's

🍴 ♿   Tel: (01925) 758759
0.1 miles east along the A50, on the right. Open; Sat-Thurs; 06.30-00.00hrs, Fri/Sat; 06.30-01.00hrs.

3 Public Toilets

WC ♿
0.1 miles east along the A50, on the right.

4 PCS Poplar 2000 Service Station (Total)

🚗 ♿ WC 24   Tel: (01925) 750810
0.1 miles east along the A50, on the right. Access, Visa, Overdrive, All Star, Switch, Dial Card, Mastercard, Amex, Diners Club, Delta, AA Paytrak, BP Supercharge, Total Cards.

5 Poplar 2000 Diesel Service Station

🚗 WC 24 £   Tel: (01925) 757777
0.1 miles east along the A50, on the right. Diesel Fuel Only. Access, Visa, Overdrive, All Star, Switch, Dial Card, Mastercard, Routex, AA Paytrak, UK Fuelcard, Securicor Fuelserv, IDS, Keyfuels, AS24.

PLACES OF INTEREST

Walton Hall & Gardens

Walton Lea Road, Higher Walton, Warrington WA4 6SN Tel: (01925) 601617 Follow the M56 west to Junction 11 and take the A56 north (Signposted 9.1 Miles)

Set amidst beautiful Cheshire countryside, there is something for everyone at Walton Hall & Gardens. Apart from the formal gardens there is a Heritage Centre, a Children's Zoo, Pitch'n'Putt and Crazy Golf courses and a bowling green all set amidst the extensive grounds. Gift Shop. Coffee Shop.

Catalyst, The Museum of the Chemical Industry

Mersey Road, Widnes, Cheshire WA8 0DF Tel: (0151) 420 1121 e-mail; info@catalyst.org.uk Follow the M56 west to Junction 12 (Signposted 11.1 Miles)

Catalyst is where science and technology really come alive through interactive and hands-on displays. Over 100 exhibits help the visitor explore the impact of chemicals on every day life, through scenes from the past, hands-on displays and multi-media programmes which summon up the sights, sounds and even the smells of yesterday. Apart from the exciting science exhibition, there is a roof top observatory with which to view Cheshire and there is a nature trail through the adjacent Spike Island Water Park with its thriving wildlife. Cafe. Gift Shop. Disabled access.

Lymm Dam

Off Crouchley Lane, Lymm, Nr Warrington WA13 0AN Tel: (01925) 758195 Follow the B5158 north and turn right along the A56 (2.5 Miles)

Lymm Dam was constructed in 1824 with the creation of a turnpike road from Warrington

M6

to Stockport and, with the remains of a slitting mill, forms a delightful country park. Sited within the Mersey Forest and adjacent to the Trans Pennine Trail, it is comprised of a variety of woodland and meadow settings and is the home for a wide range of wildlife. Picnic areas.

M6 | JUNCTION 21

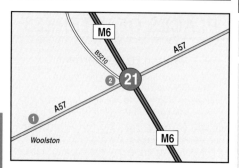

NEAREST A&E HOSPITAL

Warrington Hospital, Lovely Lane, Warrington WA5 1QG Tel: (01925) 635911

Take the A57 west to Warrington and continue along the A49. At the roundabout junction of the A49, A570 and the A50 turn right along the A570 and proceed through the town centre towards Prescot. At the junction with the B5210, the first major roundabout, turn right into Lovely Lane and the hospital is on the right. (Distance Approx 5.2 miles)

FACILITIES

1 The Rope & Anchor

🏨 🍴 ♿   Tel: (01925) 814996

0.9 miles west along the A57, in Woolston, on the right. (Scottish & Newcastle) Open all day. Lunch; Mon-Sat; 12.00-14.30hrs, Sun; 12.00-16.00hrs, Dinner; Mon-Fri; 17.30-20.30hrs, Sat; 17.30-20.00hrs.

2 The Garden Court Holiday Inn

🏨 🏨 🍴 ♿ 🎠   Tel: (01925) 838779

0.1 miles north along the B5210, on the left. Restaurant Open; Lunch; Mon-Fri; 12.00-14.00hrs, Sat & Sun; Closed, Dinner; 18.30-21.45hrs daily. Bar Snacks available throughout the day.

M6 | JUNCTION 21A

JUNCTION 21A IS A MOTORWAY INTERCHANGE ONLY WITH THE M62 AND THERE IS NO ACCESS TO ANY FACILITIES

M6 | JUNCTION 22

THERE ARE NO FACILITIES WITHIN ONE MILE OF THIS JUNCTION

NEAREST A&E HOSPITAL

Warrington Hospital, Lovely Lane, Warrington WA5 1QG Tel: (01925) 635911

Take the A49 south and continue along the A49 towards Warrington. At the first set of traffic lights, the junction with the A574, turn right into Kerfoot Street, continue along Folly Lane and the hospital is on the left just past the railway bridge. (Distance Approx 4.3 miles)

PLACES OF INTEREST

Gullivers World

Warrington, Cheshire WA5 5YZ
Information Line Tel: (01925) 444888
Follow the A49 west and continue along the A574 (Signposted along route, 4.6 Miles)

Specifically designed for families with children between the age of two and thirteen years all 40 rides and attractions are aimed to cater for their specific needs. Set within a beautiful parkland there is a huge variety of attractions including themed displays and live shows. Catering for the younger market, great emphasis is placed on making the park as safe as possible. There are a number of cafes and restaurants. Disabled Access.

M6 JUNCTION 23

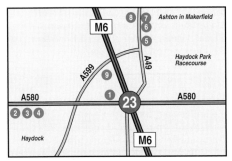

NEAREST NORTHBOUND A&E HOSPITAL

Royal Albert Edward Infirmary, Wigan Lane, Wigan WN1 2NN Tel: (01942) 244000
Proceed to Junction 25 and take the A49 north. At the first roundabout turn right along the B5238. Bear right at the junction with the A49 and turn left along the B5375. After about 0.5 miles bear right along the B5376 and the hospital is on the right along this road. (Distance Approx 6.9 miles)

NEAREST SOUTHBOUND A&E HOSPITAL

Warrington Hospital, Lovely Lane, Warrington WA5 1QG Tel: (01925) 635911
Proceed south to Junction 22 and take the A49 south and continue along the A49 towards Warrington. At the first set of traffic lights, the junction with the A574, turn right into Kerfoot Street, continue along Folly Lane and the hospital is on the left just past the railway bridge. (Distance Approx 7.3 miles)

FACILITIES

1 Haydock Island (Shell)

Tel: (01925) 293690
0.1 miles west along the A580, on the right. Access, Visa, Overdrive, All Star, Switch, Dial Card, Mastercard, Amex, Diners Club, Delta, UK Fuelcard, Shell Cards, BP Supercharge, Smart Card

2 New Boston Service Station (Elf)

Tel: (01942) 716227
1 mile west along the A580, on the left. Access, Visa, Overdrive, All Star, Switch, Dial Card, Mastercard, Amex, Diners Club, Delta, AA Paytrak, UK Fuelcard, Elf Cards, Total Fina Cards, Securicor Fuelserv, Keyfuels.

3 Little Chef

Tel: (01942) 272048
1 mile west along the A580, on the left. Open; 07.00-22.00hrs daily.

4 Travelodge St.Helens

Tel: (01942) 272055
1 mile west along the A580, on the left

5 Posthouse Haydock

Tel: 0870 400 9039
0.4 miles north along the A49, on the right. Restaurant Open; Breakfast Mon-Fri; 06.30-09.30hrs, Sat & Sun; 07.30-10.30hrs, Lunch; Mon-Fri; 12.00-14.30hrs, Dinner; 18.30-22.30hrs daily.

6 The Bay Horse

Tel: (01942) 725032
0.9 miles north along the A49, on the right. (Greenall's) Open all day. Meals served; Sun-Thurs; 12.00-21.30hrs, Fri & Sat; 12.00-22.00hrs.

7 Premier Lodge

Tel: (01942) 725032
0.9 miles north along the A49, on the right

8 The Angel

Tel: (01942) 728704
1 mile north along the A49, on the left. (Burtonwood) Open all day on Sundays. Meals served; Mon-Sat; 11.30-14.30hrs & 17.30-21.30hrs, Sun; 12.00-21.30hrs.

9 Haydock Thistle Hotel

Tel: (01942) 272000
0.6 miles west along the A599, on the left. Restaurant Open; Lunch;12.00-14.00hrs Daily, Dinner; Mon-Sat; 19.00-22.00hrs, Sun; 19.00-21.00hrs.

M6

| Pets Welcome | Cash Dispenser | 24 Hour Facilities | Toilets | Alcoholic Drinks | Food and Drink | Fuel |

M6 JUNCTION 24

THIS IS A RESTRICTED ACCESS JUNCTION

Vehicles can only exit from the southbound lanes. Vehicles can only enter the motorway along the northbound lanes.

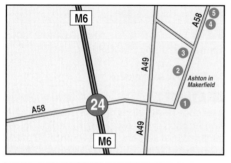

NEAREST NORTHBOUND A&E HOSPITAL

Royal Albert Edward Infirmary, Wigan Lane, Wigan WN1 2NN Tel: (01942) 244000

Proceed to Junction 25 and take the A49 north. At the first roundabout turn right along the B5238. Bear right at the junction with the A49 and turn left along the B5375. After about 0.5 miles bear right along the B5376 and the hospital is on the right along this road. (Distance Approx 5.6 miles)

NEAREST SOUTHBOUND A&E HOSPITAL

Warrington Hospital, Lovely Lane, Warrington WA5 1QG Tel: (01925) 635911

Proceed south to Junction 22 and take the A49 south and continue along this road towards Warrington. At the first set of traffic lights, the junction with the A574, turn right into Kerfoot Street, continue along Folly Lane and the hospital is on the left just past the railway bridge. (Distance Approx 8.6 miles)

FACILITIES

1 KFC

Tel: (01942) 723146
0.6 miles east along the A58, in Ashton in Makerfield, on the right. Open; 11.00-0.00hrs daily.

2 Sir Thomas Gerard

Tel: (01942) 713519
0.7 miles east along the A58, in Ashton in Makerfield, on the left (Wetherspoon) Open all day. Meals served all day Mon-Sat; 11.00-22.00hrs, Sun; 12.00-21.30hrs.

3 The Robin Hood

Tel: (01942) 721560
0.8 miles east along the A58, in Ashton in Makerfield, on the left. (Scottish & Newcastle) Open all day. Meals served; Mon-Fri; 12.00-14.30hrs, Sat; 12.00-16.30hrs. Bar Snacks served all day Thurs-Sat; 11.00-0.00hrs.

4 The Rockleigh Hotel

Tel: (01942) 727156
1 mile east along the A58, in Ashton in Makerfield, on the right. Grill Room Open; 19.00-21.00hrs daily

5 Caledonian Service Station (Esso)

Tel: (01942) 718352
1 mile east along the A58, in Ashton in Makerfield, on the right. Access, Visa, Overdrive, All Star, Switch, Dial Card, Mastercard, Amex, Diners Club, Delta, Shell Agency, BP Supercharge, Esso Cards. Open; 06.30-23.00hrs Daily.

M6 JUNCTION 25

THIS IS A RESTRICTED ACCESS JUNCTION

Vehicles can only exit from the northbound lanes. Vehicles can only enter the motorway along the southbound lanes.

| | Large Hotel | | Small Hotel | | Motel | | Guest House/ Bed & Breakfast | | Disabled Facilities | | Childrens Facilities |

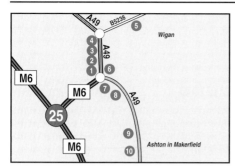

Wigan

Ashton in Makerfield

NEAREST A&E HOSPITAL

Royal Albert Edward Infirmary, Wigan Lane, Wigan WN1 2NN Tel: (01942) 244000
Take the A49 north and at the first roundabout turn right along the B5238. Bear right at the junction with the A49 and turn left along the B5375. After about 0.5 miles bear right along the B5376 and the hospital is on the right along this road. (Distance Approx 3.5 miles)

FACILITIES

1   Travel Inn

🛏️M ♿   Tel: (01942) 493469
0.2 miles north along the A49, on the left

2   Wheat Lea Park

🍴 🍽️ ♿ 🍷   Tel: (01942) 493469
0.2 miles north along the A49, on the left. (Whitbread) Open all day. Meals served; 11.30-22.00hrs daily

3   McDonald's

🍽️ ♿   Tel: (01942) 230139
0.5 miles north along the A49, on the left. Open; 07.30-0.00hrs Daily.

4   Marus Bridge Service Station (BP)

🍴 24   Tel: (01942) 242437
0.6 miles north along the A49, on the left. Access, Visa, Overdrive, All Star, Switch, Dial Card, Mastercard, Amex, Diners Club, Delta, Routex, AA Paytrak, Shell Agency, BP Cards, Mobil Cards.

5   Wigan Service Station (Esso)

🍴 ♿ WC 24   Tel: (01942) 707154
1 mile north along the B5288, on the left. Access, Visa, Overdrive, All Star, Switch, Dial Card, Mastercard, Amex, Diners Club, Delta, AA Paytrak, UK Fuelcard, Shell Gold, BP Supercharge, Mobil Cards, Esso Cards.

6   Goose Green (Shell)

🍴 ♿ WC 24   Tel: (01942) 823900
0.2 miles north along the A49, on the right. Access, Visa, Overdrive, All Star, Switch, Dial Card, Mastercard, Amex, Diners Club, AA Paytrak, Shell Cards, BP Agency, Smart Card.

7   The Red Lion

🍷 🍽️ ♿ 🍴   Tel: (01942) 831089
0.1 miles south along the A49, on the right. (Greenalls) Open all day. Meals served all day; Sun-Fri; 12.00-21.30hrs, Sat; 12,.00-22.00hrs.

8   The Cranberry Hotel

🛏️ 🍷 🍽️ 🍴 🍴   Tel: (01942) 243519
0.1 miles south along the A49, on the right. Bar Snacks available 19.00-21.00hrs daily

9   The Park Hotel

🍷 🍽️ ♿ 🍴   Tel: (01942) 270562
0.3 miles south along the A49, on the right. (Burtonwood) Open all day. Meals served;12.00-22.00hrs Daily.

10   The Bath Springs

🍷 🍽️ 🍴   Tel: (01942) 202716
0.8 miles south along the A49, on the right. (Free House) Open all day. Lunch; Mon-Fri;12.00-14.30hrs.

PLACES OF INTEREST

Wigan Pier

Wallgate, Wigan WN3 4EU Tel: (01942) 323666
Follow the A49 north (Signposted 3.1 Miles)

Set in an 8.5 acre site alongside the Leeds-Liverpool Canal, Wigan Pier offer hours of entertainment, education and interaction for all the family. For over a 100 years the pier, at the canal basin, was the hub of industrial Wigan with its sidings, warehouses, loading bays and barges all busy with the movement of coal and cotton. With the decline of both the waterways and these traditional industries the pier became derelict, but in the mid 1980's it was totally renovated and reopened as a heritage centre. Today the exhibitions include the "Way We Were", Opie's Museum of Memories, the Trencherfield Mill Engine and the Machinery Hall. There are waterbus trips, available throughout the year, and canal boat cruises in the summer. Gift Shops. Cafe. Disabled access.

Visa, Overdrive, All Star, Switch, Dial Card, Mastercard, Amex, Diners Club, Delta, Shell Gold, BP Supercharge, Esso Cards.

---

3   The Priory Wood

Tel: (01942) 211516

0.1 miles west along the A577, on the right. (Beefeater) Open all day. Meals served; Sun-Thurs; 12.00-22.30hrs, Sat & Sun; 12.00-23.00hrs

---

4   Travel Inn

Tel: (01942) 211516

0.1 miles west along the A577, on the right

M6   JUNCTION 26

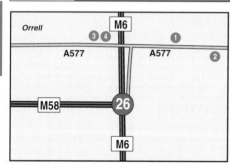

NEAREST A&E HOSPITAL

Royal Albert Edward Infirmary, Wigan Lane, Wigan WN1 2NN Tel: (01942) 244000

Take the A577 east into Wigan and proceed north along the A49 towards Preston. Turn left along the B5375 and bear right along the B5376. The hospital is on the right hand side of this road. (Distance Approx 4.5 miles)

FACILITIES

1   KFC

Tel: (01942) 216849

0.6 miles east along the A577, on the left. Open; 11.00-0.00hrs Daily.

---

2   Pemberton Service Station (Esso)

Tel: (01942) 706307

1 mile east along the A577, on the right. Access,

PLACES OF INTEREST

Wigan Pier

Wallgate, Wigan WN3 4EU Follow the A577 east (Signposted 2.9 Miles)

For details please see Junction 25 information.

M6   JUNCTION 27

(map showing Wrightington, B5250, M6, A5209, B5206, B5239)

NEAREST NORTHBOUND A&E HOSPITAL

Chorley & South Ribble District Hospital, Preston Road, Chorley PR7 1PP Tel: (01257) 261222

Take the A5209 east towards Standish and continue along the B5239. Turn left at the end along the A5106 and after about 3.5 miles continue into the A6 through Chorley and the hospital is on the left at Hartwood. (Distance Approx 7 miles)

  H Large Hotel **h** Small Hotel   **M** Motel   **B** Guest House/ Bed & Breakfast   Disabled Facilities   Childrens Facilities

NEAREST SOUTHBOUND A&E HOSPITAL

Royal Albert Edward Infirmary, Wigan Lane, Wigan WN1 2NN Tel: (01942) 244000

Take the A5209 east towards Standish and turn right along the A49. Follow this route into Wigan and the hospital is on the right hand side. (Distance Approx 3.8 miles)

FACILITIES

1   Crow Orchard Filling Station (BP)

Tel: (01257) 423983

0.1 miles west along the A5209, on the right. Access, Visa, Overdrive, All Star, Switch, Dial Card, Mastercard, Amex, Diners Club, Delta, Routex, UK Fuelcard, Shell Agency, BP Cards, Mobil Cards.

2   The Wrightington Hotel & Country Club

Tel: (01257) 425803

0.4 miles west along the A5209, on the right. Restaurant Open; Mon-Sat; 19.00-21.30hrs.

3   The Tudor Inn

Tel: (01257) 424143

0.6 miles north along the B5250, on the right. (Greenalls) Open all day. Lunch; Mon-Fri; 12.00-14.30hrs, Meals served all day Sat & Sun; 12.00-21.00hrs.

4   Wigan Standish Moat House

Tel: (01257) 499988

0.6 miles east along the A5209, on the left. Restaurant Open; Breakfast; 07.00-10.00hrs Daily, Lunch, Mon-Sat; 12.00-14.30hrs, Sun; 12.00-15.30hrs, Dinner; Mon-Thurs; 19.00-21.30hrs, Fri & Sat; 19.00-22.00hrs, Sun; 19.00-21.00hrs.

5   The Charnley Arms

Tel: (01257) 424619

0.6 miles east along the A5209, on the right. (Scottish & Newcastle) Open all day. Meals served; Sun-Fri; 12.00-21.00hrs, Sat; 12.00-22.00hrs. (Organic menus)

6   Premier Lodge

Tel: (01257) 424619

0.6 miles east along the A5209, on the right

7   Total Fina Standish Service Station

Tel: (01257) 473660

0.7 miles east along the A5209, on the right. Access, Visa, Overdrive, All Star, Switch, Dial Card, Mastercard, Amex, Diners Club, Delta, BP Supercharge, Elf Cards, Total Fina Cards.

PLACES OF INTEREST

Haigh Country Park

Haigh, Wigan WN2 1PE Tel: (01942) 831107
Follow the A5209 east (Signposted 5.6 Miles)

There are a number of walks and nature trails, marked out and of varying lengths and difficulty. Rangers are on hand to help and advise and at certain times also lead guided walks. For the visually impaired there are Audio Trails for which a tape player, Braille map and guide can be borrowed from the Information Centre whilst there is a route suitable for pushchairs and wheelchair users. Adventure activities, such as rock climbing, archery and an obstacle course can be pre-booked. Model Village, Gardens, Crazy Golf. Gift Shop. Cafe. Craft Studio. Some disabled access.

Martin Mere Wildlife & Wetland Centre

Fish Lane, Burscough, Lancashire L40 0TA
Tel: (01704) 895181 Follow the A5209 west (Signposted 8.8 Miles)

A huge number and wide variety of species of tame swans, geese, ducks and flamingos are on view within the 40 acres of wildfowl gardens and 300 acres of wild refuge. There is a children's area where some of the birds can be hand fed, spacious hides for bird observation, picnic area and art and craft gallery. Gift Shop. Cafe.

Camelot Theme Park and Rare Breeds Farm

Park Hall Road, Charnock Richard, Chorley PR7 5LP Tel: (01257) 453044
Follow the A5209 west and turn right along the B5250 to Heskin Green. (Signposted 3.6 Miles)

M6

| Pets Welcome | £ Cash Dispenser | 24 24 Hour Facilities | WC Toilets | Alcoholic Drinks | Food and Drink | Fuel |

A theme park containing over 100 rides, shows and attractions, including live jousting and Merlin's Magic Show. Gift Shop. Cafe. Some disabled Access.

SERVICE AREA

BETWEEN JUNCTIONS 27 AND 28

Charnock Richard (Northbound) (Welcome Break) Tel: (01257) 791494 Granary Restaurant, La Brioche Doree French Cafe, Welcome Lodge & Shell Fuel

Charnock Richard (Southbound) (Welcome Break) Tel: (01257) 791494 Granary Restaurant & Shell Fuel.

KFC, Burger King & Red Hen Restaurant on Footbridge connecting both sites.

M6   JUNCTION 28

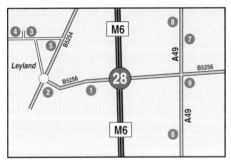

NEAREST NORTHBOUND A&E HOSPITAL

Royal Preston Hospital, Sharoe Green Lane North, Fullwood, Preston PR2 9HT
Tel: (01772) 716565

Proceed north to Junction 29 and take the A6 west exit. Follow the road through Preston town centre and turn right along the B6241. The hospital is on this road. (Distance Approx 7.9 miles)

NEAREST SOUTHBOUND A&E HOSPITAL

Chorley & South Ribble District Hospital, Preston Road, Chorley PR7 1PP
Tel: (01257) 261222

Take the B5256 east exit and turn right along the A49. After about 1.7 miles turn left along the B5248 and turn right at the end along the A6. The hospital is about 0.2 miles on at the junction of the A6 and B5252. (Distance Approx 4.3 miles)

FACILITIES

LEYLAND TOWN CENTRE IS WITHIN ONE MILE OF THIS JUNCTION.

1 Jarvis Leyland Hotel

Tel: (01772) 422922

0.1 miles west along the B5256, on the left. Restaurant Open; Breakfast; Mon-Fri; 07.00-09.30hrs, Sat & Sun; 08.00-10.00hrs, Lunch; Sun-Fri; 12.00-14.00hrs, Dinner; 19.00-21.30hrs Daily.

2 Leyland Elf Service Station

Tel: (01772) 452552

0.4 miles west along the B5256, in Leyland, on the left. Access, Visa, Overdrive, All Star, Switch, Dial Card, Mastercard, Amex, Diners Club, Delta, Elf Cards, Total Fina Cards, Solo, Electron. Open; Mon-Sat; 07.00-22.00hrs, Sun; 09.00-22.00hrs.

3 The Original Ship

Tel: (01772) 456674

1 mile along Westgate, in Leyland, on the right. (Scottish & Newcastle) Open all day. Meals served; Sun-Thurs; 12.00-21.00hrs, Fri & Sat; 12.00-17.00hrs.

4 Leyland Garage (Texaco)

Tel: (01772) 455414

1 mile along Westgate, in Leyland, on the right. Access, Visa, Overdrive, All Star, Switch, Dial Card, Mastercard, Amex, Diners Club, Delta, AA Paytrak, UK Fuelcard, BP Supercharge, Texaco Cards, Securicor Fuelserve, Keyfuels. Open; Mon-Fri; 06.30-21.00hrs, Sat & Sun: 08.00-19.30hrs.

5 The Queens

Tel: (01772) 421164

0.8 miles north along Chapel Brow (B5254), in Leyland, on the left. (Scottish & Newcastle) Open all day. Lunch; Mon-Sat;12.00-14.00hrs.

6   Shell Leyland

⬛ ♿ WC 24 £   Tel: (01772) 450940

0.5 miles south along the A49, on the right. Access, Visa, Overdrive, All Star, Switch, Dial Card, Mastercard, Amex, Diners Club, Delta, AA Paytrak, UK Fuelcard, Shell Cards, Smart, BP Agency, Esso Europe.

7   Rydal Petroleum

⬛ WC   Tel: (01772) 455253

0.5 miles north along the A49, on the right. Access, Visa, Overdrive, All Star, Switch, Dial Card, Mastercard, Delta, Open; Mon-Fri: 06.30-22.00hrs, Sat; 07.00-22.00hrs, Sun; 08.00-22.00hrs.

8   Royal Shah Indian Cuisine Restaurant

⬛ 🍽 ♿ 🅿   Tel: (01772) 622616

0.5 miles north along the A49, on the left. Open; Mon-Sat; 17.00-0.00hrs, Sun; 16.00-23.00hrs.

9   The Hayrick

⬛ 🍽 ♿ 🅿   Tel: (01772) 434668

0.2 miles east along the B5256, on the left. (Scottish & Newcastle) Open all day. Meals served; Mon-Sat; 12.00-14.00hrs & 17.00-20.00hrs, Sun; 12.00-20.00hrs.

M6   JUNCTION 29

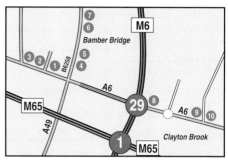

NEAREST NORTHBOUND A&E HOSPITAL

Royal Preston Hospital, Sharoe Green Lane North, Fullwood, Preston PR2 9HT
Tel: (01772) 716565

Take the A6 west exit and follow the road through Preston town centre. Turn right along the B6241 and the hospital is on this road. (Distance Approx 5.9 miles)

NEAREST SOUTHBOUND A&E HOSPITAL

Chorley & South Ribble District Hospital, Preston Road, Chorley PR7 1PP
Tel: (01257) 261222

Take the A6 south exit and the hospital is at the junction with the B5252. (Distance Approx 4.2 miles)

PLACES OF INTEREST

British Commercial Vehicle Museum

King Street, Leyland, Nr Preston PR5 1LE
Tel: (01772) 451011 Follow the B5356 west. (Signposted 0.7 Miles)

Sited on the former Leyland South Works, where for many years commercial vehicles were manufactured, it is a museum devoted to that genre with exhibits dating from the horse-drawn era to the present day, including the famous Popemobile. Imaginative displays utilize sound and lighting effects to bring the exhibits to life. Gift Shop. Cafe. Disabled access to ground floor only.

FACILITIES

1   Sainsbury's Petrol Station

⬛ £   Tel: (01772) 627762

0.6 miles west along the A6, on the right. Access, Visa, Overdrive, All Star, Switch, Dial Card, Mastercard, Amex, Delta, AA Paytrak. Open; Mon-Wed; 06.00-22.00hrs, Thurs-Fri; 06.00-22.15hrs, Sat; 06.00-22.00hrs, Sun; 07.00-22.00hrs. Toilets available in adjacent store.

2   The Poachers

⬛ 🍽 ♿ 🅿   Tel: (01772) 324100

0.7 miles west along the A6, on the right. (Scottish & Newcastle) Open all day. Meals served; Mon-Sat; 12.00-22.00hrs, Sun; 12.00-21.00hrs.

| Pets Welcome | £ Cash Dispenser | 24 24 Hour Facilities | WC Toilets | Alcoholic Drinks | Food and Drink | Fuel |

3   Lodge Inn

▛M ⓑ   Tel: (01772) 324100
0.7 miles west along the A6, on the right

4   Ye Olde Hob Inn

▜ ⓘ ⓑ ▞   Tel: (01772) 336863
0.5 miles north along the B6528, in Bamber
Bridge, on the right. (Free House) Open all day.
Meals served; Mon-Sat; 12.00-14.00hrs & Tues-
Sat; 18.00-21.00hrs, Sundays; 12.00-21.00hrs.

5   Bamber Bridge Service Station (Elf)

▛   Tel: (01772) 337807
0.6 miles north along the B6258, in Bamber
Bridge, on the right. Access, Visa, Overdrive,
All Star, Switch, Dial Card, Mastercard, Amex,
Diners Club, Delta, AA Paytrak, Elf Cards, Total
Fina Cards. Open; Mon-Sat; 07.00-22.00hrs,
Sun; 08.00-22.00hrs.

6   The Black Bull

▜ ⓘ ▞ ▛B   Tel: (01772) 335703
0.7 miles north along the B6258, in Bamber
Bridge, on the right. (Unique Pub Co) Open
all day. Lunch & Evening Meals.

7   The Mackenzie

▜ ⓘ ⓑ ▞ ▛B   Tel: (01772) 463902
0.8 miles north along the B6258, in Bamber
Bridge, on the right. (Whitbread) Open all day.
Meals served; 12.00-14.00hrs & 17.00-19.00hrs.
Bar Snacks served all day, Mon-Sat; 11.00-
23.00hrs, Sun; 12.00-22.00hrs.

8   Novotel Hotel

▛H ▜ ⓘ ⓑ   Tel: (01772) 313331
0.1 miles east along the A6, on the left.
Restaurant Open; 06.00-0.00hrs daily

9   Clayton Brook Service Station (Esso)

▛ ⓑ WC 24 £   Tel: (01772) 336064
0.7 miles east along the A6, on the left. Access,
Visa, Overdrive, All Star, Switch, Dial Card,
Mastercard, Amex, Delta, Routex, Shell Gold,
Esso Cards.

10   Brook House Hotel

▛H ▜ ⓘ ⓑ ▞   Tel: (01772) 336403
1 mile east along the A6, on the left. Restaurant
Open for Evening Meals; Mon-Sat.

PLACES OF INTEREST

Hoghton Tower

Hoghton, Preston, Lancashire PR5 0SH
Tel: (01254) 852986 Follow the A6 south towards
Chorley and turn left along the B5256. (5.3 Miles)

Ancestral home of the de Hoghton family since
Norman times, the house was rebuilt in 1565
and restored during the late 19thC. Relics of
James I's visit in 1617 are preserved here and it
was in the magnificent Banqueting Hall that
he famously knighted the beef "Sir Loin".
William Shakespeare started his working life
here and the house contains some fine 17thC
panelling. Cafe. Gift Shop. No Disabled access.

M6   JUNCTION 30

JUNCTION 30 IS A MOTORWAY
INTERCHANGE ONLY WITH THE M61 AND
THERE IS NO ACCESS TO ANY
FACILITIES

M6   JUNCTION 31

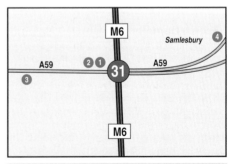

NEAREST A&E HOSPITAL

Royal Preston Hospital, Sharoe Green Lane
North, Fullwood, Preston PR2 9HT Tel: (01772)
716565

Take the A59(T) west exit and at the roundabout
turn right along the A5085. At the crossroads with

the A6063 and a minor road, turn right along the minor road and left at the end along the B6241. The hospital is along this road. (Distance Approx 3.2 miles)

FACILITIES

1   Tickled Trout Service Station (BP)

🛏️ 🚿 WC 24 £   Tel: (01772) 877656

0.1 miles west along the A59, on the right. Access, Visa, Overdrive, All Star, Switch, Dial Card, Mastercard, Amex, Diners Club, Delta, Routex, Shell Agency, BP Cards.

2   Tickled Trout Hotel

🏨 🍴 🍽️   Tel: (01772) 877671

0.1 miles west along the A59, on the right. Restaurant Open; Lunch; Sun-Fri; 12.00-14.00hrs, Sat; Closed, Dinner; 19.00-21.45hrs daily

3   New Hall Lane Filling Station (BP)

🍴 24 £   Tel: (01772) 794212

0.8 miles west along the A59, on the left. Access, Visa, Overdrive, All Star, Switch, Dial Card, Mastercard, Amex, Diners Club, Delta, Routex, AA Paytrak, UK Fuelcard, Shell Agency, BP Cards, Securicor Fuelserv.

4   Samlesbury Service Station (Esso)

🍴   Tel: (01772) 877204

1 mile west along the A59, on the left. Access, Visa, Overdrive, All Star, Switch, Dial Card, Mastercard, Amex, Diners Club, Delta, AA Paytrak, Shell Gold, Shell UK, BP Supercharge. Open; 06.00-0.00hrs Daily.

PLACES OF INTEREST

Harris Museum & Art Gallery

Market Square, Preston PR1 2PP
Tel: (01772) 258248. Follow the A59 west into Preston (Signposted 3.2 Miles)

Designed by James Hibbert in the Greek revival style, this magnificent listed building was opened in 1893. Funded by a local successful businessman and reminiscent of the British Museum, as well as the fine collection of paintings and watercolours by major 19thC British artists there is also an excellent exhibition of the story of Preston. Gift Shop. Cafe. Disabled access.

M6   JUNCTION 31A

THIS IS A RESTRICTED ACCESS JUNCTION

Vehicles can only exit from the northbound lanes. Vehicles can only enter the motorway along the southbound lanes.

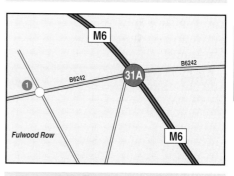

NEAREST A&E HOSPITAL

Royal Preston Hospital, Sharoe Green Lane North, Fullwood, Preston PR2 9HT Tel: (01772) 716565

Proceed north to Junction 32 and take the M55 west. At Junction 1 take the A6 south exit and at the crossroads turn left along the B6241. The hospital is along this road. (Distance Approx 5 miles)

FACILITIES

1   The Anderton Arms

🍴 🍽️ 🚿 ♿   Tel: (01772) 797605

0.6 miles west along the B6242, on the right. (Bass) Open all day. Meals served; Mon-Sat;12.00-22.00hrs, Sun; 12.00-21.00hrs

M6

M6 — JUNCTION 32

JUNCTION 32 IS A MOTORWAY INTERCHANGE ONLY WITH THE M55 AND THERE IS NO ACCESS TO ANY FACILITIES

SERVICE AREA

BETWEEN JUNCTIONS 32 AND 33

Forton Services (Northbound) (Granada)

Tel: (01524) 791775 Little Chef, Burger King, Travelodge and Shell & BP Fuel

Forton Services (Southbound) (Granada)

Tel: (01524) 791775 Self-Service Restaurant, Coffee Shop and Shell & BP Fuel

Footbridge connection between sites.

M6 — JUNCTION 33

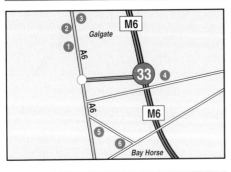

NEAREST NORTHBOUND A&E HOSPITAL

Royal Lancaster Infirmary, Ashton Road, Lancaster LA1 4RT Tel: (01524) 65944

Take the A6 north and after about 3 miles turn left at a crossroads along a minor road towards Scotforth. Turn right at the end along the A588 and the hospital is along this road. (Distance Approx 3.9 miles)

NEAREST SOUTHBOUND A&E HOSPITAL

Royal Preston Hospital, Sharoe Green Lane North, Fullwood, Preston PR2 9HT Tel: (01772) 716565

Proceed south to Junction 32 and take the M55 west. At Junction 1 take the A6 south exit and at the crossroads turn left along the B6241. The hospital is along this road. (Distance Approx 16.3 miles)

FACILITIES

1 The Plough Inn

Tel: (01524) 751337

0.7 miles north along the A6, in Galgate, on the left. (Inn Partnerships) Open all day. Meals served; Mon-Fri; 12.00-14.30hrs & 17.30-20.30hrs, Sat & Sun; 11.00-20.30hrs.

2 Green Dragon

Tel: (01524) 751062

0.9 miles north along the A6, in Galgate, on the left. (Thwaites) Open all day except Wednesday. Meals served; Thurs-Tues; 12.00-14.00hrs & 17.00-20.00hrs, Sat & Sun; 12.00-21.00hrs.

3 New Inn

Tel: (01524) 751643

0.9 miles north along the A6, in Galgate, on the right. (Free House) Meals served; Tues-Sun; 12.00-15.00hrs & 17.00-20.00hrs

4 Hampson House Hotel

Tel: (01524) 751158

0.4 miles east along Hampson Lane, on the left. Restaurant Open; Mon-Sat; 12.00-14.00hrs & 18.30-21.30hrs, Sun; 12.00-15.30hrs & 18.30-21.30hrs.

5 Salt Oke South B&B

Tel: (01524) 752313

0.3 miles south along the A6, on the left.

6 Bay Horse Inn

Tel: (01524) 791204

0.9 miles south along the Abbeystead Road, on the right. (Mitchells) Closed all day on Monday. Lunch; Tues-Sat; 12.00-14.00hrs, Evening Meals; Tues-Sat; 19.00-21.30hrs, Meals served all afternoon on Sundays; 12.00-17.00hrs.

PLACES OF INTEREST

Lancaster

Lancaster Tourist Information Office, 29 Tower Hill,
Lancaster LA1 1YN. Tel: (01524) 32878
Follow the A6 north (Signposted 5.4 Miles)

As early as the 10thC Athelstan had lands in the area, but it was the arrival of Roger of Pitou, cousin of William the Conqueror, who established Lancaster as his base to supervise the large areas of Lancashire given to him by the king. He built Lancaster Castle to keep out the marauding Scots and this was strengthened by John of Gaunt, Duke of Lancaster in the 15thC. Unusually, it is still in use today, as a prison, so it can mostly only be viewed from the outside although visitors can enter the 18thC Shire Hall and Crown Court. As with many ancient cities there are many fine period buildings still to be seen, including the Priory Church of St Mary, established in 1094 and rebuilt in the 14th & 15thC's, the Priory Tower, rebuilt in 1759 as a navigational landmark, and the Judge's Lodgings in Church Street, dating from the 1620's and now housing two museums; The Museum of Childhood and the Gillow & Town House Museum (Tel: 01524-32808) containing many examples of the fine workmanship of Gillows, the famous Lancaster cabinet makers. The city's wealth during the 17th and 18thC's and until its decline in the nineteenth was undoubtedly based upon its maritime operations and this period is celebrated at St George's Quay which, with its great stone warehouses and superb Custom House form the Maritime Museum (Tel: 01524-64637)

Within the city ....

Ashton Memorial & Williamson Park

Quernmore Road, Lancaster LA1 1UX
Tel: (01524) 33318

The great green copper dome of the memorial can be seen for miles around and is a magnificent viewpoint from which Morecambe Bay, the Lakeland Hills and the Forest of Bowland are all visible. It was built by John Belcher and completed in 1909 and is believed that the pre-stressed concrete dome is the first known example of such a construction. Today it houses exhibitions and multi-screen presentations about the life and Edwardian times of John Williamson, later Lord Ashton, who was the town's largest employer and the Liberal MP for many years. Williamson Park was his idea for providing employment for local people caused by the cotton famine crisis in the textile industry during the American Civil War of the 1860's. Constructed on old quarry workings, it opened in 1896, and there is now a Butterfly House in the restored Edwardian Palm House and the Conservation Garden and Wildlife Pool which opened in 1991. Gift Shop. Tea Room. Disabled Access.

M6   JUNCTION 34

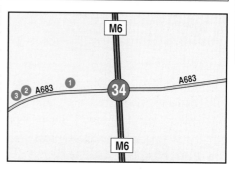

NEAREST A&E HOSPITAL

Royal Lancaster Infirmary, Ashton Road,
Lancaster LA1 4RT Tel: (01524) 65944

Take the A683 exit west and continue along the A6 south. Turn right at the roundabout along the A588 and the hospital is along this road. (Distance Approx 3.1 miles)

FACILITIES

1   Posthouse Lancaster

Tel: (01524) 65999
0.2 miles west along the A683, on the right. Traders Bar & Restaurant Open; Breakfast; Mon-Fri; 06.30-09.30hrs, Sat & Sun; 07.30-10.30hrs, Lunch; 12.00-14.00hrs daily, Dinner; 17.30-22.30hrs daily. Bar menu available all day.

 Pets Welcome   £ Cash Dispenser   24 24 Hour Facilities   WC Toilets   Alcoholic Drinks   Food and Drink   Fuel

M6

2   Lancaster Town House

B   Tel: (01524) 65527
0.9 miles west along the A683, on the right

3   Shell Lancaster

** WC 24**   Tel: (01524) 590900
1 mile west along the A683, on the right.
Access, Visa, Overdrive, All Star, Switch, Dial
Card, Mastercard, Amex, Diners Club, Delta,
Smart, Shell Cards, BP Agency.

PLACES OF INTEREST

Lancaster

Lancaster Tourist Information Office, 29 Tower Hill,
Lancaster LA1 1YN. Tel: (01524) 32878
Follow the A683 west. (Signposted 2 Miles)
For details please see Junction 33 information

Within the city ....

Ashton Memorial & Williamson Park

Quernmore Road, Lancaster LA1 1UX
For details please see Junction 33 information

M6   JUNCTION 35 & JUNCTION 35A

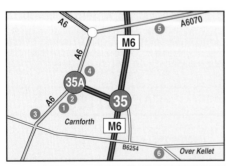

NEAREST NORTHBOUND A&E HOSPITAL

**Westmorland General Hospital, Burton Road,
Kendal LA9 7RG Tel: (01539) 732288**

Proceed north to Junction 36, take the A590 north
exit and continue into the A591. Continue along
the A6 and at the Onal Park roundabout turn south

east along the A65 and the hospital is along this
road. (Distance Approx 14.9 miles)

NEAREST SOUTHBOUND A&E HOSPITAL

**Royal Lancaster Infirmary, Ashton Road,
Lancaster LA1 4RT Tel: (01524) 65944**

Proceed to Junction 34 and take the A683 exit west.
Continue along the A6 south and turn right at the
roundabout along the A588. The hospital is along
this road. (Distance Approx 7.2 miles)

FACILITIES

1   Truckhaven (Esso)

** WC 24**   Tel: (01524) 736699
0.2 miles west along the A683, on the right.
Access, Visa, Overdrive, All Star, Switch, Dial
Card, Mastercard, Amex, Diners Club, Delta,
AA Paytrak, Shell Gold, BP Supercharge, Esso
Cards, Securicor Fuelserv, Keyfuels

2   Truckhaven Motel & Restaurant

M   Tel: (01524) 736699
0.2 miles west along the A683, on the right.
Restaurant Open; Mon-Fri; 05.30-23.00hrs
daily

3   Norjac Garage (Shell)

** WC 24**   Tel: (01524) 732208
1 mile south along the A6, on the left. Access,
Visa, Overdrive, All Star, Switch, Dial Card,
Mastercard, Amex, Diners Club, Delta, Shell
Cards, BP Agency, Smart.

4   Pine Lake Resort

       Tel: (01524) 736190
On north side of the roundabout adjacent to
Junction 35A. Morgano's French Restaurant;
Open; 09.00-23.00hrs daily

5   The Longlands Hotel

      B   Tel: (01524) 781256
1 mile east along the A6070, on the right. (Bass)
Open all day. Lunch; Mon-Fri:12.00-14.00hrs,
Sat & Sun: 12.00-14.30hrs, Evening Meals;
Mon-Fri; 17.30-21.00hrs, Sat; 17.00-21.30hrs
Sun; 17.00-21.00hrs.

6   The Eagles Head

▮ ▮ ▮ ▮   Tel: (01524) 732457

1 mile east along the B6254, in Over Kellet, on the right. (Mitchells) Lunch; 12.00-14.00hrs Daily, Evening Meals; 18.30-21.00Hrs Daily [Meals not served on Mondays between December and March]

SERVICE AREA

BETWEEN JUNCTIONS 35 AND 36

Burton in Kendal Services (Northbound Only) (Granada)

Tel: (01524) 781234 Burger King, Country Kitchen Self-Service Restaurant, Travelodge & BP Fuel.

M6   JUNCTION 36

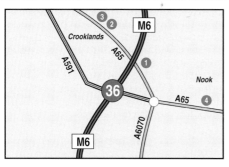

NEAREST NORTHBOUND A&E HOSPITAL

Westmorland General Hospital, Burton Road, Kendal LA9 7RG Tel: (01539) 732288

Take the A590 north exit and continue into the A591. Continue along the A6 and at the Onal Park roundabout turn south east along the A65 and the hospital is along this road. (Distance Approx 7.1 miles)

NEAREST SOUTHBOUND A&E HOSPITAL

Royal Lancaster Infirmary, Ashton Road, Lancaster LA1 4RT Tel: (01524) 65944

Proceed to Junction 34 and take the A683 exit west. Continue along the A6 south and turn right at the roundabout along the A588. The hospital is along this road. (Distance Approx 14.8 miles)

FACILITIES

1   Canal Garage (Shell)

▮ WC 24   Tel: (01539) 567280

0.6 miles north along the A65, on the right. Access, Visa, Overdrive, All Star, Switch, Dial Card, Mastercard, Amex, Diners Club, Delta, AA Paytrak, Shell Cards, Smart, Securicor Fuelserv, UK Fuelcard, Keyfuels.

2   Crooklands Motor Company

▮ WC   Tel: (01539) 567414

0.9 miles north along the A65, on the right. Eurocard, Mastercard, Delta, Visa, Switch, Access. Open; 08.00-18.00hrs Daily. Attended Service.

3   Crooklands Hotel & Restaurant

▮H ▮ ▮ ▮   Tel: (01539) 567432

1 mile north along the A65, on the right. Boskins Restaurant Open; 12.00-14.00hrs & 19.00-21.00hrs Daily. Hayloft Restaurant Open; Tues-Sat; 19.00-21.00hrs

4   Beckside B&B, 4 Nook Cottages

▮B   Tel: (01539) 567387

1 mile east along the A65, in Nook, on the left

PLACES OF INTEREST

Kendal

Follow the A591 north (Signposted 7.5 Miles)

For details please see Junction 37 information.

Sizergh Castle (NT)

Sizergh Nr Kendal LA8 8AE Tel: (01539) 60070. Follow the A591 north. (4.6 Miles)

Home of the Strickland family for over 750 years, the castle reflects the turbulent history of this part of the country. The 14thC pele tower contains some exceptional Elizabethan carved wooden chimney-pieces and there is a good collection of contemporary oak furniture and portraits. The castle is surrounded by handsome gardens, including a beautiful rock garden, and the estate has flower-rich limestone pastures and ancient woodland. Gift Shop. Tea Room. Limited disabled access.

Kirkby Lonsdale

Kirkby Lonsdale Tourist Information Office,
24 Main Street, Kirkby Lonsdale LA6 2AE
Tel: (015242) 71437
Follow the A65 east. (Signposted 5.4 Miles)

Now by-passed by the main roads to the north it has retained its character and remains a very traditional and handsome market town where life still revolves around the market place and the 600 year old cross. Thought to have been settled by marauding seafarers there is, in the origin of its name, more than a suggestion that Kirkby Lonsdale has links with the Danes. Visitors today will find lovely Georgian buildings crowding along the winding main street with interesting alleyways and courtyards to discover, excellent shops to browse in, and some wonderful tea shops.

SERVICE AREA

BETWEEN JUNCTIONS 36 AND 37

Killington Lake Services (Southbound Only)
(RoadChef)

Tel: (01539) 620739   Food Fayre Self-Service Restaurant, Wimpy Bar, RoadChef Lodge & BP Fuel

M6   JUNCTION 37

THERE ARE NO FACILITIES WITHIN ONE MILE OF THIS JUNCTION

NEAREST A&E HOSPITAL

Westmorland General Hospital, Burton Road, Kendal LA9 7RG Tel: (01539) 732288

Take the A684 exit west and after about 1 mile turn left along an unclassified road. Turn right at the end along the B6254 and the hospital is along this route. (Distance Approx 6.1 miles)

PLACES OF INTEREST

Kendal

Kendal Tourist Information Office, Town Hall, Highgate, Kendal LA9 4DL. Tel: (01539) 725758.
Follow the A684 west (Signposted 6.3 Miles)

Sited in the valley of the River Kent, it was once one of the most important woollen textile centres of northern England. The Kendal woollen industry was founded in 1331 by John Kemp, a Flemish weaver, and it flourished until it was overtaken in the 19thC by the huge West Riding of Yorkshire mills. The town was also famous for its Kendal Bowmen, skilled archers who were instrumental in the defeat of the Scots at the Battle of Flodden Field in 1513. Once a bustling town, major traffic now by-passes it, the narrow streets of Highgate, Stramongate and Stricklandgate are still busy and the fine stagecoaching inns of the 17th & 18thC's, to which Prince Charles Edward is said to have retreated after his abortive 1745 rebellion, still line the streets. The numerous alleyways throughout the town are known as yards, a distinctive feature of Kendal inasmuch as they formed a series of defences for the townspeople against the threat of raids by the Scots. The ruins of Kendal Castle, high on a hill overlooking the town, stand on the site of one of the original Roman camps that guarded the route to the Scottish border, and Katherine Parr, the last of Henry VIII's six wives, lived at this castle in the 16thC before she was married. An exhibition at Kendal Museum of Natural History & Archaeology on Station Road (Tel: 01539-721374) tells the castle's story with computer interactives and reconstructions.

Sedbergh

Sedbergh Tourist Information Office,
72 Main Street, Sedbergh LA10 5AD
Tel: (01539) 620125
Follow the A684 west (Signposted 4.6 Miles)

This old market town, with its cobbled streets, was first granted a charter in 1251 to hold fairs and a market. The old norse name for the settlement here was "Setberg" meaning flat-topped hill which perfectly describes Sedbergh's location on top of the Howgill Fells. The old major routes, from Lancaster to Newcastle and Kendal to York, passed through the town and, during the 19thC, the town bustled with the constant stream of stage coaches. Today it is a focal point for the surrounding rural communities and much of the heart of the town, where many of the older buildings still survive, has been deemed a Conservation Area. Sedbergh lies within the boundaries of the Yorkshire Dales National Park and is a major National Park Centre.

| | Large Hotel | | Small Hotel | | Motel | | Guest House/ Bed & Breakfast | | Disabled Facilities | | Childrens Facilities |

M6 JUNCTION 38

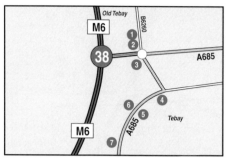

NEAREST NORTHBOUND A&E HOSPITAL

Penrith New Hospital, Bridge Lane, Penrith
CA11 8HX Tel: (01768) 245300

Proceed north to Junction 40 and take the A66 exit east. Turn left at the roundabout and the hospital is on the right. (Distance Approx 17 miles)

NEAREST SOUTHBOUND A&E HOSPITAL

Westmorland General Hospital, Burton Road,
Kendal LA9 7RG Tel: (01539) 732288

Proceed to Junction 37 and take the A684 exit west. After about 1 mile turn left along an unclassified road and then turn right at the end along the B6254. The hospital is in this road. (Distance Approx 14.5 miles)

FACILITIES

1 M6 Diesel Services

⛽ 24   Tel: (01539) 624336

0.1 miles north along the B6260, on the left. Access, Visa, Overdrive, All Star, Switch, Dial Card, Mastercard, Diners Club, Delta, Routex, Securicor Fuelserv, Morgan Fuelcard, UK Fuelcard, IDS, Keyfuels, AS24.

2 Junction 38 Cafe

🍴 🍽 ♿ 🔌   Tel: (01539) 624505

Open 24 hours Daily (NB. Closed between Sat 17.30 & Sun 07.00hrs and Sun 22.30 & Mon 07.00hrs during winter season)

3 Primrose Cottage B&B

🛏 🐾   Tel: (01539) 624791

0.1 miles south along the A685, in Tebay, on the right

4 Public Toilets

WC

0.2 miles south along the A685, in Tebay, on the left

5 Carmel Guest House

🛏   Tel: (01539) 624651

0.3 miles south along the A685, in Tebay, on the left

6 Cross Keys Inn

🍽 🍴 🔌 🛏   Tel: (01539) 624240

0.4 miles south along the A685, in Tebay, on the right. (Free House) Open all day Fri, Sat & Sun. Lunch 12.00-14.30hrs, Evening Meals 18.00-21.00hrs.

7 Barnaby Rudge Tavern and B&B

🍽 🍴 ♿ 🔌 🛏   Tel: (01539) 624328

0.9 miles south along the A685, in Tebay, on the right. (Free House) Lunch 12.00-14.00hrs and Evening Meals 19.00-21.00 hrs Daily.

PLACES OF INTEREST

Kirkby Stephen

Kirkby Stephen Tourist Information Office,
Market Street, Kirkby Stephen CA17 4QN
Tel: (017683) 71199
Follow the A685 east (Signposted 11.7 Miles)

Set amidst hills and heath-clad moors, intersected by upland valleys with small streams and on the pretty River Eden, it was the Norsemen who first established a village here. The Vikings named it "Kirke and bye", meaning churchtown and although it is essentially part of the Eden Valley, Kirkby Stephen has a strong Yorkshire Dales feel about it. The northern end of this delightful little town is dominated by the 13thC St Stephen's Church and between here and the market place stand the cloisters which served for a long time as a butter market. It gained some importance during the expansion of the railways in the 19thC, with the Settle and Carlisle line passing on the western side and the North Eastern Railway establishing an engine shed and large station within the town on its Darlington to Tebay line. Indeed much of today's A685 was built on the trackbed of the latter following its total closure in the 1970's.

SERVICE AREA

BETWEEN JUNCTIONS 38 AND 39

Tebay Services (Northbound) (Westmorland Ltd)

Tel: (01539) 624511 Restaurant, Coffee Shop, Westmorland Hotel, Coffee Bar & BP Fuel

Tebay Services (Southbound) (Westmorland Ltd)

Tel: (01539) 624511 Restaurant & BP Fuel

A unique concept in motorway services with a philosophy based on home cooking and giving the traveller that something special. This independent company utilizes local produce, including some organic vegetables, and home ground spices to produce delicious English food as well as traditional Indian curries. The food is cooked to order outside of normal hours. Cakes and bread are freshly baked on the premises and barbecued steaks and burgers are a speciality.

M6 | JUNCTION 39

THERE ARE NO FACILITIES WITHIN ONE MILE OF THIS JUNCTION

NEAREST NORTHBOUND A&E HOSPITAL

Penrith New Hospital, Bridge Lane, Penrith CA11 8HX Tel: (01768) 245300

Proceed north to Junction 40 and take the A66 exit east. Turn left at the roundabout and the hospital is on the right. (Distance Approx 12 miles)

NEAREST SOUTHBOUND A&E HOSPITAL

Westmorland General Hospital, Burton Road, Kendal LA9 7RG Tel: (01539) 732288

Proceed to Junction 37 and take the A684 exit west. After about 1 mile turn left along an unclassified road and then turn right at the end along the B6254. The hospital is in this road. (Distance Approx 19.9 miles)

M6 | JUNCTION 40

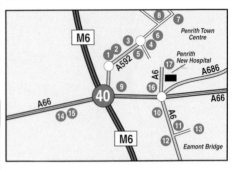

NEAREST A&E HOSPITAL

Penrith New Hospital, Bridge Lane, Penrith CA11 8HX Tel: (01768) 245300

Take the A66 exit east, turn left at the roundabout along the A6 and the hospital is on the right. (Distance Approx 0.7 miles)

FACILITIES

PENRITH TOWN CENTRE IS WITHIN ONE MILE OF THE JUNCTION

1 Mark Johns Motors

⛽ WC   Tel: (01768) 892906

0.2 miles north along the A592, on the left. Access, Visa, Overdrive, All Star, Switch, Dial Card, Mastercard, Amex, Diners Club, Delta, BP Supercharge. Open; Mon-Fri; 07.00-21.00hrs, Sat; 07.00-20.00hrs, Sun; 08.00-20.00hrs.

2 Red Rooster Road Stop & Takeaway

🍴 ♿   Tel: (01768) 892906

0.2 miles north along the A592, on the left. Open; Mon-Fri; 07.00-21.00hrs, Sat; 07.00-20.00hrs, Sun; 08.00-20.00hrs.

3 Davidsons Junction 40 (Esso)

⛽ WC 24   Tel: (01768) 867101

0.2 miles north along the A592, on the left. All pumps card operated. Bunkering of Diesel Fuel for Commercial Vehicles Only.

| H Large Hotel | h Small Hotel | M Motel | B Guest House/ Bed & Breakfast | Disabled Facilities | Childrens Facilities |

4 Agricultural Hotel

▮▮▮▮▮B   Tel: (01768) 862622
0.6 miles north along the A592, on the right.
(Jennings) Open all day. Meals served; 12.00-
14.00hrs & 18.00-20.30hrs daily

5 The Station House

▮B ▮   Tel: (01768) 866714
0.6 miles north along the A592, on the right
(Free House)

6 Safeway Penrith Petrol Station

▮ ▮ WC   Tel: (01768) 867631
0.7 miles north along the A592, on the right.
Access, Visa, Overdrive, All Star, Switch, Dial
Card, Mastercard, Amex, Diners Club, Delta,
BP Cards, Electron. Open; Mon-Sat; 07.00-
22.00hrs, Sun; 08.00-20.00hrs.

7 Glen Cottage Hotel & Restaurant

▮ ▮ ▮ ▮h   Tel: (01768) 862221
1 mile north along the A592, in Penrith, on
the right. Villa Bianca Restaurant (Italian)
Open; Mon-Sat; 12.00-14.00hrs & 18.00-
21.30hrs, Sun; Closed. Disabled facilities in
Restaurant.

8 The Royal Hotel

▮ ▮ ▮ ▮   Tel: (01768) 862670
1 mile north along the A592, in Penrith, on
Wilson Row. (Free House) Open all day on Sun.
Lunch 12.00-14.00hrs, Evening Meals 18.00-
21.00 hrs.

9 North Lakes Shire Inns

▮H ▮ ▮ ▮ ▮   Tel: (01768) 868111
Adjacent to north side of the roundabout at
the junction. Restaurant Open; Breakfast; Mon-
Fri; 07.30-09.30hrs, Sat & Sun; 08.00-10.00hrs,
Lunch; Mon-Fri; 12.30-14.00hrs, Sat & Sun;
Closed, Dinner; Mon-Sat; 19.00-21.30hrs,
Sun;19.00-20.45hrs

10 Eamont House B&B

▮B
0.9 miles south along the A6, in Eamont Bridge,
on the right

11 The Beehive Inn

▮ ▮ ▮ ▮   Tel: (01768) 862081
1 mile south along the A6, in Eamont Bridge,
on the left. (Pubmaster) Open all day Sat &
Sun. Meals served; Mon-Sat; Lunch; 12.00-
15.00hrs, Evening Meals; 18.00-21.00hrs, Sun;
11.00-21.00hrs.

12 The Crown

▮ ▮ ▮ ▮ ▮B   Tel: (01768) 892092
1 mile south along the A6, in Eamont Bridge,
on the right. (Free House) Open all day. Meals
served; Mon-Fri; 11.00-14.30hrs and 18.00-
21.00hrs, Sat & Sun; 11.00-21.00hrs.

13 Lowther Glen B&B

▮B   Tel: (01768) 864405
1 mile south along the A6, in Eamont Bridge,
in Lowther Glen.

14 Little Chef

▮ ▮   Tel: (01768) 868303
0.4 miles west along the A66, on the left. Open;
07.00-22.00hrs daily

15 Travelodge Penrith

▮M ▮ ▮   Tel: (01768) 866958
0.4 miles west along the A66, on the left

16 Bridge Lane Esso Service Station

▮ ▮ WC 24   Tel: (01768) 899303
0.8 miles north along the A6, on the left.
Access, Visa, Overdrive, All Star, Switch, Dial
Card, Mastercard, Amex, Diners Club, Delta,
Shell Gold, BP Supercharge, Esso Cards.

17 Shell Penrith

▮ ▮ WC 24   Tel: (01768) 212900
1 mile north along the A6, on the right. Shell
Cards, Access, Visa, Overdrive, All Star, Switch,
Dial Card, Mastercard, Amex, Diners Club,
Delta, Smart, BP Agency.

PLACES OF INTEREST

Penrith

Penrith Tourist Information Office, Robinson's School, Middlegate, Penrith CA11 7PT
Tel: (01768) 867466. Follow the A592 east (Signposted 1 Mile)

The capital of the Kingdom of Cumbria in the 9th & 10thC's, it was ransacked by Scottish raiders in the 14thC and the evidence of its former vulnerability can still be seen in the charming mixture of narrow streets and wide open spaces, such as Great Dockray and Sandgate, into which cattle were herded during the raids. Penrith has a splendid Georgian church, St Andrew's, of Norman origin but extensively rebuilt between 1719 and 1772, surrounded by a number of interesting buildings. The ruins of Penrith Castle bear witness to the town's importance in defending the surrounding countryside. It was built around 1399 and enlarged for the Duke of Gloucester, later Richard III, when he was Lord Warden of the Marches and responsible for keeping the peace along the borders. The castle has been in ruins since 1550 but still remains an impressive monument.

In the town centre...

Penrith Museum

Robinson's School, Middlegate, Penrith CA11 7PT
Tel: (01768) 212228

Set in a 300 year old school, the museum records the local history of Penrith and its surrounding area.

The Alpaca Centre

Snuff Mill Lane, Stainton, Penrith CA11 0HA
Tel: (01768) 891440. Follow the A6 west, turn left along the A592 towards Ullswater and then second right (1.6 Miles)

This unique centre is located in rolling countryside near to Lake Ullswater and has been developed to expand the knowledge of the Alpaca, its products and viability as a farm animal for fibre production. The animals can be viewed from the edge of the paddock or from within the adjacent tea room. The "Spirit of Andes" shop has a wide variety of Alpaca goods and clothing for sale. Disabled Access.

Rheged

Redhills, Penrith CA11 0DQ Tel: (01768) 868000.
website; www.rheged.com
e-mail; enquiries@rheged.com
Follow the A66 west (Signposted 0.6 Miles)

The Rheged Discovery Centre is Cumbria - The Lake District's newest, largest and most dramatic visitor attraction is due to open in the summer of 2000. The centrepiece is the first all British large format film, shown on a six storey high cinema screen, which will take the visitor on a dramatic journey back in time to reveal 2000 years of Cumbria's history, mystery and magic. Built on seven spectacular levels it also includes a Cumbrian showcase of rather special shops, restaurants combining local delicacies with the panoramic views, a superb sound and light show and a truly inspiring journey through Cumbria, all within Britain's largest earth covered building. Even Mountain Hall with its babbling brooks and massive limestone crags, makes the visitor feel as though they are in the very heart of Cumbria.

Lowther Leisure & Wildlife Park

Hackthorpe, Penrith CA10 2HG
Tel: (01931) 712523. Follow the A6 south to Hackthorpe (4.7 Miles)

Enjoy a whole day of family entertainment in the natural beauty of Lowther Park with all weather attractions for every age group including Stevenson's Crown Circus. Restaurant. Cafe. Picnic Areas. Disabled Access.

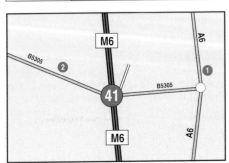

M6   JUNCTION 41

NEAREST A&E HOSPITAL

Penrith New Hospital, Bridge Lane, Penrith
CA11 8HX Tel: (01768) 245300

Proceed south to Junction 40 and take the A66 east.
Turn left at the roundabout and the hospital is on
the right. (Distance Approx 3.9 miles)

FACILITIES

1 Stoneybeck Inn

🎌 🍴 ♿ ⚡ Tel: (01768) 862369
1 mile east along the B5305, at the junction
with the A6. (Free House) Open all day. Lunch;
11.30-14.20hrs, Evening Meals; 18.00-21.00hrs
daily.

2 2M's Texaco Petrol Station

📋 WC Tel: (01768) 864019
0.3 miles west along the B5305, on the right.
Access, Visa, Overdrive, Switch, Dial Card,
Mastercard, Delta, Texaco Cards. Open; Mon-
Fri; 08.30-18.00hrs, Sat; 09.00-13.00hrs, Sun;
Closed. Attended service if required.

PLACES OF INTEREST

Hutton-in-the-Forest

Penrith CA11 9TH Tel: (017684) 84449
Follow the B5305 towards Wigton. (Signposted
from motorway, 2.1 Miles)

The romantic and historic home of Lord and
Lady Inglewood, dating from the 14thC when
it was originally built as a pele tower.
Alterations and additions over the centuries,
both inside and out, show a varying range of
architectural styles with the long gallery,
unusual in a north of England home, dating
from the 1630's whilst the hall of c1680 is
dominated by the Cupid Staircase. Many of the
upstairs rooms date from the mid-18th and
19thC's and the house features fine collections
of furniture, tapestries, portraits and ceramics.
The grounds, set in the mediaeval Forest of
Inglewood, include a beautiful 18thC walled
garden, 17thC topiary terraces and an extensive
Victorian woodland walk. Tea Room. Gift Shop.
Limited disabled access.

SERVICE AREA

BETWEEN JUNCTIONS 41 AND 42

Southwaite Services (Northbound) (Granada)

Tel: (01697) 473476 Burger King, Harry
Ramsden's, Fresh Express Self-Service Restaurant,
Travelodge & Esso Fuel

Southwaite Services (Southbound) (Granada)

Tel: (01697) 473476 Little Chef, Burger King, Harry
Ramsden's, Fresh Express Self-Service Restaurant,
Travelodge & Esso Fuel

Footbridge and Vehicle bridge connection between
sites.

M6 | JUNCTION 42

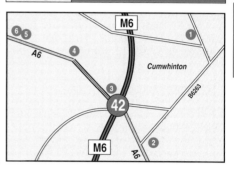

NEAREST A&E HOSPITAL

Cumberland Infirmary, Newtown Road, Carlisle
CA2 7HY Tel: (01228) 523444

Take the A6 north into Carlisle, bear right at the
end following the route to the A7 north and at the
roundabout turn left along the A595 towards
Cockermouth. After about 0.6 miles bear right along
the B5307 and the hospital is on the right hand
side of this road. (Distance Approx 5 miles)

FACILITIES

1 Lowther Arms

🎌 🍴 ♿ ⚡ 🛏 Tel: (01228) 560905
1 mile north along the B6263, in Cumwhinton,
on the left in Garlands Road. (Free House) Open
all day. Lunch & Evening Meals served daily.

2   Golden Fleece Services & Transport Cafe (BP)

🛏 WC 🍴 24   Tel: (01228) 542766

0.1 miles south along the A6, on the left. Access, Visa, Overdrive, All Star, Switch, Dial Card, Mastercard, Amex, Diners Club, Delta, Routex, Shell Agency, BP Cards, UK Fuels, Securicor Fuelserv, IDS, Keyfuels, Allied Card. Food is available in the Transport Cafe; 07.00-19.00hrs daily.

3   Carrow House Milestone Pub & Restaurant

🛏 🍴 ♿ 🎠   Tel: (01228) 532073

0.1 miles north along the A6, on the left. (Banks's) Open all day. Bar Snacks served; Mon-Fri;12.00-18.00, Sat; 12.00-22.00, Sun; 12.00-21.00hrs. Restaurant Open; Mon-Thurs; 12.00-14.00hrs & 17.00-22.00hrs, Fri & Sat; 12.00-14.00hrs & 17.00-22.30hrs, Sun; 12.00-21.00hrs.

4   Carleton Filling Station (BP)

🛏 WC   Tel: (01228) 592479

0.5 miles north along the A6, on the right. Access, Visa, Overdrive, All Star, Switch, Dial Card, Mastercard, Amex, Diners Club, Delta, Routex, UK Fuelcard, Shell Agency, BP Cards, Mobil Cards, Securicor Fuelserv. Open; Mon-Sat; 07.00-22.00hrs, Sun; 08.00-22.00hrs.

5   The Green Bank

🛏 🍴 ♿ 🎠   Tel: (01228) 528846

0.9 miles north along the A6, on the right. (Scottish & Newcastle) Open all day. Lunch served; 12.00-14.00hrs daily, Evening meals; Mon-Sat; 18.00-21.00hrs, Sun; 19.00-21.00hrs.

6   Dhaka Tandoori Restaurant

🛏 🍴 ♿   Tel: (01228) 523855

0.9 miles north along the A6, on the right. Open; Mon-Sat; 18.00-0.00hrs, Sun; 12.00-0.00hrs

PLACES OF INTEREST

Carlisle

Carlisle Tourist Information Centre, Old Town Hall, Greenmarket, Carlisle CA3 8JH
Tel: (01228) 625600.
Follow the A6 north (Signposted 4.1 Miles)

A great Roman centre, and the military base for the Petriana Regiment, Luguvalium, as Carlisle was known during this period, also became a major civilian settlement with fountains, mosaics, statues and centrally heated homes. There is an extensive collection of Roman remains, from both the city and the Cumbrian section of Hadrians Wall in the Tullie House Museum & Art Gallery in Castle Street (Tel: 01228-534781). There has been a castle at Carlisle since at least 1092 when William Rufus first built a palisaded fort but there was almost certainly a fortress prior to this, and Roman times, as the name of the city comes from the Celtic "Caer Lue", hill fort. Like many great mediaeval cities, Carlisle was surrounded by walls and the best place to view these 11thC structures is in a street called West Walls, at the bottom of Sally Port Steps and near the Tithe Barn. Many fine buildings still remain, including the Guildhall of 1407, now a museum (Tel: 01228-534781), Carlisle Cathedral, founded in 1122 (Tel: 01228-548151) and the Old Town Hall of the 17thC and now in use as the Tourist Information Centre.

Within the city ...

Carlisle Castle

Castle Way, Carlisle CA3 8UR Tel: (01228) 591922.

During the Scottish occupation in the 12thC, King David laid out a new castle with stone removed from Hadrian's Wall and the keep can still be seen, enclosed by massive inner and outer walls. The present castle is entered though a 14thC gatehouse, complete with portculis, and contains a maze of vaulted passages, chambers, staircases, towers and dungeons. It was beseiged for eight months during the Civil War by the Parliamentarians under General Leslie and after it was captured the castle was repaired and rebuilt utilizing stone from the adjacent Cathedral. Indeed some six out of the eight bays were used

M6

rendering Carlisle Cathedral to one of the smallest in England. Today the castle also contains the King's Own Royal Border Regiment Museum. Gift Shop. Limited disabled access.

M6 JUNCTION 43

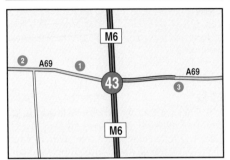

NEAREST A&E HOSPITAL

Cumberland Infirmary, Newtown Road, Carlisle CA2 7HY Tel: (01228) 523444

Take the A69 west into Carlisle, turn right at the end and left at the roundabout along the A595 towards Cockermouth. After about 0.6 miles bear right along the B5307 and the hospital is on the right hand side of this road. (Distance Approx 3 miles)

FACILITIES

1 Tesco Carlisle Petrol Station

▐ 24   Tel: (01228) 600400
0.2 miles west along the A69, on the right. Access, Visa, Overdrive, All Star, Switch, Dial Card, Mastercard, Amex, Delta. Toilets and cash machines available in adjacent store (Open 24 hours)

2 Brunton Park Service Station (Esso)

▐ WC   Tel: (01228) 528715
0.7 miles west along the A69, on the right. Access, Visa, Overdrive, All Star, Switch, Dial Card, Mastercard, Amex, Diners Club, Delta, Shell Gold, Esso Cards. Open; 06.00-0.00hrs Daily

3 The Waterloo

▐ ▐ ▐ ▐   Tel: (01228) 513347
0.6 miles east along the A69, on the right. (Inn Partnership) Meals served; 12.00-14.30hrs & 19.00-20.45 daily.

PLACES OF INTEREST

Carlisle

Follow the A69 west (Signposted 2.3 Miles)

For details please see Junction 42 information

Within the city ...

Carlisle Castle

Castle Way, Carlisle CA3 8UR

For details please see Junction 42 information

M6 JUNCTION 44

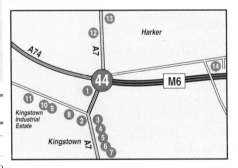

NEAREST A&E HOSPITAL

Cumberland Infirmary, Newtown Road, Carlisle CA2 7HY Tel: (01228) 523444

Take the A7 south into Carlisle, cross the river and turn right along the A595. After about 0.6 miles bear right along the B5307 and the hospital is on the right hand side of this road. (Distance Approx 3 miles)

FACILITIES

1 Posthouse Carlisle

▐H ▐ ▐ ▐ ▐   Tel: (01228) 531201
0.1 miles south along the A7, on the right.

| Pets Welcome | £ Cash Dispenser | 24 24 Hour Facilities | WC Toilets | Alcoholic Drinks | Food and Drink | Fuel |

(Entrance in Parkhouse Lane). Restaurant Open; 12.00-22.00hrs daily

2 BP Kingstown Filling Station

Tel; (01228) 523031

0.3 miles south along the A7, on the right. Access, Visa, Overdrive, All Star, Switch, Dial Card, Mastercard, Amex, Diners Club, Delta, Routex, AA Paytrak, Shell Agency, BP Cards. Open; 06.00-22.00hrs daily. Staff Toilets only, but available for customer use at discretion of staff.

3 Kingstown Hotel

Tel: (01228) 515292

0.5 miles south along the A7, on the left

4 The Coach & Horses

Tel: (01228) 525535

0.5 miles south along the A7, on the left. (Scottish & Newcastle) Open all day. Meals served; Mon-Sat; 12.00-14.00hrs & 18.30-21.00hrs, Sun; 12.00-14.30hrs & 18.30-21.00hrs.

5 Newfield Grange Hotel

Tel: (01228) 819926

0.7 miles south along the A7, in Newfield Drive, on the left

6 Gosling Bridge

Tel: (01228) 515294

0.9 miles south along the A7, on the left. (Greenalls) Open all day. Meals served all day; Sun-Thurs; 12.00-21.30hrs, Fri & Sat; 12.00-22.00hrs.

7 Premier Lodge

Tel: (01228) 515294

0.9 miles south along the A7, on the left

8 McDonald's

Tel: (01228) 512701

0.3 miles west along Parkhouse Road, on the left. Open; Sun-Thurs; 07.00-21.00hrs, Fri & Sat; 07.00-23.00hrs.

9 BP Truckstop Motel

Tel: (01228) 534192

0.4 miles west along Parkhouse Road, in Kingstown Industrial Estate, on the left

10 BP Truckstop

Tel (01228) 534192

0.4 miles west along Parkhouse Road, in Kingstown Industrial Estate, on the left. Diesel Fuel only

11 Asda Carlisle Petrol Station

Tel: (01228) 526550

0.6 miles west along Parkhouse Road, on the left. Access, Visa, Overdrive, All Star, Switch, Dial Card, Mastercard, Amex, Diners Club, Delta, AA Paytrak, Asda Cards. Toilets available in adjacent store during opening hours. Open; Mon-Sat; 07.00-22.00hrs, Sun; 09.00-16.30hrs.

12 Harker Service Station

Tel: (01228) 674274

0.6 miles north along the A7, on the left. UK Fuels, Securicor, Keyfuels, IDS, Visa, Mastercard, Access, Switch, AS24, Delta. Open; Mon-Fri; 07.30-20.00hrs, Sat; 08.00-18.00hrs, Sun; 09.00-18.00hrs. Diesel Fuel on 24 hour card operated pumps.

13 The Hill Cottage B&B

0.8 miles north along the A7, on the right

14 The Steadings B&B

Tel: (01228) 523019

0.9 miles east along the A689, on the right

PLACES OF INTEREST

Carlisle

Follow the A7 south (Signposted 2.1 Miles)

For details please see Junction 42 information

M6

Within the city ...

Carlisle Castle

Castle Way, Carlisle CA3 8UR

For details please see Junction 42 information

Solway Aviation Museum

Carlisle Airport, Carlisle CA6 4NW
Tel: (01228) 573641
Follow the A689 east (Signposted 6.4 Miles)

The outdoor collection includes a Vulcan B2 bomber, a Meteor night fighter, an English Electric Lightning and a Westland helicopter. The indoor museum has displays of a wartime airfield and an air raid shelter amongst numerous exhibits. Gift Shop. Cafe and Restaurant. Picnic Area. Disabled Access.

MOTORWAY ENDS

(TOTAL LENGTH OF MOTORWAY 225.2 MILES)

M6

  Pets Welcome   Cash Dispenser   24 Hour Facilities   Toilets   Alcoholic Drinks Food and Drink   Fuel

M6

OFF THE MOTORWAY
M40

M40

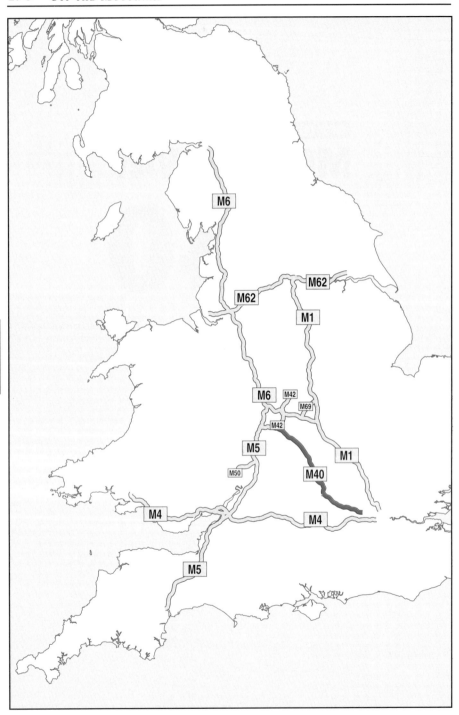

The M40 Motorway

The motorway forms a second major route between London, Birmingham and the North West, through the Chilterns, Oxfordshire and Warwickshire. Commencing near Uxbridge the motorway heads north westwards, passing Beaconsfield, the centre of an area known as the Chiltern Hundreds and which occupies a niche in Parliamentary history as one of the "rotten boroughs", and the town of Wycombe which is by-passed along the south side. Light planes can be seen crossing the carriageways as they take off and land at the adjacent Wycombe Air Park, home of the Booker-Blue Max Collection of historical aircraft. Continuing north westwards, the 300m high BT Tower comes into view at Stokenchurch and just north of here is where some of the major engineering works involved in the construction of the motorway were undertaken with the carriageways grading steeply through a deep cutting made in the western edge of the Chiltern Hills. For those travelling west there is a panoramic view across to the Cotswold Hills with the cooling towers of Didcot Power Station clearly visible to the south.

The motorway starts to head in a more northerly direction as it skirts around the eastern side of the historic city of Oxford, which dates back to Saxon times, and Ot Moor, several square miles of very poorly drained flatland, virtually impossible to farm and a home for a wide variety of wildlife. Further north an unusual landmark, a redundant water tower originally built in 1909 to supply water to Bicester, is visible alongside the carriageways before the motorway passes Banbury, the cross of which is immortalized in the children's nursery rhyme, on the west.

Between Banbury and the mediaeval town of Warwick, which is passed on the north side there are two more landmarks, both to the east, the 14thC beacon in Burton Hills Country Park and the unique 17thC Chesterton Windmill, a cylindrical building constructed in stone and surmounted on arches. The motorway continues north westwards and, just before it ends and merges with the M42, an obelisk commemorating victory in the Battle of Waterloo is visible on the north side of the carriageway.

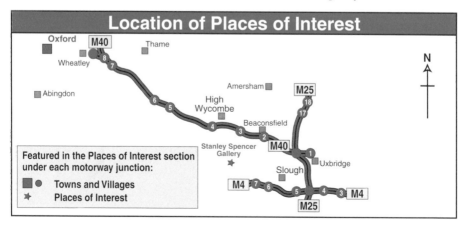

M40

Location of Places of Interest

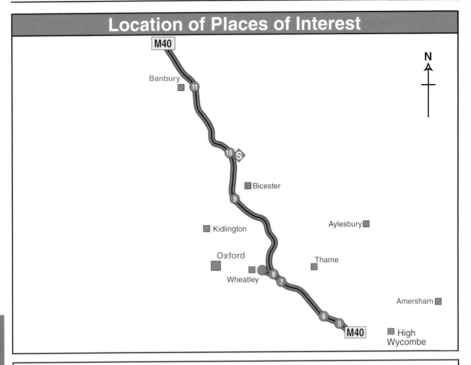

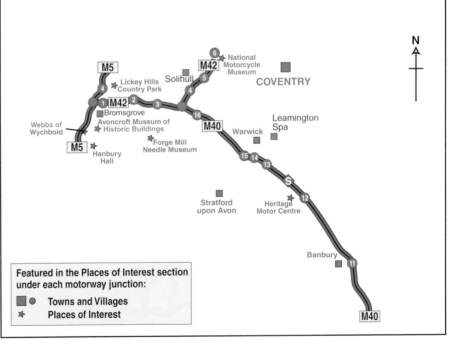

Featured in the Places of Interest section
under each motorway junction:

■ ● Towns and Villages
✷ Places of Interest

M40 JUNCTION 1A

THIS JUNCTION IS A MOTORWAY INTERCHANGE WITH THE M25 ONLY AND THERE IS NO ACCESS TO ANY FACILITIES

M40 JUNCTION 2

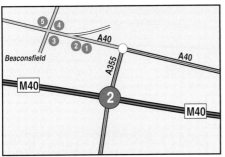

NEAREST NORTHBOUND A&E HOSPITAL

Wycombe General Hospital, Queen Alexandra Road, High Wycombe HP11 2TT
Tel: (01494) 526161

Proceed to Junction 4 and take the A404 north towards High Wycombe. Turn first left at the bottom of the hill and the Hospital is on the left. (Distance Approx 7.8 miles)

NEAREST SOUTHBOUND A&E HOSPITAL

Wexham Park Hospital, Wexham Road, Slough SL2 4HL. Tel: (01753) 633000

Take the A355 south towards Slough and turn left at the first roundabout (at Farnham Royal) along the B412. Turn right at the end of this road and left to Stoke Green. Turn left at the roundabout along Wexham Street and the hospital is on the right. (Distance Approx 6.5 miles)

FACILITIES

1 The White Horse

Tel: (01494) 673946
0.9 miles north along the A40, in Beaconsfield, on the left (Free House) Open all day. Meals served 12.00-14.30hrs daily.

2 The Old White Swan

Tel: (01494) 673800
0.9 miles north along the A40, in Beaconsfield, on the left (Whitbread) Open all day. Lunch; 12.00-14.30hrs daily

3 The Royal Saracens Head

Tel: (01494) 674119
1 mile north along the A40, in Beaconsfield, on the left (Whitbread) Open all day. Meals served 12.00-22.30hrs daily.

4 White Hart Hotel

Tel: (01494) 671211
1 mile north along the A40, in Beaconsfield, on the right (Toby) Open all day on Sat & Sun. Restaurant Open; Mon-Sat 12.00-14.30hrs & 18.00-21.00hrs, Sun; 12.00-21.00hrs.

5 The George Hotel

Tel: (01494) 673086
1 mile north along the A40, in Beaconsfield, on the right.

M40 JUNCTION 3

THIS IS A RESTRICTED ACCESS JUNCTION

Vehicles can only exit from the northbound lanes
Vehicles can only enter along the southbound lanes

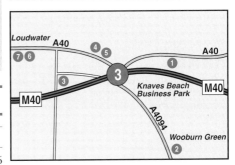

NEAREST NORTHBOUND A&E HOSPITAL

Wycombe General Hospital, Queen Alexandra Road, High Wycombe HP11 2TT
Tel: (01494) 526161

Proceed to Junction 4 and take the A404 north towards High Wycombe. Turn first left at the bottom of the hill and the Hospital is on the left. (Distance Approx 4.6 miles)

NEAREST SOUTHBOUND A&E HOSPITAL

Wexham Park Hospital, Wexham Road, Slough SL2 4HL Tel: (01753) 633000

Proceed to Junction 2, take the A355 south towards Slough and turn left at the first roundabout (at Farnham Royal) along the B412. Turn right at the end of this road and left to Stoke Green. Turn left at the roundabout along Wexham Street and the hospital is on the right. (Distance Approx 8.7 miles)

FACILITIES

1 Browns Ford Centre (Shell)

🍴 WC   Tel: (01494) 678881
0.5 miles east along the A40, on the right. Access, Visa, Overdrive, All Star, Switch, Dial Card, Mastercard, Amex, Diners Club, BP Supercharge, Shell Cards, Smartcard. Open; Mon-Fri; 07.00-22.00hrs, Sat & Sun; 07.00-21.00hrs.

2 The Rose & Crown

🍺 🍴 🛏 🚻   Tel: (01628) 520681
1 mile south along the A4034, in Wooburn Green, on the left. (Whitbread) Open all day. Meals served; 12.00-21.30hrs daily

3 The Happy Union

🍺 🍴 🚻 🛏   Tel: (01628) 520972
0.6 miles along Loudwater Road, on the left. (Whitbread) Open all day on Sat & Sun. Meals served; Mon-Sat; 12.00-14.00hrs, Sun; 12.00-15.00hrs

4 The Paper Mill

🍺 🍴 🚻 🛏   Tel: (01494) 537080
0.1 miles west along the A40, on the right. (Brewers Fayre) Open all day. Meals served; 11.30-22.00hrs daily.

5 Travel Inn

🏨M 🚻   Tel: (01494) 537080
0.1 miles west along the A40, on the right.

6 Loudwater Shell

🍴 🚻 WC   Tel: (01494) 894710
1 mile west along the A40, on the left. Access, Visa, Overdrive, All Star, Switch, Dial Card, Mastercard, Amex, AA Paytrak, Diners Club, Delta, BP Supercharge, Smartcard. Open; Mon-Fri; 07.00-23.00hrs, Sat & Sun; 08.00-22.00hrs.

7 King George V

🍺 🍴 🚻 🛏   Tel: (01494) 520928
1 mile west along the A40, on the left. (Courage) Open all day. Meals served; 11.00-15.00hrs daily

M40   JUNCTION 4

NEAREST A&E HOSPITAL

Wycombe General Hospital, Queen Alexandra Road, High Wycombe HP11 2TT
Tel: (01494) 526161

Take the A404 north exit towards High Wycombe. Turn first left at the bottom of the hill and the Hospital is on the left. (Distance Approx 1.1 miles)

FACILITIES

1 The Blacksmiths Arms

🍺 🍴 🚻 🛏   Tel: (01494) 525323
0.2 miles south along the Marlow Bottom Road, on the right. (Beefeater) Open all day. Bar Meals served 12.00-22.00hrs daily, Restaurant open; Mon-Fri; 12.00-14.30hrs & 17.00-22.30hrs, Sat;

12.00-23.00hrs, Sun; 12.00-22.30hrs.

2   Monkton Farm B&B

🚥B   Tel: (01494) 521082
1 mile south via Marlow Bottom Road and
Winchbottom Lane

3   Posthouse High Wycombe

🚥H 🍴 🍽 ♿   Tel: 0870 400 9042
0.1 miles north along the A4010, on the left.
The Junction Restaurant; Breakfast; Mon-Fri;
06.30-09.30hrs, Sat & Sun; 07.30-10.00hrs,
Lunch; 12.00-14.00hrs daily, Dinner; 19.00-
23.00hrs daily. Bar Meals available 24hrs.

4   Frankie & Benny's

🍴 🍽 ♿ 🐾   Tel: (01494) 511958
0.2 miles north along the A4010, in Crest Road,
on the left. Open; 12.00-23.00hrs daily

5   TGI Fridays

🍴 🍽 ♿   Tel: (01494) 450067
0.2 miles north along the A4010, in Crest Road,
on the left. Open; Mon-Sat; 12.00-23.30hrs,
Sun; 12.00-22.30hrs.

6   Asda Petrol Station

⛽ £ WC   Tel: (01494) 441611
0.2 miles north along the A4010, in Crest Road,
on the left. Access, Visa, Overdrive, All Star,
Switch, Dial Card, Mastercard, Amex, Diners
Club, Delta, Routex, AA Paytrak, Asda Business
Card. Open: Mon-Fri; 06.30-0.00hrs, Sat; 06.30-
22.00hrs, Sun; 08.00-18.00hrs. Toilets and Cash
Machine available in adjacent store during
opening hours.

7   Turnpike Petrol Station (Esso)

⛽ WC   Tel: (01494) 523471
0.7 miles north along the A4010, on the left.
Visa, Overdrive, All Star, Switch, Dial Card,
Mastercard, Amex, AA Paytrak, Shell Gold, BP
Supercharge, Esso Cards. Open; 07.00-23.00hrs
Daily.

8   The Turnpike

🍴 🍽 ♿ 🐾   Tel: (01494) 529419
0.7 miles north along the A4010, on the right.

(The Hungry Horse) Open all day. Meals served;
Mon-Fri; 12.00-15.00hrs & 18.00-22.00hrs, Sat;
12.00-22.00hrs, Sun; 12.00-21.00hrs.

M40   JUNCTION 5

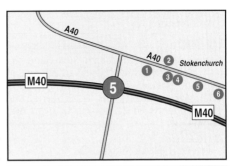

NEAREST NORTHBOUND A&E HOSPITAL

**John Radcliffe Hospital, Headley Way,
Headington, Oxford OX3 9DU
Tel: (01865) 741166**

Proceed to Junction 8 and take the A40 west to
Oxford. At the first roundabout take the second exit
towards Oxford and turn right at the second set of
traffic lights into Headley Way. The hospital entrance
is on the right. (Distance Approx 16.4 miles)

NEAREST SOUTHBOUND A&E HOSPITAL

**Wycombe General Hospital, Queen Alexandra
Road, High Wycombe HP11 2TT
Tel: (01494) 526161**

Proceed to Junction 4 and take the A404 north
towards High Wycombe. Turn first left at the bottom
of the hill and the Hospital is on the left. (Distance
Approx 8.7 miles)

FACILITIES

1   Tower Super Stop (Total)

⛽ WC £   Tel: (01494) 484382
0.4 miles east along the A40, in Stokenchurch,
on the right. Access, Visa, Overdrive, All Star,
Switch, Dial Card, Mastercard, Amex, Diners
Club, Delta, UK Fuelcard. Total Cards. Open;
06.00-22.00hrs daily

2   The Kings Arms Hotel

🚥H 🍴 🍽 ♿   Tel: (01494) 609090
0.5 miles east along the A40, in Stokenchurch,

	Pets Welcome	£	Cash Dispenser	24	24 Hour Facilities	WC	Toilets	🍴	Alcoholic Drinks	🍽	Food and Drink	⛽	Fuel

on the left. (Dhillon Hotels) Open all day. Restaurant Open; Breakfast; Mon-Sat; 07.00-09.00hrs, Sun; 08.00-10.00hrs, Lunch; Sun; 12.00-16.00hrs, Dinner; Mon-Fri; 19.00-21.30hrs. Brasserie Open; Mon-Sat; 10.00-22.00hrs, Sun; 12.00-22.00hrs.

3 Ye Fleur de Lis

Tel: (01494) 482269
0.5 miles east along the A40, in Stokenchurch, on the right. (Whitbread) Open all day on Sat & Sun. Meals served; Mon-Thurs; 12.00-14.30hrs & 18.00-21.30hrs, Fri & Sat; 12.00-14.30hrs & 18.00-22.00hrs, Sun; 12.00-18.00hrs.

4 The Four Horse Shoes

Tel: (01494) 482265
0.5 miles east along the A40, in Stokenchurch, on the right. (Carlsberg Tetley) Open all day Fri, Sat & Sun. Meals served; Mon-Fri; 12.00-14.00hrs

5 Mowchak Bar & Indian Restaurant

Tel: (01494) 485005
0.9 miles east along the A40, in Stokenchurch, on the right. Open; 12.00-14.00hrs & 17.30-23.30hrs daily

6 Stokenchurch Service Station (Murco)

Tel: (01494) 483355
1 mile east along the A40, in Stokenchurch, on the right. Access, Visa, Overdrive, All Star, Switch, Dial Card, Mastercard, Amex, Delta, Murco Cards. Open; Mon-Fri; 08.00-20.00hrs, Sat; 08.00-17.00hrs, Sun; Closed.

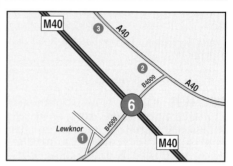

M40 JUNCTION 6

NEAREST NORTHBOUND A&E HOSPITAL

John Radcliffe Hospital, Headley Way, Headington, Oxford OX3 9DU
Tel: (01865) 741166

Proceed to Junction 8 and take the A40 west to Oxford. At the first roundabout take the second exit towards Oxford and turn right at the second set of traffic lights into Headley Way. The hospital entrance is on the right. (Distance Approx 13.9 miles)

NEAREST SOUTHBOUND A&E HOSPITAL

Wycombe General Hospital, Queen Alexandra Road, High Wycombe HP11 2TT
Tel: (01494) 526161

Proceed to Junction 4 and take the A404 north towards High Wycombe. Turn first left at the bottom of the hill and the Hospital is on the left. (Distance Approx 11.2 miles)

FACILITIES

1 Ye Olde Leathern Bottel

Tel: (01844) 351482
0.6 miles west, in Lewknor, along the High Street. (Brakspears Henley Brewery) Meals served: Lunch; 12.00-14.00hrs daily, Evening Meals; Sun-Thurs; 19.00-21.30hrs, Fri & Sat; 18.00-22.00hrs.

2 The Lambert Arms

Tel: (01844) 351496
0.4 miles north along the A40, on the left

| Large Hotel | Small Hotel | Motel | Guest House/ Bed & Breakfast | Disabled Facilities | Childrens Facilities |

3   Peel Guest House

▭B ⛟   Tel: (01844) 351310
1 mile north along the A40, on the left.

M40   JUNCTION 7

THIS IS A RESTRICTED ACCESS JUNCTION

Vehicles can only exit from the northbound lanes
Vehicles can only enter along the southbound lanes

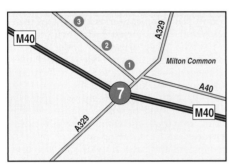

NEAREST NORTHBOUND A&E HOSPITAL

John Radcliffe Hospital, Headley Way,
Headington, Oxford OX3 9DU
Tel: (01865) 741166

Proceed to Junction 8 and take the A40 west to
Oxford. At the first roundabout take the second exit
towards Oxford and turn right at the second set of
traffic lights into Headley Way. The hospital entrance
is on the right. (Distance Approx 8.2 miles)

NEAREST SOUTHBOUND A&E HOSPITAL

Wycombe General Hospital, Queen Alexandra
Road, High Wycombe HP11 2TT
Tel: (01494) 526161

Proceed to Junction 4 and take the A404 north
towards High Wycombe. Turn first left at the bottom
of the hill and the Hospital is on the left. (Distance
Approx 16.9 miles)

FACILITIES

1   The Three Pigeons Inn

🍴 ⵎ ⛟ ▭B   Tel: (01844) 279247
0.1 miles north along the A40, in Milton
Common, on the right. (Tetley Carlsberg)

Meals served; Mon-Sat; 12.00-15.00hrs & 18.00-
21.30hrs, Sun; 12.00-15.00hrs (Carvery) &
19.00-21.30hrs

2   The Oxford Belfrey

▭H ⵎ ⵎ ⛟   Tel: (01844) 279381
0.2 miles north along the A40, in Milton
Common, on the right. The Terrace Restaurant
Open; Breakfast; Mon-Fri; 07.00-09.30hrs, Sat
& Sun; 08.00-10.00hrs, Lunch; Sun-Fri; 12.30-
14.00hrs, Dinner; Sun-Fri; 19.00-21.30hrs, Sat;
19.00-22.00hrs

3   Lantern Service Station

▯ WC   Tel: (01844) 279336
0.5 miles north along the A40, on the right.
Access, Visa, Keyfuels. Diesel Fuel Only. Open;
Mon-Sat; 09.00-18.00hrs, Sun; Closed.

M40   JUNCTION 8/8A

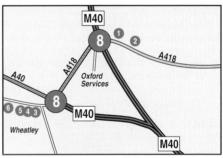

NEAREST A&E HOSPITAL

John Radcliffe Hospital, Headley Way,
Headington, Oxford OX3 9DU
Tel: (01865) 741166

Proceed along the A40 west to Oxford and at the
first roundabout take the second exit towards
Oxford. Turn right at the second set of traffic lights
into Headley Way and the hospital entrance is on
the right. (Distance Approx 6.7 miles)

 Pets Welcome   £ Cash Dispenser   24 Hour Facilities   WC Toilets   Alcoholic Drinks   Food and Drink   Fuel

FACILITIES

SERVICE AREA

Oxford Services (Welcome Break)

Tel: (01865) 877000 Red Hen, The Granary, KFC, La Brioche Doree, Burger King, Welcome Lodge & BP Fuel.

---

1   Waterstock Golf Club

🔧 🍴 ♿ 🔌   Tel: (01844) 338093

0.1 miles east along the A418, on the left. Restaurant Open; 08.00-20.00hrs daily.

---

2   Pymroyd House B&B

🛏️B   Tel: (01844) 339631

0.3 miles east along the A418, on the left

---

3   The Bridge Harvester Inn

🔧 🍴 ♿ 🔌   Tel: (01865) 875270

0.8 miles west along London Road, in Wheatley, on the left. (Bass) Open all day. Meals served; Mon-Thurs; 12.00-14.30hrs & 17.00-21.00hrs, Fri; 12.00-14.30hrs & 17.00-22.00hrs, Sat; 12.00-22.00hrs, Sun; 12.00-21.00hrs.

---

4   Travelodge Oxford East

🛏️M ♿   Tel: (01865) 875705

0.8 miles west along London Road, in Wheatley, on the left

---

5   Asda Petrol Station

⛽   Tel: (01865) 873888

0.8 miles west along London Road, in Wheatley, on the left. Access, Visa, Overdrive, All Star, Switch, Dial Card, Mastercard, Amex, Diners Club, Delta, Asda Business Card. Toilets and Cash Machine in adjacent store. Open; Mon & Tues; 07.00-19.00hrs, Wed & Thurs; 07.00-20.00hrs, Fri; 07.00-21.00hrs, Sat; 07.00-19.00hrs, Sun; 09.00-16.30hrs.

---

6   The Plough

🔧 🍴

1 mile west along London Road, in Wheatley, on the left. Meals served daily.

PLACES OF INTEREST

Oxford

Oxford Tourist Information Office, The Old School, Gloucester Green, Oxford OX1 2DA Tel: (01865) 726871. Follow the A40 west (Signposted 7.4 Miles)

Of Saxon origin, it was captured from the Danish invaders in 912 by Edward the Elder and his son, Athelstan established a mint in the town in 924. The city was taken by William the Conqueror in 1067 and he established a castle here, the remains of which adjoin the County Hall. Many historical events took place here; Maud surrendered to Stephen in 1142, in 1258 the "Mad Parliament" under Henry III passed the "Provisions of Oxford" and Ridley, Latimer and Cranmer were martyred within a few years of the founding of the see in 1542. The city and colleges abound with many of the finest period buildings in the country, dating from the 13thC through the subsequent centuries and embracing some of the best examples of the classical styles of architecture.

M40   JUNCTION 9

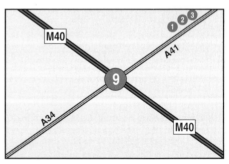

NEAREST NORTHBOUND A&E HOSPITAL

Horton General Hospital, Oxford Road (A4260), Banbury OX16 9AL Tel: (01295) 275500

Proceed to Junction 11 and take the A422 west towards Banbury and turn left at the second roundabout. Follow this route and turn left at the end along the A4260. The hospital is on the left along this road. (Distance Approx 18 miles)

NEAREST SOUTHBOUND A&E HOSPITAL

John Radcliffe Hospital, Headley Way,
Headington, Oxford OX3 9DU
Tel: (01865) 741166

Take the A34 south west towards Oxford and turn left along the A44. At the first roundabout turn left along the A40 and then follow the slip road to the B4150. Turn right at the first roundabout along the B4495 and the hospital is along this road. (Distance Approx 10.3 miles)

FACILITIES

THERE ARE NO FACILITIES WITHIN ONE MILE OF THE JUNCTION

1 Bicester Services (Esso)

🅿️ ♿ WC 24   Tel: (01869) 324451

2.5 miles north along the A41, on the left. Keyfuels, Access, Visa, Overdrive, All Star, Switch, Dial Card, Mastercard, Amex, Diners Club, Delta, BP Supercharge, Shell Gold, Esso Cards.

2 Little Chef

🍴 ♿   Tel: (01869) 248176

2.5 miles north along the A41, on the left. Open; 07.00-22.00hrs daily.

3 Burger King

🍴 ♿   Tel: (01869) 248176

2.5 miles north along the A41, on the left. Open; 10.00-22.00hrs daily.

PLACES OF INTEREST

Oxford

Follow the A34 west (Signposted 10.5 Miles)

For details please see Junction 8 information.

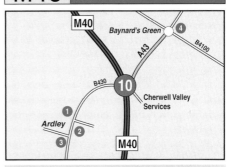

M40   JUNCTION 10

NEAREST NORTHBOUND A&E HOSPITAL

Horton General Hospital, Oxford Road (A4260),
Banbury OX16 9AL Tel: (01295) 275500

Proceed to Junction 11 and take the A422 west towards Banbury and turn left at the second roundabout. Follow this route and turn left at the end along the A4260. The hospital is on the left along this road. (Distance Approx 12.4 miles)

NEAREST SOUTHBOUND A&E HOSPITAL

John Radcliffe Hospital, Headley Way,
Headington, Oxford OX3 9DU
Tel: (01865) 741166

Proceed to Junction 9, take the A34 south west towards Oxford and turn left along the A44. At the first roundabout turn left along the A40 and then follow the slip road to the B4150. Turn right at the first roundabout along the B4495 and the hospital is along this road. (Distance Approx 15.9 miles)

FACILITIES

SERVICE AREA

Cherwell Valley Services (Granada)

Tel: (01869) 346060 Fresh Express, Burger King, Little Chef, Travelodge & Esso Fuel

1 The Corner Garage (Q8)

🅿️ WC   Tel: (01869) 345329

0.3 miles south along the B430, in Ardley, on the right. Access, Visa, Overdrive, All Star, Switch, Dial Card, Mastercard, Amex, Delta, Q8 Cards. Open; Mon-Thurs; 07.00-19.00hrs, Fri; 07.00-20.00hrs, Sat; 08.00-17.00hrs, Sun; 08.00-14.00hrs.

M40

| | Pets Welcome | £ | Cash Dispenser | 24 | 24 Hour Facilities | WC | Toilets | | Alcoholic Drinks | | Food and Drink | | Fuel |

2 Fox & Hounds

🍴 🍺 ♿ 🏠   Tel: (01869) 346883

0.3 miles south along the B430, in Ardley, on the left. (Free House) Meals served; 12.00-14.30hrs & 18.30-21.30hrs daily

3 The Old Post Office B&B

🏠B   Tel: (01869) 345958

0.5 miles south along Church Road, in Ardley, on the left.

4 Baynards Green Service Station (Esso)

🍴 WC 24   Tel: (01869) 345312

0.6 miles north along the A43, on the right. Access, Visa, Overdrive, All Star, Switch, Dial Card, Mastercard, Amex, Diners Club, Delta, Routex, AA Paytrak, UK Fuelcard, Shell Gold, BP Supercharge, Esso Cards.

M40 JUNCTION 11

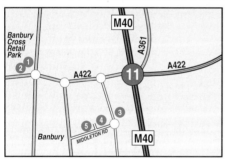

NEAREST A&E HOSPITAL

Horton General Hospital, Oxford Road (A4260), Banbury OX16 9AL Tel: (01295) 275500

Take the A422 west towards Banbury and turn left at the second roundabout. Follow this route and turn left at the end along the A4260. The hospital is on the left along this road. (Distance Approx 2.2 miles)

FACILITIES

1 Tesco Petrol Station

🍴 24   Tel: (01295) 457400

1 mile west along the A422, in Banbury Cross Retail Park, on the right. Access, Visa, Overdrive, All Star, Switch, Dial Card, Mastercard, Amex, Delta, AA Paytrak, Tesco Fuel Card. Toilets and Cash Machine available in adjacent store (Open 24 hours)

2 Burger King

🍴 ♿   Tel: (01295) 275744

1 mile west along the A422, in Banbury Cross Retail Park, on the right. Open; Mon-Thurs; 09.00-22.00hrs, Fri & Sat; 09.00-23.00hrs, Sun; 10.00-21.00hrs.

3 Ermont Way Service Station (Esso)

🍴 ♿ WC 24   Tel: (01295) 279989

0.4 miles south along Ermont Way, in Banbury, on the left. Access, Visa, Overdrive, All Star, Switch, Dial Card, Mastercard, Amex, Diners Club, Delta, Shell Gold, BP Supercharge, Esso Cards.

4 The Pepper Pot

🍴 🍺 ♿ 🏠   Tel: (01295) 261790

0.7 miles along Middleton Road, on the right. (Banks's) Open all day. Meals served; Mon-Sat; 12.00-14.00hrs & 18.00-20.30hrs, Sun; 12.00-14.00hrs.

5 Blacklock Arms

🍴 🍺 ♿ 🏠   Tel: (01295) 263079

0.8 miles along Middleton Road, on the right. (Bass) Open all day. Meals served; Mon-Sat; 12.00-14.00hrs & 17.30-21.30hrs, Sun; 12.00-20.30hrs.

PLACES OF INTEREST

Banbury

Banbury Tourist Information Centre, 8 Horsefair, Banbury OX16 0AA Tel: (01295) 259855. Follow the A422 west. (Signposted 1.8 Miles)

Banbury dates from Saxon times and it was during the Mediaeval period that the town expanded. A castle was erected in Banbury in 1125 and this was twice beseiged, latterly during the Civil War, culminating in its near total demolition. There are several old buildings surviving in the town including Ye

| 🏨 Large Hotel | 🏠 Small Hotel | 🏩 Motel | 🏠B Guest House/ Bed & Breakfast | ♿ Disabled Facilities | 🏠 Childrens Facilities |

Olde Reindere Inn from the 16thC and No.16 Market Place, a timber framed structure dating from the late 15thC. Within the town there was a strong connection with the Puritan movement and the original legendary cross of nursery rhyme fame was destroyed by them in the 17thC. A Victorian Gothic replacement, in the style of an Eleanor Cross, was installed to commemorate the wedding of Queen Victoria's eldest daughter to the Crown Prince of Prussia.

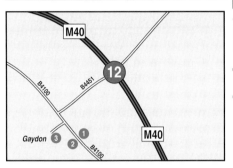

M40 JUNCTION 12

NEAREST NORTHBOUND A&E HOSPITAL

South Warwickshire Hospital, Lakin Road, Warwick CV34 5BW Tel: (01926) 495321
Proceed to Junction 13, take the A452 and then the A425 north into Warwick. Follow the A429 towards Coventry and bear left immediately after the railway bridge. The hospital is on the left side of this road. (Distance Approx 9.2 miles)

NEAREST SOUTHBOUND A&E HOSPITAL

Horton General Hospital, Oxford Road (A4260), Banbury OX16 9AL Tel: (01295) 275500
Proceed to Junction 11 and take the A422 west towards Banbury and turn left at the second roundabout. Follow this route and turn left at the end along the A4260. The hospital is on the left along this road. (Distance Approx 12.8 miles)

FACILITIES

1 Gaydon Service Station (Esso)

🅿 ♿ WC   Tel: (01926) 642278
0.7 miles south along the B4100, on the left. Access, Visa, Overdrive, All Star, Switch, Dial Card, Mastercard, Amex, Diners Club, Delta,

Shell Gold, BP Supercharge, Esso Cards. Open; 06.00-23.00hrs Daily.

2 Gaydon Inn

🅿 🍴 ♿ 🍺   Tel: (01926) 640388
0.7 miles south along the B4100, on the right. (Punch Taverns) [English & Italian menus] Lunch 12.00-14.30hrs, Evening Meals; Mon-Sat; 18.00-21.30hrs, Sun; 19.00-21.00hrs.

3 Malt Shovel

🅿 🍴 ♿ 🍺   Tel: (01926) 641221
0.8 miles west, in Gaydon Village. (Whitbread) Open all day. Meals served; 12.00-15.00hrs & 18.00-21.30hrs daily.

PLACES OF INTEREST

The Heritage Motor Centre

Banbury Road, Gaydon CV35 0BJ
Tel: (01926) 641188.
Signposted from the junction. (3.4 Miles)
This fascinating centre holds the largest exhibition of historic British cars and tells the story of the British motor industry from 1896 to the present day. Over 200 vehicles, embracing the world-famous makes of Austin, Morris, Rover, Wolseley, Riley, Standard, Triumph, MG and Austin-Healey, are on display and the 65 acre site also features a 4x4 off-road demonstration circuit. Gift Shop. Restaurant. Disabled access.

SERVICE AREA

BETWEEN JUNCTIONS 12 AND 13

Warwick Services (Northbound) (Welcome Break)
Tel: (01926) 651681 The Granary Restaurant, Burger King, Red Hen Restaurant, Welcome Lodge & BP Fuel

Warwick Services (Southbound) (Welcome Break)
Tel: (01926) 651681 The Granary Restaurant, Burger King, Red Hen Restaurant, KFC, La Brioche Doree, Welcome Lodge & BP Fuel

M40   JUNCTION 13

THIS IS A RESTRICTED ACCESS JUNCTION

Vehicles can only exit from the northbound lanes
Vehicles can only enter along the southbound lanes

NEAREST NORTHBOUND A&E HOSPITAL

South Warwickshire Hospital, Lakin Road, Warwick CV34 5BW Tel: (01926) 495321

Take the A452 and then the A425 north into Warwick. Follow the A429 towards Coventry and bear left immediately after the railway bridge. The hospital is on the left side of this road. (Distance Approx 3.8 miles)

NEAREST SOUTHBOUND A&E HOSPITAL

Horton General Hospital, Oxford Road (A4260), Banbury OX16 9AL Tel: (01295) 275500

Either;

Proceed to Junction 11 and take the A422 west towards Banbury and turn left at the second roundabout. Follow this route and turn left at the end along the A4260. The hospital is on the left along this road. (Distance Approx 18.2 miles)

Or;

Proceed to Junction 12, return to Junction 13 and follow the above Directions (Distance Approx 14.6 miles)

FACILITIES

THERE ARE NO FACILITIES WITHIN ONE MILE OF THE JUNCTION

PLACES OF INTEREST

Warwick

Warwick Tourist Information Centre, The Court House, Jury Street, Warwick CV34 4EW. Tel: (01926) 492212. Follow the A452 north (Signposted 3.2 Miles)

Standing by the River Avon, the mediaeval castle (Tel: 01926-406600) is one of the most splendid and well preserved fortresses in England. Established in Norman times, it was largely destroyed during the Barons' revolt in 1264 and most of the present structure dates from the 14thC. Although the centre of the town was rebuilt after a fire in 1694 some of the mediaeval structures including Lord Leycester's Hospital (Tel: 01926-491422), a beautiful collection of 15thC half timbered buildings enclosing a galleried courtyard, still remain amongst the elegant Queen Anne edificies.

M40   JUNCTION 14

THIS IS A RESTRICTED ACCESS JUNCTION

Vehicles can only exit from the southbound lanes
Vehicles can only enter along the northbound lanes

NEAREST NORTHBOUND A&E HOSPITAL

South Warwickshire Hospital, Lakin Road, Warwick CV34 5BW Tel: (01926) 495321

Proceed to Junction 15 and take the A429 north to Warwick. Follow this route through the town centre and bear left immediately after the railway bridge. The hospital is on the left of this road. (Distance Approx 3.8 miles)

NEAREST SOUTHBOUND A&E HOSPITAL

South Warwickshire Hospital, Lakin Road, Warwick CV34 5BW Tel: (01926) 495321

Take the A452 and then the A425 north into Warwick. Follow the A429 towards Coventry and bear left immediately after the railway bridge. The hospital is on the left side of this road. (Distance Approx 3.9 miles)

FACILITIES

THERE ARE NO FACILITIES WITHIN ONE MILE OF THE JUNCTION

PLACES OF INTEREST

Warwick

Follow the A452 south (Signposted 3.2 Miles)

For details please see Junction 13 information

 Large Hotel   Small Hotel   Motel   Guest House/ Bed & Breakfast   Disabled Facilities   Childrens Facilities

M40 JUNCTION 15

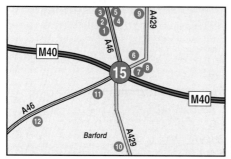

NEAREST A&E HOSPITAL

South Warwickshire Hospital, Lakin Road, Warwick CV34 5BW Tel: (01926) 495321

Take the A429 north to Warwick follow this route through the town centre and bear left immediately after the railway bridge. The hospital is on the left of this road. (Distance Approx 2.6 miles)

FACILITIES

1 Little Chef

Tel: (01926) 491764
1 mile north along the A46, on the left. Open; 07.00-22.00hrs daily.

2 Burger King

Tel: (01926) 491764
1 mile north along the A46, on the left. Open; 11.00-21.00hrs daily.

3 BP Express Shopping, Warwick North

Tel: (01926) 496862
1 mile north along the A46, on the left. Access, Visa, Overdrive, All Star, Switch, Dial Card, Mastercard, Amex, Diners Club, Delta, Routex, AA Paytrak, Shell Agency. BP Cards. Open; 06.00-22.00hrs daily.

4 Little Chef

Tel: (01926) 491560
1 mile north along the A46, on the right. Open; 07.00-22.00hrs daily.

5 BP Express Shopping, Warwick South

Tel: (01926) 499932
1 mile north along the A46, on the right. Access, Visa, Overdrive, All Star, Switch, Dial Card, Mastercard, Amex, Diners Club, Delta, Routex, AA Paytrak, Shell Agency. BP Cards. Open; 06.00-22.00hrs daily

6 Hilton National Warwick

Tel: (01926) 499555
0.1 miles north along the A429, on the left. Sonnets Restaurant Open; Breakfast; Mon-Fri; 07.00-09.30hrs, Sat & Sun; 07.30-10.00hrs, Lunch; Sun-Fri; 12.30-14.00hrs, Dinner; 18.30-22.00hrs daily.

7 Express by Holiday Inn, Warwick

Tel: (01926) 483000
0.1 miles north along the A429, on the right.

8 Porridge Pot

Tel: (01926) 401697
0.1 miles north along the A429, on the right. (Tom Cobleighs) Open all day. Meals served; Mon-Sat; 11.00-21.30hrs, Sun; 12.00-21.00hrs.

9 Stratford Road Filling Station (Jet)

Tel: (01926) 491228
1 mile north along the A429, on the left. Access, Visa, Overdrive, All Star, Switch, Dial Card, Mastercard, Amex, Diners Club, Delta, BP Supercharge, Jet Cards. Bakery and Filling Station Open; Mon-Fri; 07.00-22.00hrs, Sat; 07.00-21.00hrs, Sun; 08.00-21.00hrs.

10 The Joseph Arch

Tel: (01926) 624365
1 mile south along the A429, in Barford, on the right. (Punch Taverns) Open all day on Fri, Sat & Sun. Lunch; Mon-Sat; 12.00-14.00hrs, Sun; 12.00-15.00hrs, Evening Meals; Mon-Sat; 19.00-21.00hrs.

11 The Old Rectory Restaurant and B&B

Tel: (01926) 624562
0.2 miles south along the A46, on the left. Restaurant Open; 19.00-21.00hrs daily.

 Pets Welcome   Cash Dispenser   24 Hour Facilities   Toilets   Alcoholic Drinks   Food and Drink   Fuel

12 Hillcrest Guest House

B Tel: (01926) 624386

1 mile south along the A46, on the left

PLACES OF INTEREST

Warwick

Follow the A429 (Signposted 2.2 Miles)

For details please see Junction 13 information

Stratford upon Avon

Stratford upon Avon Tourist Information Centre, Bridgefoot, Stratford upon Avon CV37 6GW Tel: (01789) 293127. Follow the A46 south (Signposted 6.2 Miles)

On the banks of the River Avon, Stratford has many interesting features as well as being the home town of William Shakespeare. A market town dating back to at least 1196 when it was granted its charter by King John, the focal point of interest is his birthplace, a small timber framed house in Henley Street, (Tel: 01789-204016) but there are also a considerable number of buildings of the Tudor and Jacobean periods. The Shakespeare Memorial Theatre, opened on April 23rd, 1932, is the second on the site, the first was built in 1879 and destroyed by a fire in 1926. (Tel: 01789-295623)

M40 JUNCTION 16

THIS IS A RESTRICTED ACCESS JUNCTION

Vehicles can only exit from the southbound lanes
Vehicles can only enter along the northbound lanes

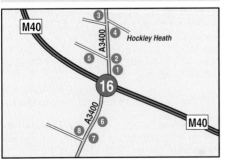

NEAREST SOUTHBOUND A&E HOSPITAL

South Warwickshire Hospital, Lakin Road, Warwick CV34 5BW Tel: (01926) 495321

Proceed to Junction 15 and take the A429 north to Warwick. Follow this route through the town centre and bear left immediately after the railway bridge. The hospital is on the left of this road. (Distance Approx 11.4 miles)

NEAREST NORTHBOUND A&E HOSPITAL

Solihull Hospital, Lode Lane, Solihull B91 2JL Tel: (0121) 711 4455

Proceed north to Junction 5 on the M42 and take the A41 north towards Birmingham. After about 1.5 miles turn left along Lode Lane (B425) and the hospital is on the left. (Distance Approx 8.5 miles)

FACILITIES

1 Old Royal Oak

Tel: (01564) 785252

0.4 miles north along the A3400, on the right. (Greene King) Open all day. Meals served; Mon-Fri; 12.00-15.00hrs & 18.00-21.45hrs, Sat & Sun; 12.00-21.45hrs.

2 Rose Cottage B&B

B Tel: (01564) 782936

0.5 miles north along the A3400, on the right

3 Wharf Tavern

Tel: (01564) 782075

1 mile north along the A3400, in Hockley Heath, on the left. (Scottish & Newcastle) Open all day. Lunch, Mon-Sat; 12.00-14.00hrs, Evening Meals, Mon-Sat; 18.30-21.30hrs, Meals served all day Sun; 12.00-20.00hrs.

4 Hockley Heath Service Station (Shell)

WC Tel: (01564) 782244

1 mile north along the A3400, in Hockley Heath, on the right. Access, Visa, Overdrive, All Star, Switch, Dial Card, Mastercard, Amex, Diners Club, AA Paytrak, UK Fuelcard, Shell Cards, BP Supercharge, Smartcard. Open; Mon-Sat; 07.00-22.30hrs, Sun; 08.00-22.30hrs.

| **H** | Large Hotel | **h** | Small Hotel | **M** | Motel | **B** | Guest House/ Bed & Breakfast | | Disabled Facilities | | Childrens Facilities |

5   Nuthurst Grange Hotel & Restaurant

Tel: (01564) 783972

0.6 miles north along Nuthurst Grange Lane, on the left. Restaurant Open; Breakfast; Mon-Fri; 07.30-09.30hrs, Lunch; Sun-Fri; 12.00-14.00hrs, Dinner; 19.00-21.30hrs daily.

6   Little Chef

Tel: (01564) 784149

0.8 miles south along the A3400, on the left. Open; 07-00-22.00hrs daily.

7   Village Garden B&B

Tel: (01564) 783553

1 mile south along the A3400, on the left

8   Ye Olde Pounde Cafe

Tel: (01564) 782970

1 mile south along the A3400, on the right. Open; Mon-Fri; 08.00-15.00hrs, Sat & Sun; Closed.

MOTORWAY MERGES WITH THE M42

(TOTAL LENGTH OF MOTORWAY 87.7 MILES)

M40

	Pets Welcome		Cash Dispenser		24 Hour Facilities		Toilets		Alcoholic Drinks		Food and Drink		Fuel
						WC							

M40

OFF THE MOTORWAY

M42

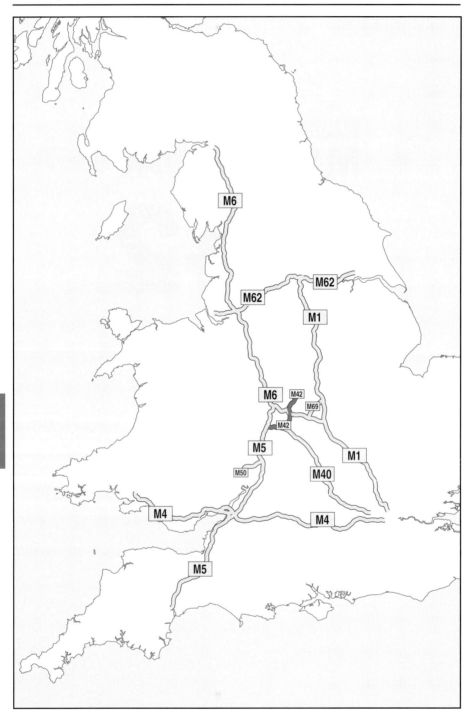

M42

The M42 Motorway

Although only a short motorway of under forty miles long, it provides a direct north east-south west connection and avoids the need for traffic to go around the north side of Birmingham.

Commencing at its junction with the M5 at Lickey End just north of the old nail-making town of Bromsgrove, the motorway heads due east, passing the small village of Alvechurch, the football team of which in 1971 took part in the longest ever FA Cup tie of 11 hours, on the south and part of the Phoenix (former BMW/Rover) Longbridge Plant is visible on top of a hill to the north.

At its junction with the M40, the motorway then bears north passing under one of the new generation of bow string style road bridges, at Blythe Valley Business Park, which are assembled alongside of the roadway and slid into place thus alleviating weeks of traffic disruption using traditional methods. The motorway continues northwards past Solihull, home of the world famous Land Rover and the Renewal Christian Centre and, just before the link with the M6, the suspended roof supports of the NEC Arena can be seen to the west and low flying aircraft landing and taking off from the adjacent Birmingham International Airport cross the carriageways.

North of the M6 the eastern suburbs of Birmingham can be seen to the west and the small town of Coleshill which dates back to 799AD is on the east. Just before the M42 ends, Tamworth, of saxon origin, is on the west side and, at Appleby Magna, the motorway has an end-on junction with the A42 to the M1 north and Nottingham.

M42

Location of Places of Interest

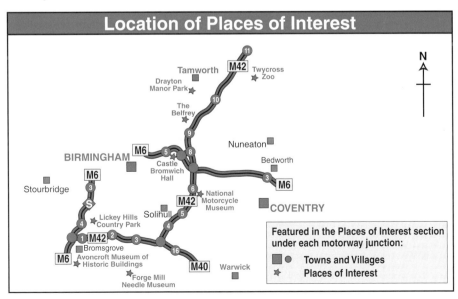

M42   JUNCTION 1

THIS IS A RESTRICTED ACCESS JUNCTION

There is no exit for eastbound vehicles.
There is no access for westbound vehicles

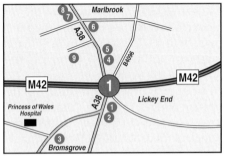

NEAREST NORTH & EASTBOUND A&E HOSPITAL

Selly Oak Hospital, Raddlebarn Road, Selly Oak, Birmingham B29 6JD Tel: (0121) 627 1627

Take the A38 north to Birmingham and into Selly Oak. The hospital is signposted within the city. (Distance Approx 9.6 miles)

NEAREST SOUTHBOUND A&E HOSPITAL

Worcester Royal Infirmary, Ronkswood Branch, Newtown Road, Worcester WR5 1HN
Tel: (01905) 763333

Proceed to Junction 6 (M5) and take the westbound exit along the A4440 and the route is signposted A&E Hospital. (Distance Approx 14.5 miles)

NEAREST MINOR INJURY UNIT

The Princess of Wales Community Hospital, Stourbridge Road, Bromsgrove B61 0BB
Tel: (01527) 488000

Follow the A38 into Bromsgrove. Signposted within the town. Opening Hours; 09.00-17.00, Monday to Friday. (Distance Approx 1 mile)

FACILITIES

1   The Forest

🍴 🛏 ⚡ ♿   Tel: (01527) 872063

On the south side of the Roundabout. (Harvester) Open all day Sat & Sun; Meals served; Lunch; Mon-Fri; 12.00-14.30hrs, Evening Meals; Mon-Thurs; 17.00-21.00hrs, Fri; 17.00-22.00hrs. Meals served all day; Sat; 12.00-22.00hrs, Sun; 12.00-21.00hrs.

2   Forest Service Station (Esso)

🍴 ♿ WC 24   Tel: (01527) 570142

0.1 miles south along the A38, on the left. Access, Visa, Overdrive, All Star, Switch, Dial Card, Mastercard, Amex, Diners Club, Delta, Shell Gold, BP Supercharge, Esso Cards,

3   Shell Bromsgrove

🍴 ♿ WC 24 £   Tel: (01527) 834830

1 mile south along the Birmingham Road, in Bromsgrove, on the left. Access, Visa, Overdrive, All Star, Switch, Dial Card, Mastercard, Amex, Diners Club, Delta, AA Paytrak, BP UK Agency, Smartcard, Shell Cards

4   Marlgrove Super Stop (Total)

🍴 WC 24 £   Tel: (01527) 579854

0.3 miles north along the A38, on the right. Access, Visa, Overdrive, All Star, Switch, Dial Card, Mastercard, Amex, Diners Club, Delta, Total Cards.

5   Marlgrove Club & Motel

🏨M 🍴 🛏   Tel: (01527) 872889

0.3 miles north along the A38, on the right. Cafe Open: 07.00-14.00hrs daily. Restaurant Open; Mon-Sat; 19.30-21.30hrs.

6   The Marlbrook

🛏 🍴 ⚡ ♿   Tel: (01527) 878060

0.6 miles north along the A38, on the right. (Punch Taverns) Open all day, Meals served 12.00-21.00hrs Daily

7   Elf Service Station

🍴 WC   Tel: (01527) 570178

0.7 miles north along the A38, on the left. Access, Visa, Overdrive, Amex, All Star, Diners Club, Dial Card, Switch. Open; Mon-Fri; 07.00-23.00hrs, Sun; 08.00-23.00hrs.

8   Stakis Country Court Hotel

🏨H 🍴 🛏 ♿   Tel: (0121) 447 7888

| Large Hotel | Small Hotel | Motel | | Guest House/ Bed & Breakfast | | Disabled Facilities | | Childrens Facilities |

0.8 miles north along the A38, on the left. Restaurant Open; Mon-Fri; 07.00-10.00hrs, 12.00-14.00hrs, 19.00-21.30hrs, Sat; 07.00-10.00hrs, 19.00-21.30hrs, Sun; 07.30-10.00hrs, 12.00-14.00hrs, 19.00-21.30hrs.

9 The Royal Oak

🍴 🍽 🐾 ♿ Tel: (01527) 870141

0.7 miles along Barley Mow Lane, on the left. (Freehouse) Open all day. Meals served; Mon-Sat, 12.00-14.00hrs & 18.00-21.00hrs, Sun; 12.00-20.15hrs.

PLACES OF INTEREST

Avoncroft Museum of Historic Buildings

Stoke Heath, Worcestershire Tel: (01527) 831886
Follow the A38 south. (Signposted 3.4 miles)

For details please see Junction 5 (M5) information

M42 JUNCTION 2

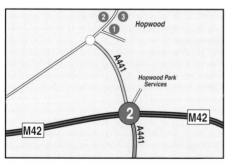

NEAREST NORTH & EASTBOUND A&E HOSPITAL

Selly Oak Hospital, Raddlebarn Road, Selly Oak, Birmingham B29 6JD Tel: (0121) 627 1627

Take the A441 north towards Birmingham. In Cotteridge bear left along the Ring Road (A4040) to Selly Oak and after about 1.5 miles turn right into Raddlebarn Road. The hospital is on the left. (Distance Approx 5.9 miles)

NEAREST SOUTH & WESTBOUND A&E HOSPITAL

Worcester Royal Infirmary, Ronkswood Branch, Newtown Road, Worcester WR5 1HN
Tel: (01905) 763333

Proceed to Junction 6 (M5) and take the westbound exit along the A4440 and the route is signposted A&E Hospital. (Distance Approx 18.9 miles)

NEAREST MINOR INJURY UNIT

The Princess of Wales Community Hospital, Stourbridge Road, Bromsgrove B61 0BB
Tel: (01527) 488000

Proceed to Junction 1 and follow the A38 into Bromsgrove. Signposted within the town. Opening Hours; 09.00-17.00, Monday to Friday. (Distance Approx 5.2 miles)

FACILITIES

SERVICE AREA

Hopwood Park Services (Welcome Break)

Tel: (0121) 447 8011 La Brioche Doree French Cafe, Granary Restaurant, Burger King, Mezzanine Restaurant and BP Fuel.

1 Petrol Express (Esso)

⛽ WC ♿ Tel: (0121) 445 6173

0.8 miles north along the A441, on the right. Access, Visa, Overdrive, All Star, Switch, Dial Card, Mastercard, Amex, Diners Club, Delta, Shell Agency, Esso Cards. Open; Mon-Fri; 06.00-22.00hrs, Sat & Sun; 07.00-22.00hrs

2 Hopwood House

🍴 🍽 🐾 ♿ Tel: (0121) 445 1716

0.8 miles north along the A441, on the left. (Banks's) Meals served; Mon-Thurs; 12.00-14.00hrs & 17.00-21.00hrs, Fri; 12.00-14.30hrs & 17.00-22.00hrs, Sat; 12.00-22.00hrs, Sun; 12.00-21.00hrs

3 Westmead Hotel & Restaurant

🛏M 🍴 🍽 🐾 Tel: (0121) 445 1202

1 mile north along the A441, on the right. The Colonial Restaurant; Open; Breakfast; Mon-Sat; 07.00-09.30hrs, Sun; 08.00-10.00hrs, Lunch; Mon-Fri; 12.00-14.00hrs, Dinner; Mon-Sat;

19.00-21.45hrs. Carvery Bar; Open; Sun; 12.00-21.30hrs.

PLACES OF INTEREST

Birmingham

Follow the A441 north (Signposted 10 Miles)

For details please see Junction 1 (M5) information.

Cadbury World

Bournville, Birmingham B30 2LD
Follow the A441 north (Brown & White tourist boards along the route) (6 Miles).

For details please see Junction 4 (M5) information.

Brindley Place & Broad Street, Birmingham

Follow the A441 north into the city centre and the route is signposted "National Indoor Arena & Convention Centre" (10 Miles)

For details please see Junction 1 (M5) information.

Forge Mill Needle Museum

Needle Mill Lane, Riverside, Redditch B98 8HY
Tel: (01527) 62509 Follow the A441 south to Redditch (3.5 Miles)
Redditch has established a reputation for manufacturing high quality springs and needles and at Forge Mill the history of the latter is described. The displays include water powered machinery and textiles. Picnic Area. Museum Shop. Cafe. Disabled access.

M42 | JUNCTION 3

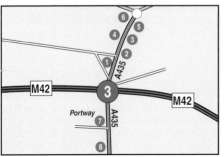

NEAREST NORTHBOUND A&E HOSPITAL

Solihull Hospital, Lode Lane, Solihull B91 2JL
Tel: (0121) 711 4455

Proceed to Junction 5 and take the A41 north towards Birmingham. After about 1.5 miles turn left along Lode Lane (B425) and the hospital is on the left. (Distance Approx 9.6 miles)

NEAREST SOUTHBOUND A&E HOSPITAL

Selly Oak Hospital, Raddlebarn Road, Selly Oak, Birmingham B29 6JD Tel: (0121) 627 1627

Proceed to Junction 2 and take the A441north towards Birmingham. In Cotteridge bear left along the Ring Road (A4040) to Selly Oak and after about 1.5 miles turn right into Raddlebarn Road. The hospital is on the left. (Distance Approx 8.7 miles)

NEAREST MINOR INJURY UNIT

The Princess of Wales Community Hospital, Stourbridge Road, Bromsgrove B61 0BB
Tel: (01527) 488000

Proceed to Junction 1 and follow the A38 into Bromsgrove. Signposted within the town. Opening Hours; 09.00-17.00, Monday to Friday. (Distance Approx 8 miles)

FACILITIES

1   Weatheroak Service Station (Total Fina)

🚻 WC 24   Tel: (01564) 824685
0.2 miles north along the A435, on the left. Access, Visa, Overdrive, All Star, Switch, Dial Card, Mastercard, Amex, Diners Club, Delta, Elf Cards, Total Fina Cards.

2   The Horse & Jockey

🍴 🛏 🚪 ♿   Tel: (01564) 822308
0.4 miles north along the A435, on the right. (Punch Retail) Open all day. Meals served; Mon-Sat; 12.00-22.00hrs, Sun; 12.00-21.30hrs.

3   Fontana Restaurant

🍴 🍴   Tel: (01564) 822824
0.7 miles north along the A435, on the right.

4   Inkford Hotel

🏨 🍴   Tel: (01564) 824330
0.9 miles north along the A435, on the left.

| Large Hotel | Small Hotel | Motel | Guest House/ Bed & Breakfast | Disabled Facilities | Childrens Facilities |

5 Star Market Wythall (Texaco)

🛒 WC ♿ 24 £ Tel: (01564) 825110
0.9 miles north along the A435, on the right.
Access, Visa, Overdrive, All Star, Switch, Dial
Card, Mastercard, Amex, Diners Club, Delta,
UK Fuelcard, Texaco Cards, Keyfuels

6 Becketts Farm Restaurant

🍴 🍽 ♿ 🐾 Tel: (01564) 823402
1 mile north along the A435, on the left. Open;
Mon-Sat; 08.00-21.30hrs, Sun; 09.00-18.00hrs.

7 The Rose & Crown

🍴 🍽 🐾 ♿ Tel: (01564) 822166
0.2 miles south along the A435, on the right.
(Greenall's) Open all day. Restaurant & Bar
Meals: Mon-Fri; 12.00-14.30hrs, 17.30-
22.00hrs, Sat 12.00-14.30hrs, 17.30-22.30hrs
[Bar Meals served all day], Sun; 12.00-21.30hrs.

8 Portway House Restaurant (Italian)

🍴 🍽 ♿ Tel: (01564) 824794
0.3 miles south along the A435, on the right.
Open; Lunch; Tues-Fri & Sun; 12.00-14.30hrs,
Evening Meals; Tues-Sat; 18.30-22.30hrs

M42   JUNCTION 3A

**THIS JUNCTION IS A MOTORWAY
INTERCHANGE ONLY WITH THE M40
AND THERE IS NO ACCESS TO ANY
FACILITIES**

M42   JUNCTION 4

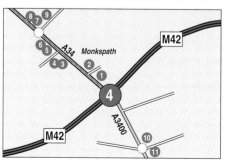

NEAREST NORTHBOUND A&E HOSPITAL

**Solihull Hospital, Lode Lane, Solihull B91 2JL
Tel: (0121) 711 4455**
Proceed to Junction 5 and take the A41 north
towards Birmingham. After about 1.5 miles turn
left along Lode Lane (B425) and the hospital is on
the left. (Distance Approx 4 miles)

NEAREST SOUTHBOUND A&E HOSPITAL

**Selly Oak Hospital, Raddlebarn Road, Selly Oak,
Birmingham B29 6JD Tel: (0121) 627 1627**
Proceed to Junction 2 and take the A441 north
towards Birmingham. In Cotteridge bear left along
the Ring Road (A4040) to Selly Oak and after about
1.5 miles turn right into Raddlebarn Road. The
hospital is on the left. (Distance Approx 14.3 miles)

NEAREST MINOR INJURY UNIT

**The Princess of Wales Community Hospital,
Stourbridge Road, Bromsgrove B61 0BB
Tel: (01527) 488000**
Proceed to Junction 1 and follow the A38 into
Bromsgrove. Signposted within the town. Opening
Hours; 09.00-17.00hrs, Monday to Friday. (Distance
Approx 13.6 miles)

FACILITIES

1 Tesco Petrol

🛒 24 Tel: (0121) 253 7500
0.2 miles north along the A34, on the right.
Access, Visa, Overdrive, All Star, Switch, Dial
Card, Mastercard, Amex, Tesco Fuel Card.
Toilets and Cash Machine available in adjacent
store during store opening hours.

2 McDonald's

🍽 🐾 ♿ Tel: (0121) 733 6327
0.3 miles north along the A34, on the right.
Open; 07.00-23.00hrs daily. "Drive Thru"open
until 23.30hrs Sun-Thurs and 0.00hrs Fri & Sat.

3 The Plough Inn

🍴 🍽 🐾 ♿ Tel: (0121) 744 2942
0.6 miles north along the A34, on the left.
(Whitbread) Open all day. Restaurant Open;
Mon-Fri; 07.00-09.00hrs, 12.00-14.30hrs &
17.00-22.30hrs, Sat & Sun; 08.00-10.00hrs &
12.00-22.15hrs. Bar Meals served; 12.00-
21.00hrs daily.

4   Travel Inn

📠M   Tel: (0121) 744 2942
0.6 miles north along the A34, on the left

5   Chez Julienne Francais

🔲 🍴 ♿   Tel: (0121) 744 7232
0.8 miles north along the A34, on the left.
Open; Lunch; Mon-Fri; 12.00-14.00hrs,
Evening Meals; Mon-Sat; 19.00-22.30hrs.

6   Jefferson's

🔲 🍴 ⛱ ♿   Tel: (0121) 733 1666
0.8 miles north along the A34, on the left. Open
all day. Meals served; Mon-Sat; 12.00-23.00hrs
and Sun; 12.00-22.30hrs.

7   Da Corrado (Italian)

🔲 🍴 ♿ ⛱   Tel: (0121) 744 1977
1 mile north along the A34, on the right. Open;
Mon-Sat; 12.00-14.00hrs & 19.00-23.00hrs,
Sun; Closed.

8   The Regency Hotel

📠H 🔲 🍴 ⛱   Tel: (0121) 745 6119
1 mile north along the A34, on the right.
Copperfields Brasserie; Open; Breakfast; 07.00-
09.30hrs daily, Lunch; Sun-Fri; 12.00-14.00hrs,
Dinner; Mon-Thurs; 19.00-21.30hrs, Fri-Sun;
19.00-22.00hrs.

9   Fatty Arbuckles

🔲 🍴 ♿ ⛱   Tel: (0121) 733 3773
1 mile north along the A34, in Monkspath
Leisure Park. Open; 12.00-23.00hrs daily

10   Boxtrees Farm B&B and Coffee Shop

📠B 🍴   Tel: (01564) 782039 & 07970 736156
0.9 miles south along the A3400, on the left.
Coffee Shop Open; Wed-Mon; 10.00-17.00hrs.
Hot meals served; 12.00-14.00hrs.

11   Boxtrees Filling Station (Total Fina)

🔲 ♿ WC   Tel: (01564) 782139
1 mile south along the A3400, on the left.
Access, Visa, Overdrive, All Star, Switch, Dial

Card, Mastercard, Amex, Diners Club, Delta,
Elf Cards, Total Fina Cards, Keyfuels. Open:
06.00-23.00hrs daily

M42	JUNCTION 5

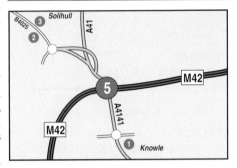

NEAREST A&E HOSPITAL

Solihull Hospital, Lode Lane, Solihull B91 2JL
Tel: (0121) 711 4455
Take the A41 north towards Birmingham. After
about 1.5 miles turn left along Lode Lane (B425)
and the hospital is on the left. (Signposted. Distance
Approx 2 miles)

FACILITIES

**THERE ARE NO FACILITIES WITHIN 1
MILE OF THIS JUNCTION**

1   Wilson Arms

🔲 🍴 ♿ ⛱   Tel: (01564) 772559
1.3 miles south along the A4141, in Knowle
on the left. (Bass) Open all day. Carvery Open;
Mon-Sat; 12.00-14.00hrs & 17.00-22.00hrs,
Sun; 12.00-22.00hrs. Bar Meals served; 12.00-
22.00hrs daily.

2   Brueton Park Filling Station (BP)

🔲 ♿ WC 🍴 £   Tel: (0121) 704 0878
1.4 miles north along the B4025, on the left,
in Solihull. Access, Visa, Overdrive, All Star,
Switch, Dial Card, Mastercard, Amex, AA
Paytrak, Diners Club, Delta, Routex, BP Cards,
Shell Agency. Open; 07.00-23.00hrs daily.
Delice de France Hot Snacks available.

Large Hotel	Small Hotel	Motel	Guest House/ Bed & Breakfast	Disabled Facilities	Childrens Facilities

3 The Golden Lion

🎯 🍴 ♿ 🐾   Tel: (0121) 704 1567

1.5 miles north along the B4025, on the right, in Solihull. (Unique Pub Co) Open all day. Meals served; Mon-Sat; 11.00-19.00hrs

PLACES OF INTEREST

Birmingham

Follow the A41 north (Signposted 9.4 Miles)

For details please see Junction 1 (M5) information.

Brindley Place & Broad Street, Birmingham

Follow the A41 north into the city centre and the route is signposted "National Indoor Arena & Convention Centre" (10 Miles)

For details please see Junction 1 (M5) information.

M42   JUNCTION 6

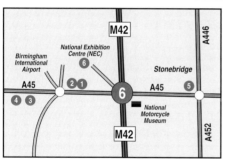

NEAREST A&E HOSPITAL

Heartlands Hospital, Bordesley Green East, Birmingham B9 5SS Tel: (0121) 766 6611

Take the A45 towards Birmingham and after about 4 miles turn right at Yardley along the A4040 (Ring Road). Follow this route for about 1 mile and turn left along Bordesley Green East (B4126). The hospital is on the left. (Distance Approx 6.9 miles)

FACILITIES

1 Arden Hotel & Restaurant

🏨 ♿ 🎯 🍴 🐾   Tel: (01675) 443221

0.5 miles west along the A45, on the right. Restaurant Open: Lunch; Sun-Fri; 12.00-14.30hrs, Sat; Closed, Dinner; Sun-Thurs; 18.30-22.00hrs, Fri & Sat; 18.30-23.00hrs.

2 Bickenhill Service Station (Esso)

⛽ 24 ♿ WC   Tel: (01675) 443485

0.6 miles west along the A45, on the right. Access, Visa, Overdrive, All Star, Switch, Dial Card, Mastercard, Amex, Diners Club, Delta, AA Paytrak, Shell Gold, BP Supercharge, Esso Cards. NB Temporarily closes between 22.35 & 23.00hrs to effect shift change

3 The Clock

🎯 🍴 ♿ 🐾   Tel: (0121) 782 3434

0.9 miles west along the A45, on the left. (Punch Taverns) Open all day. Meals served; Mon-Thurs; 12.00-21.00hrs, Fri & Sat; 12.00-22.00hrs, Sun; 12.00-20.30hrs.

4 Anne's Pantry Petrol Station (Elf)

⛽ WC £   Tel: (0121) 782 4498

1 mile west along the A45, on the left. Access, Visa, Overdrive, All Star, Switch, Dial Card, Mastercard, Amex, Diners Club, Delta, AA Paytrak, Elf Cards, Total Fina Cards. Open: 07.00-22.00hrs daily.

5 The Malt Shovel

🎯 🍴 ♿ 🐾   Tel: (01675) 442326

0.9 miles east along the A45, on the left. (Toby Carvery) Open all day. Carvery Restaurant Open; Mon-Thurs; 12.00-14.00hrs & 17.00-22.00hrs, Fri; 12.00-14.00hrs & 17.00-22.30hrs, Sat; 12.00-22.30hrs, Sun; 12.00-22.00hrs. Bar Meals served; Mon-Sat; 11.00-23.00hrs, Sun; 12.00-22.30hrs.

6 Stakis Birmingham Metropole Hotel

🏨 🎯 🍴 ♿ 🐾   Tel: (0121) 780 4242

0.6 miles, within the NEC site. Restaurant

Open; Breakfast; 07.00-10.30hrs, Lunch; 12.30-15.00hrs, Dinner; 18.00-22.30hrs daily.

into Lindrige Road and third left into Rectory Road. The hospital is on the right after 1 mile. (Distance Approx 4.9 miles)

PLACES OF INTEREST

National Motorcycle Museum

Coventry Road, Bickenhill, Solihull B92 0EJ
The entrance is adjacent to the roundabout.

For details please see Junction 4 (M6) information.

M42 JUNCTION 7 & JUNCTION 7A

THIS JUNCTION IS A MOTORWAY INTERCHANGE ONLY WITH THE M6 AND THERE IS NO ACCESS TO ANY FACILITIES

M42 JUNCTION 8

THIS JUNCTION IS A MOTORWAY INTERCHANGE ONLY WITH THE M6 AND THERE IS NO ACCESS TO ANY FACILITIES

M42 JUNCTION 9

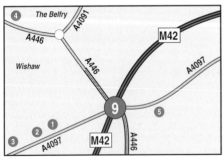

NEAREST A&E HOSPITAL

Good Hope Hospital, Rectory Road, Sutton Coldfield B75 7RR. Tel: (0121) 378 2211

Take the A449 towards Lichfield and after about 2.6 miles turn left along Holly Lane. Turn first right

FACILITIES

1   The White Horse

Tel: (01675) 470227
0.6 miles south along the A4097, on the right. (Bass) Open all day. Meals served; Sun-Tues; 12.00-21.00hrs, Wed-Sat; 12.00-22.00hrs.

2   The Old School House Hotel

Tel: (01675) 470177
0.8 miles south along the A4097, on the right.

3   The Kingsley

Tel: (01675) 470808
1 mile south along the A4097, on the right. (Beefeater) Open all day. Meals served; Mon-Fri; 12.00-14.30hrs & 17.30-23.00hrs, Sat & Sun; 12.00-23.00hrs

4   The Belfry Hotel

Tel: (01675) 470301
1 mile north along the A446, on the right. (De Vere Hotels) French Restaurant [a la carte] Open; Lunch; Mon-Sun; 12.30-14.00hrs Sat; Closed. Dinner; Mon-Sat; 19.30-21.45hrs. Sun; Closed. Garden Room [Carvery]; Open; Lunch; 12.30-14.00hrs, Dinner; 18.30-22.00hrs daily. Rileys Restaurant [Light meals] Lunch; 12.00-17.00hrs, Evening meals; 18.30-22.00hrs daily.

5   Water Park Lodge

Tel: (01675) 470533
0.3 miles north along the A4097, on the right.

PLACES OF INTEREST

Drayton Manor Park

Tamworth, Staffordshire B78 3TW
Tel: (01827) 287979
website; www.draytonmanor.co.uk
Follow the A446 north and continue along the A4091 (Signposted 4.9 Miles)

| | Large Hotel | | Small Hotel | | Motel | | Guest House/ Bed & Breakfast | | Disabled Facilities | | Childrens Facilities |

Based around the former home of Sir Robert Peel, part of the manor is utilized as a cafe, Drayton Manor Park now has over 100 dynamic and exciting rides including Stormforce 10, The Shockwave and Splash Canyon. Other attractions include a zoo and childrens farm. Shop. Cafe. Disabled access.

The Belfry

Wishaw, North Warwickshire B76 9PR
Tel: (01675) 470301 website; www.thebelfry.com
e-mail; enquiries@thebelfry.com
Follow the A446 north (Signposted 1 Mile)

Set in 500 acres of North Warwickshire countryside, the world-famous Belfry is a must for golf enthusiasts with its three championship golf courses. Casual visitors are welcome to view the scene of some of the most renowned encounters in the Ryder Cup and browse through the largest on-course Golf, Leisure and Lifestyle Shop in Europe. The complex, the venue for the Ryder Cup in 2001, also includes a 324 bedroom 4-star hotel, 5 restaurants, 8 bars, a night club and a floodlit driving range.

M42 JUNCTION 10

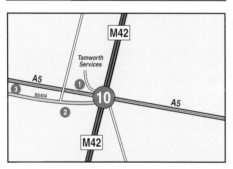

NEAREST NORTHBOUND A&E HOSPITAL

George Eliot Hospital, College Street, Nuneaton CV10 7DJ Tel: (01203) 351351

Take the A5 east towards Hinckley and at Atherstone bear left along the B4111 to Nuneaton town centre. Follow the A444 south towards Coventry and turn right into College Street. The hospital is on the left. (Distance Approx 11.6 miles)

NEAREST SOUTHBOUND A&E HOSPITAL

Good Hope Hospital, Rectory Road, Sutton Coldfield B75 7RR Tel: (0121) 378 2211

Proceed to Junction 9, take the A449 towards Lichfield and after about 2.6 miles turn left along Holly Lane. Turn first right into Lindrige Road and third left into Rectory Road. The hospital is on the right after 1 mile. (Distance Approx 10.8 miles)

FACILITIES

SERVICE AREA

Tamworth Services (Granada)

Tel: (01827 260120) Burger King, Little Chef, Travelodge and Esso Fuel.

1 Kinsall Green Garage (Murco)

🚻 **WC**   Tel: (01827) 283838

0.1 miles north along the A5, on the right. Access, Visa, Overdrive, All Star, Switch, Dial Card, Mastercard, Delta. Open; Mon-Sat; 07.00-19.00 hours. Sun; Closed.

2 Centurion Park

🍴 🍴 ♿ 🅿   Tel: (01827) 260587

0.3 miles north along the B5404, on the left. (Brewers Fayre) Open all day. Meals served all day Mon-Sat 11.30-22.00hrs, Sun; 12.00-22.00hrs.

3 The Red Lion

🍴 🍴 🅿 🛏B   Tel: (01827) 280818

0.9 miles north along the B5404, in Wilncote, on the right. (Free House) Open all day Wed-Sat. Meals served; Mon-Sat; 12.00-14.00hrs.

PLACES OF INTEREST

Drayton Manor Park

Tamworth, Staffordshire B78 3TW
Follow the A5 west (Signposted 3.3 Miles)

For details please see Junction 9 information

| Pets Welcome | £ Cash Dispenser | 24 Hour Facilities | **WC** Toilets | Alcoholic Drinks | Food and Drink | Fuel |

Tamworth

Follow the A5 west (3.4 Miles)

In the second half of the 8thC Offa, King of Mercia, built a great palace in Tamworth and fortified it by encircling the palace and town with a ditch, of which traces remain as Offa's Dyke and King's Ditch. The town was destroyed twice by the Danes and later invaded by the Scandinavians, with evidence of this occupation being revealed in some of the street names such as Gunlake. Tamworth Castle, sited on an eminence beside the River Anker and close to its junction with the River Thame, was first constructed by Ethelfreda, King Alfred's daughter, in the 10thC. The Normans built a motte and bailey castle in sandstone in the 1180's and the building, set in delightful Pleasure Grounds is open to public view (Tel: 01827-63563). The vast Parish Church was founded in 963, rebuilt by the Normans and rebuilt again after the Great Fire of Tamworth in 1345 and has a fine 15thC tower with a remarkable double staircase. The town has some fine 18thC buildings and, in front of the Town Hall of 1701, there is a bronze statue of Sir Robert Peel, Tamworth's most famous son.

M42 JUNCTION 11

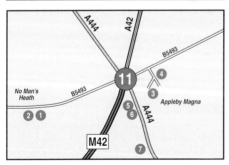

NEAREST NORTHBOUND A&E HOSPITAL

Queen's Hospital, Belverdere Road, Burton upon Trent DE13 0RB Tel: (01283) 566333

Take the A444 into Burton and at Stapenhill proceed over St.Peters Bridge and continue to the 2nd roundabout and the hospital is signposted Burton A&E from this point. (Distance Approx 11.5 miles)

NEAREST SOUTHBOUND A&E HOSPITAL

George Eliot Hospital, College Street, Nuneaton CV10 7DJ Tel: (01203) 351351

Take the A444 south into Nuneaton town centre and continue south towards Coventry. Turn right into College Street and the hospital is on the left. (Distance Approx 14.3 miles)

FACILITIES

1   The Four Counties

Tel: (01827) 830243

1 mile south along the B5493, in No Man's Heath, on the left. (Free House) Lunch; Mon-Sat; 12.00-14.00hrs, Sun; 12.00-14.30hrs, Evening Meals; Mon-Sat; 19.00-22.30hrs, Sun; 19.00-22.00hrs.

2   Four Counties Garage (Murco)

Tel: (01827) 830883

1 mile south along the B5493, in No Man's Heath, on the left. Visa, Overdrive, All Star, Switch, Dial Card, Mastercard, Amex, Delta. Open; Mon-Sat; 08.00-18.00hrs, Sun; Closed.

3   The Crown Inn

Tel: (01530) 271478

1 mile east, in Appleby Magna, in Church Street. (Marston's) Open all day. Bar meals served; Mon-Sat; 12.00-21.00hrs.

4   The Black Horse

Tel: (01530) 270588

1 mile east, in Appleby Magna, in Top Street (Banks's) Open all day Sun. Meals served; Mon-Sat; 12.00-14.00hrs & 18.00-21.30hrs, Sun; 12.00-14.00hrs.

5   McDonald's

Tel: (01530) 273195

0.1 miles south along the A444, on the right. Open; Sun-Thurs; 07.00-23.00hrs ["Drive Thru" open until 23.30hrs], Fri & Sat; 07.00-23.00hrs ["Drive Thru" open until 0.00hrs].

| | Large Hotel | | Small Hotel | | Motel | | Guest House/ Bed & Breakfast | | Disabled Facilities | | Childrens Facilities |

6 Appleby Magna Service Area (Total Fina)

▨ ▨ **WC** **24**   Tel: (01530) 273303
0.1 miles south along the A444, on the right. Access, Visa, Overdrive, All Star, Switch, Dial Card, Mastercard, Amex, Diners Club, Delta, BP Supercharge, Elf Cards, Total Fina Cards.

7 Appleby Inn Hotel

▨ ▨ ▨ ▨ ▨B   Tel: (01530) 270463
0.7 miles south along the A4444, on the left. (Free House) Open all day. Meals served all day 12.00-22.00 daily.

PLACES OF INTEREST

Twycross Zoo

Burton Road, Twycross, Nr Atherstone, Warwickshire CV9 3PX Tel: (01827) 880250
Follow the A444 south and turn left along the B4116 (4.1 Miles)

Over 1000 animals, including many endangered species, are on view at this centre for conservation and education. Other attractions include a pets corner, children's adventure playground, picnic areas and under cover soft play area. Gift Shop. Cafe. Licensed Bar.

MOTORWAY ENDS

(TOTAL LENGTH OF MOTORWAY 39.0 MILES)

| Pets Welcome | Cash Dispenser | 24 Hour Facilities | Toilets | Alcoholic Drinks | Food and Drink | Fuel |

M42

OFF THE MOTORWAY
M50

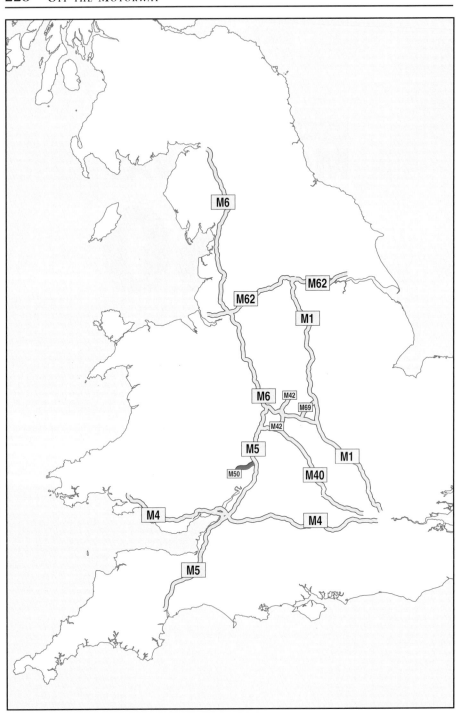

M50

The M50 Motorway

Also known as the Ross Spur, the M50 forms part of the dual carriageway link between South Wales, the Midlands and the North. Primarily a strictly rural route the only item of note is the fine view across the River Wye to Ross at the south end of the motorway. Travellers are strongly advised to ensure that they have sufficient fuel to travel the whole length as there are no facilities available between Ross and Strensham on the M5.

Location of Places of Interest

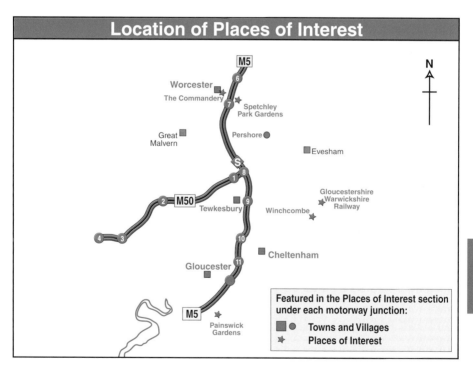

Featured in the Places of Interest section under each motorway junction:

- Towns and Villages
- Places of Interest

M50　JUNCTION 1

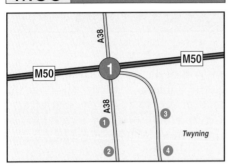

NEAREST NORTHBOUND A&E HOSPITAL

Worcester Royal Infirmary, Ronkswood Branch, Newtown Road, Worcester WR5 1HN
Tel: (01905) 763333

Proceed north to Junction 7 on the M5 and take the A44 exit west. The route is signposted A&E Hospital. (Distance Approx 13.7 miles)

NEAREST SOUTHBOUND A&E HOSPITAL

Cheltenham General Hospital, Sandford Road, Cheltenham GL53 7AN Tel: (01242) 224133

Proceed to Junction 10 on the M5 and follow the A4019 into Cheltenham. The hospital is signposted from within the town. (Distance Approx 14.6 miles)

NEAREST MINOR INJURY UNIT

Tewkesbury Hospital, Barton Road, Tewkesbury GL20 5QN. Tel: (01684) 293303.

Follow A38 south into Tewkesbury. Signposted from within the town. Open 24 hours. (Distance Approx 3.3 miles)

NEAREST WESTBOUND A&E HOSPITAL

Hereford General Hospital, Nelson Street, Hereford HR1 2PA Tel: (01432) 355444

Proceed to Junction 2, take the A417 and then the A438 to Hereford. The hospital is signposted within the city. (Distance Approx 27.9 miles)

NEAREST WESTBOUND MINOR INJURY UNIT

Ross Community Hospital, Walton Street, Ross on Wye HR9 5AD Tel: (01989) 562100

Proceed to Junction 4 and take the A449 to Ross. Continue along Broad Street, cross Market Square and go along Walford Street. Turn second left into Walton Street and the hospital is along this road. Open 24 Hours. (Distance Approx 21.5 miles)

FACILITIES

1　Stakis Puckrup Hall Hotel & Golf Club

Tel: (01684) 296200

0.4 miles south along the A38, on the right. Balharries Restaurant; Open; Breakfast; Mon-Fri; 07.00-10.00hrs, Sat & Sun; 08.00-10.00hrs, Lunch; Mon-Fri; 12.00-14.00hrs, Sun; 12.30-14.00hrs, Dinner; Mon-Sat; 19.00-22.00hrs, Sun; 19.00-21.30hrs.

2　The Crown at Shuthonger

Tel: (01684) 293714

1 mile south along the A38, in Shuthonger, on the right. (Free House) Lunch; 12.00-14.30hrs daily, Evening Meals; Tues-Sat; 19.00-21.00hrs. (Daily between February and November)

3　The Village Inn

Tel: (01684) 293500

0.8 miles south on Twyning Green, in Twyning, on the left.(Pubmaster) Meals served; Tues-Sun; 12.00-14.00hrs & 19.00-21.00hrs.

4　The Fleet Inn

Tel: (01684) 274310

1 mile south along Fleet Lane, in Twyning, on the left. (Whitbread) Open all day. Lunch; 12.00-14.30hrs daily, Evening meals; Sun-Thurs; 18.00-21.00hrs, Fri & Sat; 18.00-21.30hrs. Baguettes and Chips served between 12.00-21.00hrs daily.

M50　JUNCTION 2

THERE ARE NO FACILITIES WITHIN ONE MILE OF THE JUNCTION

NEAREST EASTBOUND A&E HOSPITAL

Worcester Royal Infirmary, Ronkswood Branch, Newtown Road, Worcester WR5 1HN
Tel: (01905) 763333

Proceed north to Junction 7 on the M5 and take the A44 exit west. The route is signposted A&E Hospital. (Distance Approx 23 miles)

NEAREST EASTBOUND MINOR INJURY UNIT

Tewkesbury Hospital, Barton Road, Tewkesbury GL20 5QN Tel: (01684) 293303

Proceed north to Junction 1 and follow the A38 south into Tewkesbury. Signposted from within the town. Open 24 hours. (Distance Approx 12.6 miles)

NEAREST WESTBOUND A&E HOSPITAL

Hereford General Hospital, Nelson Street, Hereford HR1 2PA Tel: (01432) 355444

Take the A417 and then the A438 to Hereford. The hospital is signposted within the city. (Distance Approx 18.7 miles)

NEAREST WESTBOUND MINOR INJURY UNIT

Ross Community Hospital, Walton Street, Ross on Wye HR9 5AD Tel: (01989) 562100

Proceed to Junction 4 and take the A449 to Ross. Continue along Broad Street, cross Market Square and go along Walford Street. Turn second left into Walton Street and the hospital is along this road. Open 24 Hours. (Distance Approx 12.2 miles)

M50 JUNCTION 3

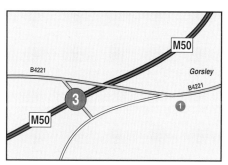

NEAREST EASTBOUND A&E HOSPITAL

Gloucestershire Royal Hospital, Great Western Road, Gloucester GL1 3PQ Tel: (01452) 528555

Take the B4215 south into Gloucester. The hospital is signposted within the town. (Distance Approx 13.5 miles)

NEAREST EASTBOUND MINOR INJURY UNIT

Tewkesbury Hospital, Barton Road, Tewkesbury GL20 5QN Tel: (01684) 293303.

Proceed north to Junction 1 and follow the A38 south into Tewkesbury. Signposted from within the town. Open 24 hours. (Distance Approx 19.5 miles)

NEAREST WESTBOUND A&E HOSPITAL

Hereford General Hospital, Nelson Street, Hereford HR1 2PA Tel: (01432) 355444

Take the B4224 north to Hereford and the hospital is signposted within the city. (Distance Approx 14.4 miles)

NEAREST WESTBOUND MINOR INJURY UNIT

Ross Community Hospital, Walton Street, Ross on Wye HR9 5AD Tel: (01989) 562100

Proceed to Junction 4 and take the A449 to Ross. Continue along Broad Street, cross Market Square and go along Walford Street. Turn second left into Walton Street and the hospital is along this road. Open 24 Hours. (Distance Approx 5.3 miles)

FACILITIES

1   The Roadmaker Inn

Tel: (01989) 720352
0.4 miles south along the B4221, on the right. (Free House) Lunch; Thurs-Sun; 11.30-14.30hrs, Evening Meals; Tues-Sat; 18.30-21.30hrs

M50 JUNCTION 4

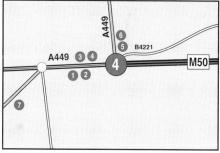

NEAREST A&E HOSPITAL

Hereford General Hospital, Nelson Street, Hereford HR1 2PA Tel: (01432) 355444

Take the A49 north to Hereford and the hospital is signposted within the city. (Distance Approx 16.3 miles)

NEAREST MINOR INJURY UNIT

Ross Community Hospital, Walton Street, Ross on Wye HR9 5AD Tel: (01989) 562100

Take the A449 to Ross, continue along Broad Street, cross Market Square and go along Walford Street. Turn second left into Walton Street and the hospital is along this road. Open 24 Hours. (Distance Approx 1.6 miles)

FACILITIES

1 The Granary Restaurant

Tel: (01989) 565301

0.4 miles south along the A449, on the left. Open; 07.00-22.00hrs daily.

2 BP Ross Spur South

WC 24 Tel: (01989) 565301

0.4 miles south along the A449, on the left. Access, Visa, Overdrive, All Star, Switch, Dial Card, Mastercard, Amex, Diners Club, Delta, Routex, Shell Agency, BP Cards.

3 Red Hen

Tel: (01989) 565301

0.4 miles south along the A449, on the right. Open; 07.00-22.00hrs daily.

4 BP Ross Spur North

WC 24 Tel: (01989) 565028

0.4 miles south along the A449, on the right. Access, Visa, Overdrive, All Star, Switch, Dial Card, Mastercard, Amex, Diners Club, Delta, Routex, Shell Agency, BP Cards.

5 The Travellers Rest

Tel: (01989) 563861

Adjacent to roundabout, on the north side. (Beefeater) Open all day. Breakfast; Mon-Fri; 07.00-09.00hrs, Sat & Sun; 08.00-10.00hrs. Food also served Sun-Thurs; 12.00-22.30hrs, Fri & Sat; 12.00-23.00hrs.

6 Travel Inn

M Tel: (01989) 563861

Adjacent to roundabout, on the north side

7 Broadlands B&B

B Tel: (01989) 563663

1 mile south along the B4234, on the left, in Ross on Wye.

MOTORWAY ENDS

(TOTAL LENGTH OF MOTORWAY 21.5 MILES)

M50

OFF THE MOTORWAY

M62

M62

The M62 Motorway

This motorway forms the strategic east-west link across northern England, connecting the ports of Liverpool, on the west coast and Kingston upon Hull on the east with Manchester, Huddersfield, Bradford and Leeds.

Travelling east from Liverpool, the glass making town of St.Helens and Clock Face are in view on the north side and the cooling towers of Fiddlers Ferry Power Station are on the horizon to the south before the motorway crosses over the site of RAF Burtonwood airfield and continues eastwards, passing Warrington on the south side before forming a junction with the M6.

As the motorway heads towards Manchester it passes the south edge of Chat Moss. This is a huge boggy area that was a major engineering problem for George Stephenson as he constructed the Liverpool & Manchester Railway in the 1820s, taking many tonf of ballast to infill and create a solid base upon which to lay the rails. East of Manchester, the motorway joins the M60 and heads due north, from which point it forms the north west portion of the new Manchester Ring Road and is re-classified as the M60 as it by-passes the city.

Continuing eastwards, as the M62, it passes between the mill towns of Rochdale and Oldham before reaching its highest point at 1221ft, some 188ft higher than Shap Summit on the M6, with Moss Moor in view on the south side and the Rishworth Moor on the north. As the motorway skirts past Booth Wood Reservoir in the valley below, it splits and Stott Hall Farm can be seen between the carriageways, one of the few, if only, instances of a dwelling occupying the central reservation.

During the construction of the M62 the opportunity was taken to construct a new reservoir, utilizing the motorway itself as a dam and, further east, the resultant Scammonden Water can be seen on the south side. As it continues eastwards panoramic views across to Elland, Halifax and the South Pennines can be seen to the north before the motorway passes between Bradford and Dewsbury and then Leeds and Wakefield, linking up with the M1. The motorway continues past Castleford to the north and Pontefract Racecourse to the south before the cooling towers of Ferrybridge Power Station come into view.

The landscape is now in complete contrast to the Pennines over which the motorway has crossed and the cooling towers of Thorpe Marsh, to the south, and Drax Power Stations, to the north are visible in the distance before it crosses over the East Coast Main Line between London (Kings Cross) and Edinburgh. The motorway links up with the M18 before it passes Goole, a port town on the confluence of the Ouse and Don, on the south side and Howden, with the tower of Howden Minster clearly visible, to the north, ending at Junction 38 where it joins end-on with the A63 which connects with Kingston upon Hull.

Location of Places of Interest

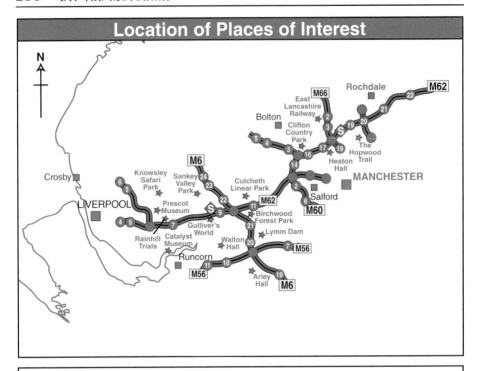

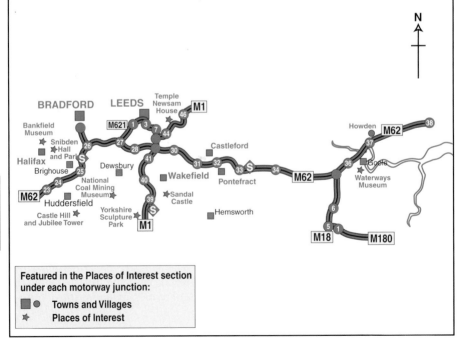

Featured in the Places of Interest section
under each motorway junction:

◼ ● Towns and Villages

☆ Places of Interest

M62

M62 — JUNCTION 4

NEAREST A&E HOSPITAL

The Royal Liverpool University Hospital, Prescot Street, Liverpool L7 8XP Tel: (0151) 706 2000

Follow the A5080 west into Liverpool (Signposted in city. Distance Approx 2.8 miles)

M62 — JUNCTION 5

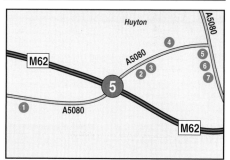

NEAREST EASTBOUND A&E HOSPITAL

Whiston Hospital, Prescot, Merseyside L35 5DL Tel: (0151) 426 1600

Proceed to Junction 6 and take the exit north to Prescot. (Signposted within town. Distance Approx 4.4 miles)

NEAREST WESTBOUND A&E HOSPITAL

The Royal Liverpool University Hospital, Prescot Street, Liverpool L7 8XP Tel: (0151) 706 2000

Proceed to Junction 4 and follow the A5080 west into Liverpool (Signposted in city. Distance Approx 4.1 miles)

FACILITIES

1 Turnpike Tavern

Tel: (0151) 738 2921
1 mile west along the A5080, on the left. (Bass) Open all day. Meals served; Mon-Sat; 12.00-19.00hrs, Sun; 12.00-17.00hrs

2 Derby Lodge

Tel: (0151) 480 4440
0.3 miles east along the A5080 on the right.

(Scottish & Newcastle) Open all day. Meals served; Mon-Fri; 07.00-09.00hrs & 12.00-22.00hrs, Sat & Sun; 07.00-10.00hrs & 12.00-22.00hrs.

3 Premier Lodge

Tel: (0151) 480 4440
0.3 miles east along the A5080 on the right

4 The Stanley Arms

Tel: (0151) 489 1747
0.6 miles east along the A5080 on the left. (Bass) Open all day. Meals served; Mon-Sat; 12.00-21.00hrs, Sun; 12.00-20.00hrs

5 The Crofters

Tel: (0151) 482 4951
0.8 miles east along the A5080 on the right. (Bass) Meals served; 12.00-15.00hrs daily

6 Repsol Service Station

Tel: (0151) 489 6246
0.8 miles east along the A5080 on the right. Access, Visa, Overdrive, All Star, Switch, Dial Card, Mastercard, Amex, Diners Club, Delta, BP Supercharge, Repsol Card. Open; 07.00-23.00hrs

7 Save Petrol Station

Tel: (0151) 481 0188
0.9 miles east along the A5080 on the right. Access, Visa, Overdrive, All Star, Switch, Dial Card, Mastercard, Amex, Diners Club, Delta, Save Card. Open; 06.00-23.00hrs

M62 — JUNCTION 6

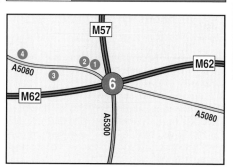

M62

| Pets Welcome | £ Cash Dispenser | 24 24 Hour Facilities | WC Toilets | Alcoholic Drinks | Food and Drink | Fuel |

NEAREST A&E HOSPITAL

**Whiston Hospital, Prescot, Merseyside L35 5DL
Tel: (0151) 426 1600**

Take the M57 exit north to Prescot and follow the A57 east. (Signposted within town. Distance Approx 1.8 miles)

FACILITIES

1 Travel Inn

[=M] [&] Tel: (0151) 480 9614
0.2 miles west along the A5080, on the right

2 Chapel Brook

[icons] Tel: (0151) 480 9614
0.2 miles west along the A5080, on the right. (Whitbread) Open all day. Meals served; 11.30-22.00hrs daily

3 Woodlands Service Station (Esso)

[icons] Tel: (0151) 480 6284
0.3 miles west along the A5080, on the left. Access, Visa, Overdrive, All Star, Switch, Dial Card, Mastercard, Amex, Diners Club, Delta, BP Supercharge, Shell Gold, Esso Cards

4 Hare & Hounds

[icons] Tel: (0151) 489 3046
0.5 miles west along the A5080, on the right. (Scottish & Newcastle) Open all day. Meals served; Mon-Sat; 12.00-14.30hrs & 17.30-20.30hrs, Sun; 12.00-20.00hrs

PLACES OF INTEREST

Knowsley Safari Park

Prescot, Merseyside L34 4AN Tel: (0151) 430 9009
website; www.knowsley.com
e-mail; safari.park@knowsley.com
Follow the M57 north and take the A58 east at Junction 2. (Signposted from junction 3.7 Miles)

A huge variety of animals are in view along the 5-mile long safari drive. Other attractions include a miniature railway, Lake farm, Seal & Parrot Shows, Reptile house, amusement park and picnic and play areas. Gift Shop. Restaurant.

The Prescot Museum

34 Church Street, Prescot, Merseyside L34 3LA
Tel: (0151) 430 7787 Follow the M57 north and take the A57 east at Junction 2. (2.6 Miles)

By 1800 Prescot had become the principal UK manufacturing base for the production of watch movements, watch and clock components and horological and precision tools, but by the end of the century the mass production of watches in the USA and the improvements made to the Swiss cottage industry of watch manufacture all but wiped it out. Not just a museum for the study of Prescot's local history, it is also a centre of national importance for the study, interpretation and display of the watchmaking craft for which the area was once famous. The exhibits include a reconstruction of a watchmakers workshop and watchmaking factory.

M62 | JUNCTION 7

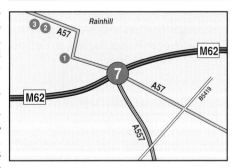

NEAREST EASTBOUND A&E HOSPITAL

Warrington Hospital, Lovely Lane, Warrington WA5 1QG Tel: (01925) 635911

Proceed to Junction 9 and follow the A49 south, at the roundabout junction of the A49, A570 and the A50 turn right along the A570 and proceed through the town centre towards Prescot. At the junction with the B5210, the first major roundabout, turn right into Lovely Lane and the hospital is on the right. (Distance Approx 8.9 miles)

NEAREST WESTBOUND A&E HOSPITAL

**Whiston Hospital, Prescot, Merseyside L35 5DL
Tel: (0151) 426 1600**

Follow the A57 north to Prescot. (Signposted within town. Distance Approx 3.1 miles)

| | Large Hotel | | Small Hotel | | Motel | | Guest House/ Bed & Breakfast | | Disabled Facilities | | Childrens Facilities |

FACILITIES

1 Stoops Filling Station (Shell)

🛗 ♿ WC 24   Tel: (0151) 426 4199
0.2 miles north along the A57, on the left.
Access, Visa, Overdrive, All Star, Switch, Dial
Card, Mastercard, Amex, Diners Club, Delta,
BP Supercharge, Shell Cards, Smartcard.

2 The Ship

🛏 🍴 ♿ 🍺   Tel: (0151) 426 4165
0.4 miles north along the A57, on the left.
(Scottish & Newcastle) Open all day. Meals
served; 12.00-22.00hrs daily

3 Premier Lodge

🏨♿   Tel: (0151) 426 4165
0.4 miles north along the A57, on the left

PLACES OF INTEREST

The Rainhill Trials Exhibition

The Library, View Road, Rainhill L35 0LE
Tel: (0151) 426 4269. Follow the A57 north into
Rainhill and View Road is on the left. (1.2 Miles)

In 1829 the directors of the Liverpool &
Manchester Railway held a competition at
Rainhill to decide whether their railway would
be powered by the preferred method of
stationary steam winding engines and cables
or by a steam locomotive capable of hauling a
load 3 times its own weight for a total distance
of 70 miles at 10miles/hr. Up to this point
steam locomotives had been crude and
unreliable but Stephenson's "Rocket" proved
to be more than adequate for the task and, once
and for all, eliminated all doubts that they
could provide the motive power on railways.
The Rainhill Trials Exhibition, located in a
railway carriage adjacent to the library,
contains many fascinating artefacts associated
with this epoch making event that shaped
transport across the world for the next 100
years.

M62 | JUNCTION 8

THERE IS NO JUNCTION 8

SERVICE AREA

BETWEEN JUNCTIONS 7 AND 9

Burtonwood (Eastbound) (Welcome Break)
Tel: (01925) 651656 Granary Restaurant, Welcome
Lodge & Shell Fuel

Burtonwood (Westbound) (Welcome Break)
Tel: (01925) 651656 Red Hen Restaurant & Shell
Fuel

Footway Tunnel connection between sites.

M62 | JUNCTION 9

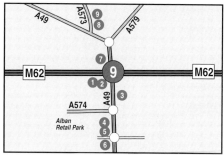

NEAREST A&E HOSPITAL

**Warrington Hospital, Lovely Lane, Warrington
WA5 1QG Tel: (01925) 635911**

Follow the A49 south and at the roundabout
junction of the A49, A570 and the A50 turn right
along the A570 and proceed through the town
centre towards Prescot. At the junction with the
B5210, the first major roundabout, turn right
into Lovely Lane and the hospital is on the right.
(Distance Approx 2.6 miles)

FACILITIES

1 Winwick Quay

🛏 🍴 ♿ 🍺   Tel: (01925) 414417
0.1 miles south along the A49, on the right.
(Whitbread) Open all day. Meals served; 11.00-
22.00hrs daily

M62

| Pets Welcome | £ Cash Dispenser | 24 24 Hour Facilities | WC Toilets | Alcoholic Drinks | Food and Drink | Fuel |

2 Travel Inn

⊨M 🔲   Tel: (01925) 414417:
0.1 miles south along the A49, on the right

3 Winwick Service Station (Texaco)

🔲 WC   Tel: (01925) 638240
0.2 miles south along the A49, on the left.
Access, Visa, Overdrive, All Star, Switch, Dial
Card, Mastercard, Amex, Delta, Shell Agency,
Fast Fuel, Texaco Cards. Open; Mon-Fri; 06.00-
22.00hrs, Sat; 08.00-22.00hrs, Sun; 08.00-
21.00hrs.

4 KFC

🔲 🔲   Tel: (01925) 419786
0.9 miles south along the A49, on the right in
Alban Retail Park. Open; 11.00-23.00hrs daily

5 Pizza Hut

🔲 🔲 🔲   Tel: (01925) 574220
0.9 miles south along the A49, on the right in
Alban Retail Park. Open; 12.00-23.00hrs daily

6 BP Longford

🔲 24 £   Tel: (01925) 633983
1 mile south along the A49, on the right.
Access, Visa, Overdrive, All Star, Switch, Dial
Card, Mastercard, Amex, AA Paytrak, Diners
Club, Delta, Routex, Shell Agency, BP Cards

7 Burger King

🔲 🔲   Tel: (01925) 573387
0.1 miles north along the A49, on the left.
Open; 10.00-22.00hrs daily. (Drive-Thru open
until; Sun-Thurs; 23.00hrs, Sat & Sun; 0.00hrs)

8 Lodge Inn

⊨M 🔲   Tel: (01925) 631416
0.6 miles north along the A573, on the right

9 The Swan

🔲 🔲 🔲 🔲   Tel: (01925) 631416
0.6 miles north along the A573, on the right.
(Scottish & Newcastle) Open all day. Meals
served; Mon-Sat; 12.00-22.00hrs, Sun; 12.00-
21.00hrs

PLACES OF INTEREST

Gullivers World

Warrington, Cheshire WA5 5YZ
(Signposted from Junction 2.9 Miles)

For details please see Junction 22 (M6)
information.

Sankey Valley Park

Bewsey Old Hall, Bewsey Farm Close, Old Hall,
Warrington WA5 5PB Tel: (01925) 571836
Follow the A49 south and turn right along the A574
(Signposted Bewsey 2.7 Miles)

Centred around Sankey Canal, opened in 1757
and claimed to be the country's first true canal,
Sankey Valley Park is part of the Mersey Forest.
Within the park, mosaics of woodlands,
grasslands, and ponds allow a variety of
wildlife, insects and flowers to flourish. There
are excellent walks, including the Sankey Canal
Trail, a childrens play area and Bewsey Old Hall,
the family seat of the Botelers, Lords of
Warrington, is open to tours. Disabled access.

M62   JUNCTION 10

**THIS JUNCTION IS A MOTORWAY
INTERCHANGE ONLY WITH THE M6 AND
THERE IS NO ACCESS TO ANY
FACILITIES**

NEAREST WESTBOUND A&E HOSPITAL

**Warrington Hospital, Lovely Lane, Warrington
WA5 1QG Tel: (01925) 635911**

Proceed to Junction 9 and follow the A49 south,
at the roundabout junction of the A49, A570 and
the A50 turn right along the A570 and proceed
through the town centre towards Prescot. At the
junction with the B5210, the first major
roundabout, turn right into Lovely Lane and the
hospital is on the right. (Distance Approx 4.3
miles)

NEAREST EASTBOUND A&E HOSPITAL

**Hope Hospital, Stott Lane, Salford M6 8HB
Tel: (0161) 789 7373**

Proceed to Junction 12 and follow the M602 east
(Signposted within city. Distance Approx 11.1
miles)

| ⊨H Large Hotel | ⊨h Small Hotel | ⊨M Motel | ⊨B Guest House/ Bed & Breakfast | 🔲 Disabled Facilities | 🔲 Childrens Facilities |

M62 JUNCTION 11

THERE ARE NO FACILITIES WITHIN ONE MILE OF THIS JUNCTION

NEAREST WESTBOUND A&E HOSPITAL

Warrington Hospital, Lovely Lane, Warrington WA5 1QG Tel: (01925) 635911

Proceed to Junction 9 and follow the A49 south, at the roundabout junction of the A49, A570 and the A50 turn right along the A570 and proceed through the town centre towards Prescot. At the junction with the B5210, the first major roundabout, turn right into Lovely Lane and the hospital is on the right. (Distance Approx 6.8 miles)

NEAREST EASTBOUND A&E HOSPITAL

Hope Hospital, Stott Lane, Salford M6 8HB Tel: (0161) 789 7373

Proceed to Junction 12 and follow the M602 east (Signposted within city. Distance 9.2 miles)

PLACES OF INTEREST

Culcheth Linear Park

Wigshaw Lane, Culcheth, Warrington WA3 4AB Tel: (01925) 765064. Follow the A574 south and then right towards Culcheth. (4 Miles)

Following part of the trackbed of the former Cheshire Lines Committee railway line between Wigan and Glazebrook, the Culcheth Linear Park offers a delightful and pleasant walk. Picnic Areas. Disabled access.

Birchwood Forest Park & Risley Moss

Ordnance Avenue, Birchwood, Warrington WA3 6QX Bichwood Forest Park Tel: (01925) 824239, Risley Moss Tel: (01925) 824339 Follow the A574 south and turn left at the first roundabout along Moss Gate. (1 Mile)

Birchwood Forest Park is the name given to all the green areas throughout Birchwood. The town is made up of three villages; Gorse Covert, Oakwood and Locking Stumps and these and the surrounding countryside are linked by a comprehensive network of footpaths creating a 550 acre parkland. There are many pleasant walks which include beautiful parkland, secluded ponds, mosses and birch woodland,

providing ideal habitats for wildlife. Playgrounds. Visitor Centre. Disabled access.

M62 JUNCTION 12

THIS JUNCTION IS A MOTORWAY INTERCHANGE WITH THE M63 AND M602 ONLY AND THERE IS NO ACCESS TO ANY FACILITIES

NEAREST A&E HOSPITAL

Hope Hospital, Stott Lane, Salford M6 8HB Tel: (0161) 789 7373

Follow the M602 east (Signposted within city. Distance 2.5 miles)

M60 JUNCTION 13

BETWEEN JUNCTIONS 13 AND 18 THE M62 MOTORWAY HAS BEEN RECLASSIFIED AS THE M60

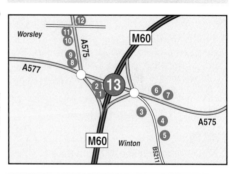

NEAREST A&E HOSPITAL

Hope Hospital, Stott Lane, Salford M6 8HB Tel: (0161) 789 7373

Proceed to Junction 12 and follow the M602 east (Signposted within city. Distance 3.5 miles)

FACILITIES

1   The John Gilbert

                     Tel: (0161) 703 7733

Adjacent to south side of west roundabout.

| Pets Welcome | Cash Dispenser | 24 Hour Facilities | Toilets | Alcoholic Drinks | Food and Drink | Fuel |

(Watling Streetings) Open all day. Meals served; 12.00-21.00hrs

2 Novotel Manchester West Hotel

Tel: (0161) 799 3535
Adjacent to south side of west roundabout. The Garden Brasserie; Open; 06.00-0.00hrs daily

3 The Bridgewater Hotel

Tel: (0161) 794 0589
0.3 miles south along the B5211, on the right. (Scottish & Newcastle) Open all day. Meals served; Mon-Thurs; 12.00-15.00hrs & 18.00-21.00hrs, Fri & Sat; 12.00-15.00hrs & 18.00-20.00hrs, Sun; 12.00-18.00hrs

4 The Barton Arms

Tel: (0161) 794 9373
0.5 miles south along the B5211, on the left. (Bass) Open all day. Meals served; Mon-Sat; 12.00-20.00hrs, Sun; 12.00-19.00hrs

5 Alder Service Station (Texaco)

Tel: (0161) 789 4665
0.8 miles south along the B5211, on the left. Access, Visa, Overdrive, All Star, Switch, Dial Card, Mastercard, Amex, Diners Club, Delta, BP Supercharge, Fastfuel, Texaco cards. Open; 07.00-22.00hrs daily

6 Tung Fong Chinese Restaurant

Tel: (0161) 794 5331
0.2 miles east along the A572, on the left. Open; Mon-Fri; 12.00-14.00hrs & 17.30-23.00hrs, Sat & Sun; 17.30-23.00hrs

7 Cafe Bar Rioja Tapas Bar

Tel: (0161) 793 6003
0.2 miles east along the A572, on the left. Open; 12.00-15.00hrs & 17.30-23.00hrs daily

8 Worsley Old Hall

Tel: (0161) 799 2960
0.3 miles north along the A575, on the left. (Brewers Fayre) Open all day. Meals served; Mon-Sat; 11.30-22.00hrs, Sun; 12.00-22.00hrs

9 Marriott Manchester Hotel & Country Club

Tel: (0161) 975 2000
0.3 miles north along the A575, on the left. Brindley's Restaurant; Breakfast; Mon-Thurs; 06.30-10.30hrs, Fri-Sun; 07.00-11.00hrs, Lunch; Sun-Fri; 12.00-14.00hrs, Dinner; 19.00-22.00hrs

10 The Cock

Tel: (0161) 790 2381
0.8 miles north along the A575, on the left. (Scottish & Newcastle) Open all day. Meals served; 12.00-22.00hrs daily

11 Worsley Service Station (Esso)

Tel: (0161) 702 0370
0.9 miles north along the A575, on the left. Access, Visa, Overdrive, All Star, Switch, Dial Card, Mastercard, Amex, AA Paytrak, Diners Club, Delta, Shell Gold, Esso Cards

12 The Willows

Tel: (0161) 790 4951
1 mile north along the A575, on the right

M60 JUNCTION 14

THIS IS A RESTRICTED ACCESS JUNCTION AND INTERCHANGE WITH THE M61

Vehicles can only exit from the southbound lanes and travel west along the A580

Vehicles can only enter the motorway along the northbound lanes from the A580 east carriageway

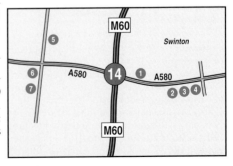

NEAREST A&E HOSPITAL

Hope Hospital, Stott Lane, Salford M6 8HB
Tel: (0161) 789 7373
Proceed to Junction 12 and follow the M602 east
(Signposted within city. Distance 4.5 miles)

FACILITIES

1 XS Superbowl

🚹 🍴 ♿ Tel: (0161) 794 3374
0.2 miles east along the A580, on the left.
Within the building: Wimpy Restaurant; Open
10.00-0.00hrs daily & Bridgewater Bar; Open
during licensing hours.

2 Deansbrook Service Station (Total)

🍴 24 Tel: (0161) 728 8800
0.9 miles east along the A580, on the right.
Access, Visa, Overdrive, All Star, Switch, Dial
Card, Mastercard, Amex, AA Paytrak, Diners
Club, Delta, BP Supercharge, Elf Card, Total
Fina Card.

3 The New Ellesmere

🛏 🍴 ♿ 🍷 Tel: (0161) 728 2791
1 mile east along the A580, on the right.
(Scottish & Newcastle) Open all day. Meals
served; 07.30-09.30 & 12.00-21.30hrs daily

4 Premier Lodge

🛏M ♿ Tel: (0161) 728 2791
1 mile east along the A580, on the right.

5 The Willows

🍺B 🍴 ♿ Tel: (0161) 790 4951
1 mile north along the A575, on the right

6 Worsley Service Station (Esso)

🍴 ♿ WC £ 24 Tel: (0161) 702 0370
0.9 miles south along the A575, on the right.
Access, Visa, Overdrive, All Star, Switch, Dial
Card, Mastercard, Amex, AA Paytrak, Diners
Club, Delta, Shell Gold, Esso Cards

7 The Cock

🛏 🍴 ♿ 🍷 Tel: (0161) 790 2381
1 mile south along the A575, on the right.
(Scottish & Newcastle) Open all day. Meals
served; 12.00-22.00hrs daily

M60 JUNCTION 15

**JUNCTION 15 [M60] IS A MOTORWAY
INTERCHANGE ONLY WITH THE M61 AND
THERE IS NO ACCESS TO ANY
FACILITIES**

M60 JUNCTION 16

**THIS IS A RESTRICTED ACCESS
JUNCTION**

Vehicles can only exit from the southbound lanes.
Vehicles can only enter the motorway along the
northbound lanes.

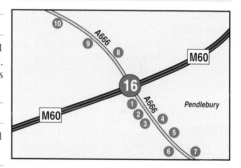

NEAREST WESTBOUND A&E HOSPITAL

Hope Hospital, Stott Lane, Salford M6 8HB
Tel: (0161) 789 7373
Proceed to Junction 12 and follow the M602 east
(Signposted within city. Distance 6.3 miles)

NEAREST EASTBOUND A&E HOSPITAL

**North Manchester General Hospital, Delauneys
Road, Crumpsall, Manchester M8 5RB**
Tel: (0161) 795 4567
Proceed to Junction 17, follow the A56 south and
turn left along the A576 and right along the
A6010 (Signposted along this route. Distance
Approx 7.5 miles)

M62

	Pets Welcome	£	Cash Dispenser	24	24 Hour Facilities	WC	Toilets		Alcoholic Drinks	Food and Drink

 Fuel

FACILITIES

1 The Robin Hood

Tel: (0161) 794 2906
Adjacent to south side of motorway junction. (Punch Taverns) Open all day. Meals served; 12.00-21.45 hrs daily

2 Golden Lion

Tel: (0161) 794 3016
0.1 miles south along the A666, on the right. (Whitbread) Open all day. Meals served; Mon-Sat; 11.30-22.00hrs, Sun; 12.00-22.00hrs

3 Oddfellows Arms

Tel: (0161) 794 5691
0.4 miles south along the A666, on the right. (Freehouse) Open all day Fri-Sun. Meals served; Mon-Fri 12.00-14.00hrs. NB Accommodation is Bed & Evening Meal.

4 Pendle Hill Service Station (Esso)

Tel: (0161) 281 7989
0.7 miles south along the A666, on the left. Access, Visa, Overdrive, All Star, Switch, Dial Card, Mastercard, Amex, AA Paytrak, Diners Club, Delta, Shell Gold, Esso Cards

5 Dilash Balti House

Tel: (0161) 728 5333
0.8 miles south along the A666, on the left. Open; Sun-Thurs; 17.00-0.00hrs, Fri/Sat & Sat/Sun; 17.00-01.00hrs

6 Royal Oak

Tel: (0161) 794 4517
0.9 miles south along the A666, on the right. (Freehouse) Open all day. Bar meals served Mon-Sat; 12.00-16.00hrs, Sun Lunch; 12.00-17.30hrs.

7 McDonald's

Tel: (0161) 794 7415
1 mile south along the A666, on the left. Open; Mon-Thurs; 07.00-23.00hrs, Fri-Sun; 07.00-0.00hrs

8 Clifton Park Hotel

Tel: (0161) 794 3761
0.4 miles north along the A666, on the right. Restaurant Open; Mon-Sat; 18.00-23.00hrs, Sun; 15.00-17.00hrs & 18.00-23.00hrs

9 Save Service Station

Tel: (0161) 794 4063
0.8 miles north along the A666, on the left. Access, Visa, Overdrive, All Star, Switch, Dial Card, Mastercard, Delta, Save Card. Open: 07.00-23.00hrs daily.

10 Unity Brook

Tel: (01204) 797831
1 mile north along the A666, on the left. (Bass) Open all day. Meals served; Mon-Sat; 12.00-19.45hrs, Sun; 12.00-16.45hrs

PLACES OF INTEREST

Clifton Country Park

Clifton House Road, Swinton M27 6NG Tel: (0161) 793 4219. Follow the A666 north (0.7 Miles)

Set within 80 acres of countryside, this beautiful park features a wide variety of different wildlife habitats; woodland, open grassland, wetland areas, a pond and a lake. Gift Shop.

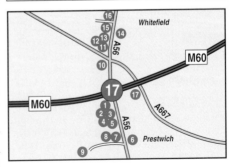

M60   JUNCTION 17

M62

NEAREST A&E HOSPITAL

North Manchester General Hospital, Delauneys Road, Crumpsall, Manchester M8 5RB
Tel: (0161) 795 4567
Follow the A56 south and turn left along the A576 and right along the A6010 (Signposted along this route. Distance Approx 5.3 miles)

FACILITIES

1 Paddock Filling Station (BP)

Tel: (0161) 773 9081

0.1 miles south along the A56, on the right. Keyfuels, Access, Visa, Overdrive, All Star, Switch, Dial Card, Mastercard, Amex, AA Paytrak, Diners Club, Delta, Routex, Shell Agency, BP Cards. Open; 06.00-23.00hrs daily

2 Travel Inn

Tel: (0161) 798 0827

0.1 miles south along the A56, on the right

3 TGI Fridays

Tel: (0161) 798 7125

0.1 miles south along the A56, on the right. Open; Mon-Fri; 11.00-23.00hrs, Sat & Sun; 11.30-23.30hrs

4 Tesco Petrol

Tel: (0161) 910 9400

0.2 miles south along the A56, on the right. Access, Visa, Overdrive, All Star, Switch, Dial Card, Mastercard, Amex, Delta, BP Supercharge, Tesco Fuelcard. Toilets and cash machine available in adjacent store (Open 24 hours)

5 Grapes of Prestwich

Tel: (0161) 773 2570

0.4 miles south along the A56, on the right. (Inn Partnership) Open all day. Meals served; 12.00-14.30hrs daily

6 Red Lion

Tel: (0161) 773 6163

0.7 miles south along the A56, on the left. (Joseph Holt) Open all day. Meals served; 12.00-14.00hrs daily

7 Royal Bengal Tandoori Restaurant

Tel: (0161) 773 6311

0.7 miles south along the A56, on the right. Open; 18.00-01.00hrs daily

8 The Cottage Cafe

Tel: (0161) 773 6022

0.7 miles south along the A56, on the right, in Church Lane. Open; Mon-Tues & Thurs-Fri; 08.00-15.00hrs, Wed & Sat; 08.00-13.30hrs

9 The Church

Tel: (0161) 798 6727

0.7 mile south along the A56, on the right in Church Lane. (John Smiths) Open all day. Meals served; Mon-Fri; 12.00-14.30hrs, Sat & Sun; 12.00-16.00hrs

10 McDonald's

Tel: (0161) 767 9731

0.3 miles north along the A56, on the left. Open; 07.30-23.00hrs (Drive-Thru until 0.00hrs)

11 Khan Saab

Tel: (0161) 766 2148

0.4 miles north along the A56, on the left. Open; Sun-Thurs; 18.00-23.30hrs, Fri & Sat; 18.00-0.00hrs

12 Akrams

Tel: (0161) 796 0403

0.5 miles north along the A56, on the left. Open; Mon-Thurs; 18.00-23.30hrs, Fri & Sat; 18.00-0.00hrs, Sun; 12.00-23.00hrs

13 The Masons

Tel: (0161) 766 2713

0.7 miles north along the A56, on the left. (Whitbread) Open all day. Meals served; 12.00-15.30hrs daily

14 Roma Coffee Lounge & Restaurant

Tel: (0161) 766 2941

0.8 miles north along the A56, on the right. Open; Mon-Fri; 09.15-18.45hrs, Sat; 09.00-16.50hrs

M62

	Pets Welcome	£	Cash Dispenser	24	24 Hour Facilities	WC	Toilets		Alcoholic Drinks		Food and Drink		Fuel

15 Bulls Head

🎿 🍽 ♿ 📶 Tel: (0161) 766 5968

0.9 miles north along the A56, on the left. (Whitbread) Meals served; Mon-Fri; 11.30-14.00hrs

16 Total Whitefield

🍴 WC Tel: (0161) 767 9624

1 mile north along the A56, on the left. Access, Visa, Overdrive, All Star, Switch, Dial Card, Mastercard, Amex, Diners Club, Delta, BP Supercharge, Total Fina Card. Open; Mon-Sat; 07.00-23.00hrs, Sun; 08.00-22.00hrs

17 Kirkhams Service Station (Esso)

🍴 ♿ WC 24 £ Tel: (0161) 773 2486

0.7 miles east along the A667, on the right. Access, Visa, Overdrive, All Star, Switch, Dial Card, Mastercard, Amex, Diners Club, Delta, Shell Gold, Esso Cards

PLACES OF INTEREST

Heaton Hall

Heaton Park, Prestwich, Manchester M25 5SW
Tel: (0161) 773 1231
Follow the A56 south and turn left along Scholes Lane (A6044) (Signposted 2 miles)

Built by James Wyatt in 1722 and set in a large parkland, it was a former residence of the Earls of Wilton. The Hall contains one of the few remaining Etruscan rooms, with painted walls and ceiling by Biagio Rebecca. Other contents include an organ built by Samuel Green in 1790, 17thC Dutch paintings, furniture and paintings of the 18thC and the Assheton Bennett Collection of English silver. Cafe. Disabled access.

East Lancs Railway

Bolton Street Station, Bury BL9 0EY
Tel: (0161) 764 7790
website; www.east-lancs-rly.co.uk
Follow the A56 north to Bury (4.1 miles)

An 8 mile steam railway linking Bury and Rawtenstall via the pretty Irwell Valley. Opens at weekends and Bank Holidays. Cafe, Gift Shop, Disabled access.

M62   JUNCTION 18

THIS JUNCTION IS A MOTORWAY INTERCHANGE ONLY WITH THE M66 NORTH AND M60 SOUTH, AND THERE IS NO ACCESS TO ANY FACILITIES

NEAREST A&E HOSPITAL

North Manchester General Hospital, Delauneys Road, Crumpsall, Manchester M8 5RB
Tel: (0161) 795 4567

Follow the M60 south to Junction 19, take the A576 exit west and turn left along Blackley New Road (Signposted along this route. Distance Approx 4.1 miles)

SERVICE AREA

BETWEEN JUNCTIONS 18 AND 19

Birch Services (Westbound) (Granada)

Tel: (0161) 643 0911 Self Service Restaurant, Burger King, Travelodge & Esso Fuel

Birch Services (Eastbound) (Granada)

Tel: (0161) 643 0911 Self Service Restaurant, Burger King, Travelodge & Esso Fuel

Footbridge Connection between sites

M62   JUNCTION 19

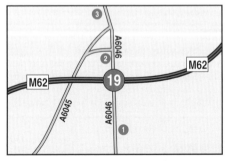

NEAREST WESTBOUND A&E HOSPITAL

Royal Oldham Hospital, Rochdale Road, Oldham OL1 2JH Tel: (0161) 624 0420

Proceed to Junction 20, follow the A627 (M) south and turn left along the A6048. (Distance Approx 5.8 Miles)

M62 (vertical, left margin)

| Large Hotel | Small Hotel | Motel | Guest House/ Bed & Breakfast | Disabled Facilities | Childrens Facilities |

NEAREST EASTBOUND A&E HOSPITAL

North Manchester General Hospital,
Delauneys Road, Crumpsall,
Manchester M8 5RB Tel: (0161) 795 4567

Proceed to Junction 18, follow the M66 south,
take the A576 exit west and turn left along
Blackley New Road (Signposted along this route.
Distance Approx 6.9 miles)

FACILITIES

1 Hollin Service Station (BP)

🖷 24   Tel: (0161) 653 0709

0.8 miles south along the A6046, on the left.
Access, Visa, Overdrive, All Star, Switch, Dial
Card, Mastercard, Amex, AA Paytrak, Diners
Club, Delta, Routex, Shell Agency, BP Cards

2 Hopwood Service Station (Texaco)

🖷 WC 24   Tel: (01706) 692100

0.4 miles north along the A6046, on the left.
Access, Visa, Overdrive, All Star, Switch,
Mastercard, Amex, Diners Club, Delta, Fast
Fuel, Texaco Cards.

3 Starkey Arms

🍴 🍺 ♿   Tel: (01706) 622301

0.9 miles north along the A6046, on the left.
(Joseph Holt) Open all day. Meals served 12.00-
22.00hrs daily

PLACES OF INTEREST

The Hopwood Trail

Rochdale Road, Middleton M24 2GL
Tel: (01706) 356592
Follow the A6046 south into Middleton and turn left
along the A664 (2.5 Miles)

Following a circular route of nearly 2 miles
through the former Hopwood Estate, the trail
consists of 12 easily identified points and passes
through oak and birch woodlands, ancient clay
pits and coal mining areas.

M62   JUNCTION 20

THIS JUNCTION IS A MOTORWAY INTERCHANGE ONLY WITH THE A627 (M) AND THERE IS NO ACCESS TO ANY FACILITIES

NEAREST A&E HOSPITAL

Royal Oldham Hospital, Rochdale Road,
Oldham OL1 2JH Tel: (0161) 624 0420

Proceed to Junction 20, follow the A627 (M)
south and turn left along the A6048. (Distance
Approx 3.6 Miles)

PLACES OF INTEREST

Rochdale

Rochdale Tourist Information Cente, The Clock
Tower, Rochdale OL16 1AB. Tel: (01706) 356592.
Follow the A627 (M) north (Signposted 2.4 Miles)

Lying in a shallow valley formed by the little
River Roch, the town is surrounded to the
north and east by the Pennines and, with its
origins in mediaeval times, it expanded with
the booming cotton industry. Rochdale was the
birthplace of the Co-operative Movement and
the first Co-op shop, opened on December 21st,
1844 and sited in Toad Lane, still exists as the
Rochdale Pioneers Museum (Tel: 01706-
524920). The town has some famous sons and
daughters and these include John Bright, the
renowned 19thC political thinker and Gracie
Fields, the celebrated singer.

M62   JUNCTION 21

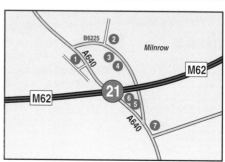

NEAREST A&E HOSPITAL

Rochdale Infirmary, Whitehall Street, Rochdale
OL12 0NB Tel: (01706) 377777

Follow the A640 north and turn right along the
A671.(Signposted within town. Distance Approx
3.3 Miles)

FACILITIES

---

1   Poachers Pocket, Milnrow

Tel: (01706) 355402
0.3 miles north along the A640, on the left, in
Harbour Lane. (Banks's) Open all day. Meals
served 12.00-22.00hrs daily

---

2   Tim Bobbin

Tel: (01706) 658992
0.6 miles east along the B6225, on the left, in
Milnrow. (Unique Pub Co) Open all day. Meals
served Mon-Sat; 12.00-19.00hrs, Sun; 12.00-
15.00hrs

---

3   Milnrow Balti Restaurant

Tel: (01706) 353651
0.7 miles east, on the right in Milnrow. Open;
Mon-Sat; 17.00-0.00hrs, Sun; 15.00-20.30hrs

---

4   BP Milnrow

Tel: (01706) 641132
1 mile east, on the right in Milnrow. Access,
Visa, Overdrive, All Star, Switch, Dial Card,
Mastercard, Amex, AA Paytrak, Diners Club,
Routex, Shell Agency, BP Cards. Open; Mon-
Sat; 06.30-22.15hrs, Sun; 07.30-22.15hrs.

---

5   The John Milne

Tel: (01706) 299999
0.4 miles north along the (B6225), on the left
in Milnrow. (Brewers Fayre) Open all day. Meals
served; Mon-Fri; 07.00-09.00hrs & 11.30-
22.00hrs, Sat; 08.00-10.00hrs & 11.30-22.00hrs,
Sun; 08.00-10.00hrs & 12.00-22.00hrs.

---

6   Travel Inn

Tel: (01706) 299999
0.4 miles north along the (B6225), on the left
in Milnrow

---

7   Waggon & Horses

Tel: (01706) 844248
0.5 miles south along the A640, on the left (JW
Lees)

PLACES OF INTEREST

Rochdale

Follow the A640 north (Signposted 3.1 Miles)
For details please see Junction 20 information

M62   JUNCTION 22

THERE ARE NO FACILITIES WITHIN ONE
MILE OF THIS JUNCTION

NEAREST WESTBOUND A&E HOSPITAL

Rochdale Infirmary, Whitehall Street, Rochdale
OL12 0NB Tel: (01706) 377777

Proceed to Junction 21, follow the A640 north
and turn right along the A671.(Signposted within
town. Distance Approx 7.8 Miles)

NEAREST EASTBOUND A&E HOSPITAL

Huddersfield Royal Infirmary, Acre Street,
Lindley, Huddersfield HD3 3EA Tel: (01484)
422191

Proceed to Junction 23 and follow the A640 east
(Signposted. Distance Approx 9 miles)

M62   JUNCTION 23

THIS IS A RESTRICTED ACCESS
JUNCTION

Vehicles can only exit from the eastbound lanes.
Vehicles can only enter the motorway along the
westbound lanes.

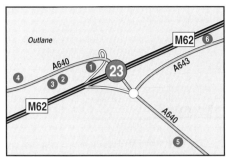

FACILITIES

1   The Swan Inn

Tel: (01422) 374419
0.3 miles west along the A640, on the left, in
Outlane (Enterprise Inns) Meals served daily

2   The Old Golf House Hotel

Tel: (01422) 379311
0.5 miles west along the A640, on the left, in
Outlane. The Grangemoor Restaurant; Open;
12.00-14.00hrs & 19.00-21.30hrs

3   Waggon & Horses

0.6 miles west along the A640, on the left, in
Outlane. Meals served daily

4   The Highlander

Tel: (01422) 370711
1 mile west along the A640, on the right, in
Outlane. (Freehouse) Meals served; Tues-Sun;
12.00-14.00hrs & 18.00-21.00hrs

5   Salendine Service Station (Fina)

Tel: (01484) 460267
0.9 miles south along the A640, on the right.
Access, Visa, Overdrive, All Star, Switch, Dial
Card, Mastercard, Amex, AA Paytrak, Diners
Club, Delta, Elf Card, Total Fina Cards.

6   The Wappy Spring Inn

Tel: (01422) 372324
0.8 miles east along the A643, on the left.
(Tetley) Meals served; Tues-Sat; 12.00-13.30hrs
& 18.30-21.20hrs, Sun; 16.30-20.00hrs.

PLACES OF INTEREST

Castle Hill & Jubilee Tower

Off Lumb Lane, Almondbury,
Huddersfield HD4 6SZ Tel: (01484) 223830
Follow the A640 east, turn right along the A62 and
right along the A616 (6 Miles)

Considered to be one of the most important
archaeological sites in Yorkshire, the hill, a high
moorland ridge overlooking the Colne and
Holme Valleys, has been occupied as a place of
defence since c20,000BC, by what are believed
to be Neolithic herdsmen from mainland
Europe, and the magnificent ramparts of an
Iron Age Fort, built in 600BC and later
destroyed by fire, can still be seen here. In 1147
the Normans restored the earthworks, building
a motte and bailey castle and the hill was used
as a beacon during the times of the Armada
and the Napoleonic Wars. In 1897 the Jubilee
Tower was built to commemorate the 60th
anniversary of Queen Victoria's reign and, at a
height of 1000ft above sea level, the top of the
tower affords splendid panoramic views. An
exhibition tracing the hill's history is contained
within the tower.

M62   JUNCTION 24

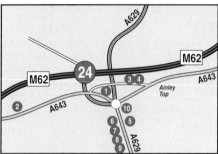

NEAREST A&E HOSPITAL

**Huddersfield Royal Infirmary, Acre Street,
Lindley, Huddersfield HD3 3EA
Tel: (01484) 422191**

Follow the A629 south (Signposted within town.
Distance Approx 1.7 miles)

FACILITIES

1   Hilton National, Huddersfield

Tel: (01422) 375431
0.1 miles west along the A643, on the right.
Restaurant Open; Breakfast; Mon-Fri; 07.00-
10.00hrs, Sat & Sun; 07.30-10.0hrs, Lunch; Sun-
Fri; 12.30-14.00hrs, Dinner; 19.00-22.00hrs
daily.

2   The Wappy Spring Inn

Tel: (01422) 372324
1 mile west along the A643, on the right.
(Tetley) Meals served; Tues-Sat; 12.00-13.30hrs
& 18.30-21.20hrs, Sun; 16.30-20.00hrs.

3   Nags Head Country Carvery

Tel: (01422) 373758
1 mile east along the A643, in New Hay Road,
Ainley Top. Restaurant Open; Mon-Sat; 12.00-
14.00hrs and 18.00-21.30hrs. Meals served all
day on Sundays; 12.00-21.00hrs. Bar meals
available, Mon-Sat; 12.00-21.30hrs, Sun; 12.00-
21.00hrs.

4   Lodge Inns

Tel: (01422) 373758
1 mile east along the A643, in New Hay Road,
Ainley Top.

5   Birchencliffe BP Petrol Station

Tel: (01484) 451362
0.3 miles south along the A629, on the left.
Access, Visa, Overdrive, All Star, Switch, Dial
Card, Mastercard, Amex, AA Paytrak, Diners
Club, Routex, BP Cards, Shell Cards.

6   Briar Court Hotel

Tel: (01484) 519902
0.3 miles south along the A629, on the right

7   Da Sandro Pizzeria Restaurant

Tel: (01484) 519902
0.3 miles south along the A629, on the right.
Lunch; Mon-Sat; 12.00-14.00hrs, Dinner; Mon;
18.00-22.30hrs, Tues-Fri; 18.00-23.00hrs, Sat;
18.00-23.30hrs. Meals served all day on
Sundays; 12.30-21.45hrs.

8   Monza BP Petrol Station

Tel: (01484) 435152
0.6 miles south along the A629, on the right.
Access, Visa, Overdrive, All Star, Switch, Dial
Card, Mastercard, Amex, Diners Club, Delta,
Routex, AA Paytrak, Shell Agency, BP Cards

9   The Cavalry Arms

Tel: (01484) 530812
0.7 miles south along the A629, on the right.
(Pubmaster) Lunch; Mon-Sat; 12.00-14.00hrs,
Evening Meals; 17.00-20.00hrs. Meals served
all day on Sundays; 12.00-20.00hrs.

10   The Ainley Top

Tel: (01422) 374360
Adjacent to east side of roundabout (Brewers
Fayre) Meals served; Mon-Sat; 11.30-22.00hrs,
Sun; 12.00-22.00hrs

M62

PLACES OF INTEREST

Halifax

Halifax Tourist Information Centre, Piece Hall,
Halifax HX1 1RE. Tel: (01422) 368725
Follow the A629 north (Signposted 5.6 Miles)

In the valley of the River Hebble and built upon a rise that ascends sharply towards a range of hills, Halifax has one of Yorkshire's most impressive pieces of municipal architecture, the large and beautiful Piece Hall. Built in 1779 the classically styled edifice consists of colonnades and balconies surrounding a large quadrangle and originally housed 315 merchant's rooms for the selling of cloth or pieces. A replica of a guillotine can be found in Gibbet Street, an old thoroughfare where executions took place, and the Church of St John the Baptist which dates from the 12th and 13thC's with most of the present building of 15thC origin, is the largest Parish Church in England.

Also in the town centre ...

Eureka!

Discovery Road, Halifax HX1 2NE
Tel: (01422) 330069

Britain's first interactive museum designed especially for children aged between 3 and 12 years. Over 400 exhibits of hands-on displays that allow visitors to touch, listen and smell as well as look are available. Gift Shop. Cafe. Disabled access.

Bankfield Museum

Akroyd Park, Boothtown Road, Halifax HX3 6HG
Tel: (01422) 354823 or (01422) 352334
Follow the A629 north into Halifax and take the
A647 north (5.9 Miles)

Set in a Victorian millowner's house, Bankfield is a centre for textiles and contemporary craft. The collection includes textiles and weird and wonderful objects and displays include the Toy Gallery and the Duke of Wellington's Regimental Museum. Disabled access

Shibden Hall & Park

Listers Road, Halifax HX3 6AG
Tel: (01422) 352246
Follow the A629 north into Halifax and take the
A58 east (Signposted along A58, 6.2 Miles)

The Old Hall lies in a valley on the outskirts of Halifax and is situated in 90 acres of parkland. The distinctive timber framed house dates from 1420 and is furnished to reflect the various periods of its history. The 17thC barn, behind the hall, houses a fine collection of horse drawn vehicles and the original buildings have been transformed into a 19thC village centre with a pub, estate worker's cottage and saddler's, blacksmith's, wheelwright's and potter's workshops. Other attractions include children's rides, a miniature railway, pitch and putt and a boating lake. Gift Shop. Cafe.

Castle Hill & Jubilee Tower

Off Lumb Lane, Almondbury
Huddersfield HD4 6SZ
Follow the A640 east, turn right along the A62 and
right along the A616 (6 Miles)

For details please see Junction 23 information.

M62 JUNCTION 25

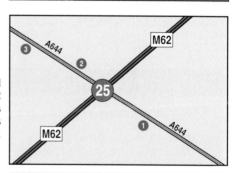

NEAREST WESTBOUND A&E HOSPITAL

Huddersfield Royal Infirmary, Acre Street,
Lindley, Huddersfield HD3 3EA
Tel: (01484) 422191

Proceed to Junction 24 and follow the A629 south (Signposted within town. Distance Approx 5.5 miles)

NEAREST EASTBOUND A&E HOSPITAL

Bradford Royal Infirmary, Duckworth Lane,
Bradford BD9 6RJ Tel: (01274) 542200

Proceed to Junction 26, follow the M606 north, continue along the A6177 and turn right along the A641 (Signposted along route. Distance Approx 10 Miles)

 Pets Welcome £ Cash Dispenser 24 24 Hour Facilities WC Toilets Alcoholic Drinks Food and Drink Fuel

FACILITIES

1   The Corn Mill

🛏️ 🍴 ♿ ✂️   Tel: (01484) 400069
0.5 miles east along the A644, on the right.
(Independent) Meals served; 12.00-20.00hrs
daily

2   Posthouse Leeds/Brighouse

🏨 🛏️ 🍴 ✂️ ♿ 🛏️   Tel: (01484) 400400
0.2 miles west along the A644, on the right.
Breakfast; Mon-Fri; 06.30-09.30hrs, Sat & Sun;
07.00-10.00hrs, Lunch; Mon-Fri; 12.00-
15.00hrs, Sat & Sun; 12.00-15.30hrs, Dinner;
18.30-22.30hrs daily.

3   Crown Service Station (Shell)

⛽ WC   Tel: (01484) 720371
0.8 miles west along the A644, on the left.
Keyfuels, Access, Visa, Overdrive, All Star,
Switch, Dial Card, Mastercard, Amex, AA
Paytrak, Diners Club, Shell Cards, Smart, Diesel
Direct. Open; Mon-Fri; 06.30-21.30hrs, Sat &
Sun; 07.00-21.00hrs daily

SERVICE AREA

BETWEEN JUNCTIONS 25 AND 26

Hartshead Moor (Eastbound) (Welcome Break)
Tel: (01274) 876584 Days Inn, Burger King, The
Granary Restaurant & Shell Fuel
Hartshead Moor (Westbound) (Welcome Break)
Tel: (01274) 876584 KFC, The Granary Restaurant
& Shell Fuel
Footbridge Connection between sites

M62   JUNCTION 26

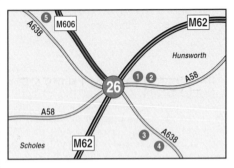

NEAREST A&E HOSPITAL

**Bradford Royal Infirmary, Duckworth Lane,
Bradford BD9 6RJ Tel: (01274) 542200**

Follow the M606 north, continue along the A6177
and turn right along the A641. (Signposted along
route. Distance Approx 6.9 Miles)

FACILITIES

1   The Hunsworth

🛏️ 🍴 ♿ ✂️   Tel: (01274) 862828
0.2 miles east along the A58, on the left.
(Brewers Fayre) Meals served; Mon-Sat; 11.00-
23.00hrs, Sun; 12.00-22.30hrs.

2   Travel Inn

🏨 ♿   Tel: (01274) 862828
0.2 miles east along the A58, on the left

3   The Horncastle

🛏️ 🍴 ✂️ 🏨   Tel: (01274) 875444
0.5 miles south along the A638, on the right.
(Free House) Meals served all day Mon-Fri;
07.30-19.00hrs, Sat & Sun; 07.30-16.00hrs.

4   The Talbot

🛏️ 🍴 ✂️ 🏨   Tel: (01924) 472086
0.8 miles south along the A638, on the right.
(Theakstons) Bar Snacks served; Lunch and
early evening daily.

M62

| Large Hotel | Small Hotel | Motel | Guest House/ Bed & Breakfast | Disabled Facilities | Childrens Facilities |

5 Richardsons Arms

🍴 🏨 ♿ 🚭   Tel: (01274) 675722
0.7 miles north along Oakenshaw Road, in
Oakenshaw, on the right. Meals served; Mon-
Sat; 07.00-14.00hrs, Sun; 08.00-12.00hrs.

PLACES OF INTEREST

Bradford
Bradford Tourist Information Centre,
Central Library, Princes Way, Bradford BD1 1NN
Tel: (01274) 753678
e-mail; public.libraries@bradford.gov.uk
Follow the M606 north (Signposted 4.6 Miles)

On a tributary of the River Aire, Bradford was
established prior to 1066 but it was during the
18thC that it developed into a major centre of
trade in worsted and wool, with the first mill
being erected in 1798. It reached its pinnacle
of wealth in the mid-19thC and many fine
buildings constructed during this period can
be seen, especially in Little Germany. Of
particular note is the Town Hall of 1873, with
an exterior ornamented with statues of English
monarchs.

Also in the city centre ...

The National Museum of Photography, Film & Television
Pictureville, Bradford BD1 1NQ
Tel: (01274) 202030 website; www.nmpft.org.uk

Recently extended, the six floors include the
Kodak and TV Heaven galleries, displays on
animation and film-making, advertising and,
in the Magic Factory, a hands-on exhibition.
There is also an IMAX cinema, shop and full
catering facilities.

Bradford Industrial & Horses at Work Museum
Moorside Mills, Moorside Road, Eccleshill,
Bradford BD2 3HP Tel: (01274) 631756
Follow the M606 north and turn right along the
A6177 (Signposted along A6177, 6.3 Miles)

An original worsted spinning mill complex
built in 1875 and now in use as a working
museum recreating life in Bradford at the turn
of the 19thC. The displays include the mill
stables, complete with Shire horses, mill

owner's house, transport gallery and back to
back cottages with working demonstrations of
horse drawn buses and trams. Cafe, Shop,
Disabled access.

M62 JUNCTION 27

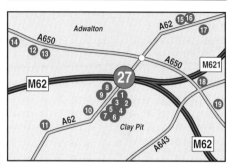

NEAREST WESTBOUND A&E HOSPITAL
Bradford Royal Infirmary, Duckworth Lane,
Bradford BD9 6RJ Tel: (01274) 542200
Proceed to Junction 26, follow the M606 north,
continue along the A6177 and turn right along
the A641 (Signposted along route. Distance
Approx 11.2 Miles)

NEAREST EASTBOUND A&E HOSPITAL
Leeds General Infirmary, Great George Street,
Leeds LS1 3EX Tel: (0113) 243 2799
Follow the M621 north, leave at Junction 2, take
the A643 north and continue along the A58.
(Signposted. Distance Approx 5.7 Miles)

FACILITIES

1 Bella Pasta (Italian)
🍴 🏨 🚭 ♿   Tel: (01924) 422374
0.3 miles east along the A62, in Centre 27, on
the left. Open; Sun-Thurs; 12.00-23.00hrs, Fri
& Sat; 12.00-23.30hrs

2 Frankie & Benny's (New York Italian)
🍴 🏨 🚭 ♿   Tel: (01924) 423747
0.3 miles east along the A62, in Centre 27, on
the left. Open; Mon-Sat; 12.00-23.00hrs, Sun;
12.00-22.30hrs.

M62

| | Pets Welcome | | Cash Dispenser | | 24 Hour Facilities | | Toilets | | Alcoholic Drinks | | Food and Drink | | Fuel |

3 Chiquito's (Mexican)

Tel: (01924) 359292

0.3 miles east along the A62, in Centre 27, on the left. Buffet Meals; Mon-Fri; 12.00-18.00hrs, Sun; 18.00-22.30hrs. A la carte Menu, Mon-Sat; 12.00-23.00hrs.

4 KFC

Tel: (01924) 422634

0.3 miles east along the A62, in Centre 27, on the left. Open; 11.00-0.00hrs Daily.

5 Pizza Hut

Tel: (01924) 420460

0.3 miles east along the A62, in Centre 27, on the left. Open; Sun-Thurs; 11.30-23.00hrs, Fri & Sat; 11.30-0.00hrs.

6 McDonald's

Tel: (01924) 456833

0.3 miles east along the A62, in Centre 27, on the left. Open; Sun-Thurs; 07.30-23.00hrs, Fri & Sat; 07.30-0.00hrs

7 Exchange Bar

Tel: (01924) 422120

0.3 miles east along the A62, in Centre 27, on the left. Food served 11.30-23.00hrs daily.

8 TGI Friday's

Tel: (01924) 475000

0.3 miles east along the A62, in The West Yorkshire Retail Park, on the right. Open; Mon-Sat; 12.00-23.00hrs, Sun; 12.00-22.00hrs

9 Repsol Service Station

Tel: (01924) 422077

0.3 miles east along the A62, in The West Yorkshire Retail Park, on the right. Keyfuels, Access, Visa, Overdrive, All Star, Switch, Dial Card, Mastercard, Amex, AA Paytrak, Diners Club, Delta, BP Cards. Open; Mon-Fri; 06.00-23.00hrs, Sat; 07.00-23.00hrs, Sun; 08.00-22.00hrs.

10 Little Chef

Tel: (01924) 440621

0.6 miles east along the A62, on the right. Open; 07.00-22.00hrs

11 The Pheasant

Tel: (01924) 473022

0.8 miles east along the A62, on the right. (Whitbread) Meals served; Mon-Thurs; 12.00-15.00hrs & 17.00-19.00hrs, Fri & Sat; 12.00-15.00hrs, Sun; 12.00-16.00hrs.

12 The Old Brickworks

Tel: (0113) 287 9132

0.6 miles west along the Wakefield Road, in Drighlington, on the left. (Whitbread) Meals served; 12.00-22.00hrs daily.

13 Travel Inn

Tel: (0113) 287 9132

0.6 miles west along the Wakefield Road, in Drighlington, on the left.

14 The New Inn

Tel: (0113) 285 2447

1 mile west along the Wakefield Road, in Drighlington, on the left. (Free House) Meals served; Mon-Fri; 17.00-20.15hrs, Sat; 12.00-14.30hrs, Sun; 12.00-18.00hrs

15 Gildersome Lodge

Tel: (0113) 252 1828

0.6 miles north along the A62, in Gildersome, on the left. (Tetley) Meals served; Mon-Thurs; 18.00-21.00hrs

16 Gildersome Service Station (Total Fina)

Tel: (0113) 238 3343

0.6 miles north along the A62, in Gildersome, on the left. Access, Visa, Overdrive, All Star, Switch, Dial Card, Mastercard, Amex, Diners Club, Delta, AA Paytrak, BP Supercharge, Elf Card, Total Fina Cards.

M62

 Large Hotel
 Small Hotel
 Motel
 Guest House/ Bed & Breakfast
 Disabled Facilities
 Childrens Facilities

17 The Mill House Bar & Family Restaurant

🔲 🍴 ♿ 🔌   Tel: (0113) 238 3810
0.7 miles north along the A62, in Gildersome, on the right. (Mill House) Meals served; 12.00-21.00hrs daily

18 The Angel

🔲 🍴 ♿ 🔌   Tel: (0113) 253 3115
0.7 miles east along the A650, on the left. (Punch Taverns) Meals served; 12.00-22.00hrs daily

19 Victoria Filling Station (Shell)

⛽ ♿ WC   Tel: (0113) 252 7538
0.9 miles east along the A650, on the left. Keyfuels, Access, Visa, Overdrive, All Star, Switch, Dial Card, Mastercard, Amex, AA Paytrak, Diners Club, Delta, BP Agency, Shell Cards. Open; Mon-Fri; 06.00-21.45hrs, Sat; 07.00-20.15hrs, Sun; 08.00-20.15hrs.

PLACES OF INTEREST

Leeds

Leeds Tourist Information Centre, The Arcade, Leeds City Station, Leeds LS1 1PL
Tel: (0113) 242 5242
Follow the M621 north (Signposted 5 Miles)

On the River Aire, there were primitive lake dwellings here and the site was in use by Norman times. In the reign of Edward III Flemish emigrants arrived here and brought with them their trade of cloth making, establishing it as a major industry within the area. The city developed rapidly in the early 19thC as the inland port on the Leeds to Liverpool and Aire & Calder Navigation canals, forming a central link between Liverpool and Hull, from where goods were exported world wide. The Canal Basin, an integral part of this system which provided extensive wharves, warehouses, boat building yards and wet and dry docks, fell into disuse as other modes of transport took over but it has, recently, been sympathetically restored and designated as a Conservation Area. The city centre contains a number of buildings from varying periods throughout its history but most of the outstanding edifices are of the classical baroque style favoured in the Victorian era.

Within the city centre ...

Royal Armouries

Armouries Drive, Leeds LS10 1LT
Tel: (0113) 220 1999 24hr Information Line
Tel: 0990 1066 66
website; www.armouries.org.uk
e-mail; enquiries@armouries.org.uk

3000 years of history is displayed within five galleries themed on War, Tournament, The Orient, Self-Defence and Hunting. Over 8000 exhibits are on display including Henry VIII's tournament armour and 16thC Indian elephant armour. There are displays, dramatic interpretations, live action events, interactive technology and a continuous daily showing of 42 specially commissioned films. Shops. Restaurants. Bars. Disabled access.

Thackray's Medical Museum

Beckett Street, Leeds LS9 7LN
Tel: (0113) 245 7084
Follow the M621 and A653 into Leeds and take the A58 east (Signposted along A58, 6.7 Miles)

Housed in the old Leeds Union Workhouse and utilizing state of the art interactive exhibits, the museum traces the history of medicine and its impact on every day life. Displays include Victorian slum life, surgical operations and a fascinating collection of objects and gadgets. Gift Shop. Cafe. Disabled access.

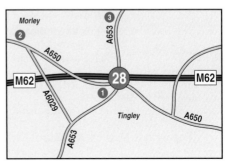

NEAREST A&E HOSPITAL

Leeds General Infirmary, Great George Street, Leeds LS1 3EX Tel: (0113) 243 2799
Follow the A653 north, bear left along the A6110, continue along the A643 north and the A58. (Signposted. Distance Approx 5.5 Miles)

FACILITIES

1   The White Bear

🍴 🛏 ⚡   Tel: (0113) 253 2768

Adjacent to south side of the roundabout. (Beefeater) Meals served; Mon-Thurs; 12.00-14.30hrs & 17.00-22.30hrs, Fri-Sun; 12.00-23.00hrs.

2   Tingley Bar Grill & Fish Restaurant

🍴 🛏 ⚡ ♿   Tel: (0113) 253 3774

0.8 miles west along the A650, on the right. Fish Restaurant; Open; 11.30-22.00hrs daily. Bar Grill; Open; Mon-Sat; 12.00-14.00hrs & 17.00-21.00hrs, Sun; 12.00-21.00hrs.

3   White Rose

🍴 🛏 ⚡ ♿   Tel: (0113) 252 3720

1 mile north along the A653, on the left. (Brewers Fayre) Meals served; Mon-Sat; 11.00-22.00hrs, Sun; 12.00-22.00hrs.

PLACES OF INTEREST

Leeds

Follow the A653 and M621 north
(Signposted 4.8 Miles)

For details please see Junction 27 information

Within the city centre ...

Royal Armouries

Armouries Drive, Leeds LS10 1LT

For details please see Junction 27 information

Thackray's Medical Museum

Beckett Street, Leeds LS9 7LN Follow the A653, M621 and A653 into Leeds and take the A58 east (Signposted along A58, 6.5 Miles)

For details please see Junction 27 information

M62   JUNCTION 29

THIS IS A MOTORWAY INTERCHANGE ONLY AND THERE IS NO ACCESS TO ANY FACILITIES

NEAREST WESTBOUND A&E HOSPITAL

Leeds General Infirmary, Great George Street, Leeds LS1 3EX Tel: (0113) 243 2799

Proceed north to Junction 7 (M621) and take the A61 north into Leeds city centre. The hospital is signposted in the city. (Distance Approx 6.8 miles)

NEAREST EASTBOUND A&E HOSPITAL

Pinderfield Hospital, Aberford Road, Wakefield WF1 4DG Tel: (01924) 201688

Proceed south to Junction 41 (M1), take the A650 exit east and continue along the A61. At the junction of the A642 and A61 turn left along the A642 and the hospital is along this road. (Distance Approx 4.6 miles)

PLACES OF INTEREST

Leeds

Follow the M1 and M621 north
(Signposted 6 Miles)

For details please see Junction 27 information

Within the city centre ...

Royal Armouries

Armouries Drive, Leeds LS10 1LT

For details please see Junction 27 information

Thackray's Medical Museum

Beckett Street, Leeds LS9 7LN Follow the M1 north, take the A61 north exit at Junction 7 (M621) and continue along the A58 (5.7 Miles)

For details please see Junction 27 information

| Large Hotel | Small Hotel | Motel | Guest House/ Bed & Breakfast | | Disabled Facilities | Childrens Facilities | | |

M62   JUNCTION 30

THERE ARE NO FACILITIES WITHIN 1 MILE OF THIS JUNCTION

NEAREST A&E HOSPITAL

Pinderfield Hospital, Aberford Road, Wakefield
WF1 4DG Tel: (01924) 201688

Follow the A642 south. (Distance Approx 2.8 miles)

PLACES OF INTEREST

Wakefield

Follow the A 642 south (Signposted 3.7 Miles)

For details please see Junction 40 (M1) information

Within the city centre ...

The Cathedral Church of All Saints

Northgate, Wakefield WF1 1HG

For details please see Junction 40 (M1) information

The Chantry Chapel of St Mary

on Wakefield Bridge

For details please see Junction 40 (M1) information

M62   JUNCTION 31

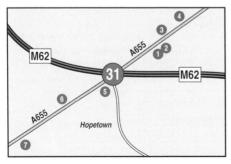

NEAREST WESTBOUND A&E HOSPITAL

Pinderfield Hospital, Aberford Road, Wakefield
WF1 4DG Tel: (01924) 201688

Proceed to Junction 30 and follow the A642 south. (Distance Approx 5.9 miles)

NEAREST EASTBOUND A&E HOSPITAL

Pontefract General Infirmary, Friarwood Lane,
Pontefract WF8 1PL. Tel: (01977) 600600

Proceed to Junction 32 and follow the A639 south (Signposted in town. Distance Approx 4.1 miles)

FACILITIES

1   Little Red Lion

Tel: (01977) 665400

0.5 miles north along the A655, on the right. (Premier Lodge) Meals served 18.00-22.00hrs daily.

2   Trading Post, Castleford

Tel: (01977) 519587

0.5 miles north along the A655, on the right. (Mansfield's) Open; Mon-Sat; 11.00-23.00hrs, Sun; 12.00-22.30hrs

3   Rising Sun

Tel: (01977) 554766

0.6 miles north along the A655, on the left. (Free House) Meals served; Mon-Sat; 12.00-14.30hrs & 17.30-21.30hrs, Sun; 12.00-21.30hrs.

4   Q8 Garage, Castleford

Tel: (01977) 553549

0.9 miles north along the A655, on the left. Access, Visa, Overdrive, All Star, Switch, Dial Card, Mastercard, Amex, Diners Club, Delta, Q8 Cards.

5   The Village Motel

Tel: (01924) 897171

0.2 miles along the Normanton Road, on the left. (Mill House) Meals served; Mon-Thurs; 12.00-22.00hrs, Fri-Sun; 12.00-12.00hrs.

6   Prospect Garage (Jet)

Tel: (01924) 895824

0.5 miles along the Normanton Road, on the right. Access, Visa, Overdrive, All Star, Switch, Dial Card, Mastercard, Amex, Jet Cards. Open;

M62

| Pets Welcome | £ Cash Dispenser | 24 24 Hour Facilities | WC Toilets | Alcoholic Drinks | Food and Drink | Fuel |

Mon-Sat; 07.00-22.00hrs, Sun; 09.00-22.00hrs.

7 Q8 Garage

🖪 Tel: (01924) 893579
1 mile along the Normanton Road, on the left. Keyfuels, Access, Visa, Overdrive, All Star, Switch, Dial Card, Mastercard, Amex, Diners Club, Delta, Q8 Cards. Open: Mon-Sat; 06.00-23.00hrs, Sun; 07.00-23.00hrs.

PLACES OF INTEREST

Castleford

Follow the A655 east (Signposted 2.3 Miles)

Sited on the River Aire the Romans established a crossing here and built a notable fort and settlement known as Lagentium or Legioleum. Archaeological finds from this period can be seen at the Library Museum. Known as the birthplace of the famous sculptor Henry Moore, the town is renowned for its excellent 18thC pottery and glassware.

M62 JUNCTION 32

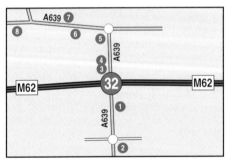

NEAREST A&E HOSPITAL

Pontefract General Infirmary, Friarwood Lane, Pontefract WF8 1PL. Tel: (01977) 600600
Follow the A639 south (Signposted in town. Distance Approx 1.6 miles)

FACILITIES

1 Parkside Inne

�)h 🍴 🚻 ♿ -ᴧ- Tel: (01977) 709911
0.2 miles south along the A639, on the left. (Regal Hotel Group) Carvery Open; Mon-Sat; 12.00-14.00hrs & 19.00-21.30hrs, Sun; 12.00-15.00hrs. Restaurant Open; Mon-Sat; 19.00-21.30hrs, Sun; 12.00-14.30hrs. Bar meals served Sun; 19.00-21.00hrs.

2 McDonald's

🚻 ♿ Tel: (01977) 602919
0.9 miles south along the A639, in the Racecourse Retail Park, on the left. Open; Mon-Thurs; 07.30-23.00hrs, Fri-Sun; 07.30-0.00hrs

3 Save Petrol Station

🖪 ♿ WC Tel: (01977) 553489
0.2 miles north along the A639, on the left. Keyfuels, Access, Visa, Overdrive, All Star, Switch, Dial Card, Mastercard, Delta, Save. Open; 06.00-23.00hrs daily.

4 Singing Choker

🍴 🚻 -ᴧ- ♿ Tel: (01977) 668383
0.3 miles north along the A639, on the left. (Tom Cobleigh's) Meals Served; Mon-Sat; 12.00-21.30 hrs, Sun; 12.00-21.00hrs.

5 Castleford Service Station (Texaco)

🖪 WC 24 Tel: (01977) 552718
0.4 miles north along the A639, on the left. Access, Visa, Overdrive, All Star, Switch, Dial Card, Mastercard, Amex, Diners Club, Delta, BP Supercharge, Fast Fuel, Texaco Cards.

6 The Royal Oak

🍴 🚻 -ᴧ- ♿ Tel: (01977) 553610
0.7 miles north along the A639, on the left. (Punch Taverns) Meals served; Mon-Sat; 12.00-20.00hrs, Sun; 12.00-16.00hrs

7 Glasshoughton Service Station (Total Fina)

🖪 WC 24 Tel: (01977) 516578

| | Large Hotel | | Small Hotel | | Motel | | Guest House/ Bed & Breakfast | | Disabled Facilities | | Childrens Facilities |

0.7 miles north along the A639, on the right. Access, Visa, Overdrive, All Star, Switch, Dial Card, Mastercard, Amex, Diners Club, Delta, Total Fina Cards.

---

8   Mirage Hotel

                    Tel: (01977) 553428

1 mile north along the A639, on the left

PLACES OF INTEREST

Castleford

Follow the A639 north (Signposted 1.7 Miles)

For details please see Junction 31 information.

Pontefract

Follow the A639 south (Signposted 1.6 Miles)

The town of Pontefract, or "Pomfret" as it was known and so called in Shakespeare, was an important crossing point over the River Aire. Pomfret Castle, constructed by the Normans just after 1069, was a Royal stronghold and used to imprison Richard II who died here. In 1649, during the Civil War, the castle was finally taken by the Parliamentary forces and they subsequently demolished it just leaving some underground passages, dungeons and a keep which is now utilized as a museum (Tel: 01977-723440). The mixture of spacious precincts and narrow streets reflects its history with the Town Hall of 1785, the Buttercross and timber framed building providing an attractive setting in the town centre. The town is renowned for a variety of sweets known as Pontefract Cakes, made from locally grown liquorice.

M62   JUNCTION 33

SERVICE AREA

Ferrybridge Services (Granada)

Tel: (01977) 672767 Self Service Restaurant, Burger King, Little Chef, Travelodge, Esso Fuel.

NEAREST A&E HOSPITAL

Pontefract General Infirmary, Friarwood Lane, Pontefract WF8 1PL. Tel: (01977) 600600

Follow the A1 north and turn left along the A628. (Signposted in town. Distance Approx 2.5 miles)

PLACES OF INTEREST

Pontefract

Follow the A1 north and turn left along the A628. (Signposted 2.5 Miles)

For details please see Junction 32 information.

M62   JUNCTION 34

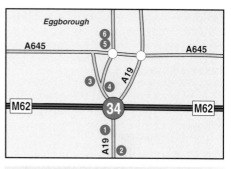

NEAREST WESTBOUND A&E HOSPITAL

Pontefract General Infirmary, Friarwood Lane, Pontefract WF8 1PL Tel: (01977) 600600

Proceed to Junction 33, follow the A1 north and turn left along the A628. (Signposted in town. Distance Approx 7 miles)

NEAREST EASTBOUND A&E HOSPITAL

Goole & District Hospital, Woodland Avenue, Goole DN14 6RX Tel: (01724) 282282

Proceed to Junction 36 and follow the A614 east (Signposted. Distance Approx 12.6 miles)

FACILITIES

1   George & Dragon

                    Tel: (01977) 661319

0.6 miles south along the A19, on the right. (Independent) Meals served; Lunch; 12.00-

14.00hrs daily. Evening Meals; Sun-Fri; 19.00-21.00hrs, Sun; 19.00-21.30hrs.

2   Whitley Bridge Service Station

🍴 WC   Tel: (01977) 663040

1 mile south along the A19, on the left. Keyfuels, Access, Visa, Overdrive, All Star, Switch, Dial Card, Mastercard, Amex, Delta, BP Cards. Open; 06.00-22.00hrs daily.

3   The Jolly Miller

🍴 🍴 ♿ 🅿   Tel: (01977) 661348

0.4 miles north along the Selby Road, in Eggborough, on the left. (Pubmaster) Meals served all day.

4   Station Garage

🍴 WC   Tel: (01977) 661256

0.4 miles north along the Selby Road, in Eggborough, on the left. Access, Visa, Overdrive, All Star, Switch, Dial Card, Mastercard, Amex, AA Paytrak, Diners Club, Delta, Routex, BP Supercharge. Open; Mon-Fri; 07.30-18.30hrs, Sat; 07.30-15.00hrs, Sun; 09.00-12.30hrs.

5   The Horse & Jockey

🍴 🍴 ♿ 🅿   Tel: (01977) 661295

1 mile north along the Selby Road, in Eggborough, on the left. (Enterprise Inns) Meals served 12.00-14.00hrs daily

6   Selby Road Filling Station

🍴 WC   Tel: (01977) 661651

1 mile north along the Selby Road, in Eggborough, on the left. Access, Visa, Switch, Mastercard, Delta, BP Supercharge. Open; Mon-Fri; 08.00-18.00hrs, Sat; 08.00-17.00hrs, Sun; 09.00-13.00hrs.

M62   JUNCTION 35

THIS JUNCTION IS A MOTORWAY INTERCHANGE WITH M18 ONLY AND THERE IS NO ACCESS TO FACILITIES

NEAREST WESTBOUND A&E HOSPITAL

Pontefract General Infirmary, Friarwood Lane, Pontefract WF8 1PL. Tel: (01977) 600600

Proceed to Junction 33, follow the A1 north and turn left along the A628. (Signposted in town. Distance Approx 15 miles)

NEAREST EASTBOUND A&E HOSPITAL

Goole & District Hospital, Woodland Avenue, Goole DN14 6RX. Tel: (01724) 282282

Proceed to Junction 36 and follow the A614 east (Signposted. Distance Approx 4.6 miles)

M62   JUNCTION 36

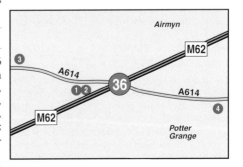

NEAREST A&E HOSPITAL

Goole & District Hospital, Woodland Avenue, Goole DN14 6RX. Tel: (01724) 282282

Follow the A614 east (Signposted. Distance Approx 1.9 miles)

FACILITIES

1   Glews Garage (Shell)

🍴 ♿ WC 24 🍴   Tel: (01405) 764525

0.2 miles west along the A614, on the left. Access, Visa, Overdrive, All Star, Switch, Dial Card, Mastercard, Amex, Diners Club, Delta, BP Supercharge, Shell Cards, Esso Cards. Food available Mon-Sat; 08.00-17.00hrs.

2   McDonald's

🍴 ♿   Tel: (01405) 766747

0.2 miles west along the A614, on the left. Open; Sun-Thurs; 07.30-23.00hrs, Fri & Sat; 07.30-0.00hrs

| Large Hotel | Small Hotel | Motel | Guest House/ Bed & Breakfast | Disabled Facilities | Childrens Facilities |

3   Woodside Cafe (Transport)

🍴   Tel: (01405) 839321
0.9 miles west along the A614, on the right.
Open; Mon-Fri; 06.20-17.00hrs

4   Rawcliffe Road Service Station (Total Fina)

🗂 24   Tel: (01405) 763445
1 mile east along the A614, in Goole, on the right. Access, Visa, Overdrive, All Star, Switch, Dial Card, Mastercard, Amex, Diners Club, Delta, BP Supercharge, Total Fina Cards.

PLACES OF INTEREST

The Waterways Museum & Adventure Centre

Dutch River Side, Goole DN14 5TB
Tel: (01405) 768730
Follow the A614 east and turn right along the A161. (Signposted 2.4 Miles)

The collection tells the story of Goole's development as a canal terminus to the Aire & Calder Navigation and as a port connecting to the North Sea. The displays include contemporary social history and interactive exhibitions.

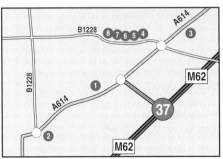

M62   JUNCTION 37

NEAREST WESTBOUND A&E HOSPITAL
Goole & District Hospital, Woodland Avenue, Goole DN14 6RX. Tel: (01724) 282282
Proceed to Junction 36 and follow the A614 east (Signposted. Distance Approx 4.7 miles)

NEAREST EASTBOUND A&E HOSPITAL
The Hull Royal Infirmary, Anlaby Road, Kingston upon Hull HU3 2JZ
Tel: (01482) 328541
Proceed to Junction 38, follow the A63 east and continue along the A164 and A1105. (Distance Approx 23.1 miles)

FACILITIES

1   Junction 37 Service Station (Rix)

🗂 ♿ WC   Tel: (01430) 430388
0.6 miles west along the A614, on the right. Keyfuels, Access, Visa, Overdrive, All Star, Switch, Dial Card, Mastercard, Delta, BP Supercharge, Securicor Fuelserv, UK Fuels. Open; Mon-Fri; 07.00-22.00hrs, Sat; 07.30-20.30hrs, Sun; 08.30-20.30hrs.

2   The Ferryboat Inn

🍴 🍴 ♿ 🐾   Tel: (01430) 430300
1 mile west along the A614, on the left. (Free House) Lunch; 12.00-14.00hrs daily, Evening Meals; Mon-Sat; 18.00-22.00hrs, Sun; 18.00-21.00hrs.

3   Brian Leighton Garages (BP)

🗂 ♿ WC 24   Tel: (01430) 430717
1 mile north along the A614, on the right. Access, Visa, Overdrive, All Star, Switch, Dial Card, Mastercard, Amex, AA Paytrak, Diners Club, Delta, Routex, BP Cards, Shell Agency, Mobil Cards.

4   The Wellington Hotel

🛏 🍴 🍴 🐾   Tel: (01430) 430258
1 mile north along the A614, in Howden, on the right. (Independent) Meals served; 12.00-14.30hrs & 18.30-22.00hrs daily

5   Bowmans Hotel

🛏 🍴 🍴 🐾   Tel: (01430) 430805
1 mile north along the A614, in Howden, on the right. (Independent) Meals served; Mon-Sat; 12.00-14.30hrs & 18.30-21.30hrs, Sun; 12.00-14.30hrs.

M62

6   The Wheatsheaf

 Tel: (01430) 432334

1 mile north along the A614, in Howden, on the right. (Unique Pub Co) Meals served; Mon; 17.00-19.00hrs, Tues-Thurs; 12.00-14.00hrs & 17.00-19.00hrs, Fri-Sun; 12.00-19.00hrs

7   The White Horse Inn

Tel: (01430) 430326

1 mile north along the A614, in Howden, on the right. (Whitbread) Lunch; 12.00-14.30hrs daily, Evening meals; Mon-Thurs; 17.30-19.30hrs.

8   The Station

Tel: (01430) 431301

1 mile north along the A614, in Howden, on the right (Free House)

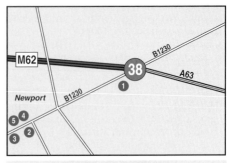

NEAREST A&E HOSPITAL

The Hull Royal Infirmary, Anlaby Road,
Kingston upon Hull HU3 2JZ
Tel: (01482) 328541

Follow the A63 east and continue along the A164 and A1105. (Distance Approx 14.7 miles)

PLACES OF INTEREST

Howden

Follow the A614 north (Signposted 1 Mile)

With its cobbled streets and fine Georgian buildings, the ancient and interesting market town of Howden is dominated by the 135ft high Bishop Skirlaw's tower of Howden Minster. Known as the Church of St Peter it was rebuilt in the 13th & 14thC's but only parts of it remain today. The chapter house of 1399, Saltmarshe Chantry-chapel and Grammar School can still be seen, as well as some remains of the Bishop's Palace. The famous mediaeval chronicler, Roger de Hoveden, was born here in 1204.

FACILITIES

1   Triangle Motors (BP)

Tel: (01430) 424119

0.1 miles east along the B1230, on the left. Access, Visa, Overdrive, All Star, Switch, Mastercard, Amex, Delta, Routex, BP Cards, Shell Cards. Open; Mon-Sat; 07.00-22.00hrs, Sun; 08.00-22.00hrs.

2   Gibson Garages (Club)

Tel: (01430) 440099

1 mile east along the B1230, in Newport, on the left. Access, Visa, Switch, Mastercard, Delta. Open; Mon-Fri; 08.00-19.00hrs, Sat & Sun; 08.00-17.00hrs.

3   The Jolly Sailor

Tel: (01430) 449191

1 mile east along the B1230, in Newport, on the left. (Free House) Lunch; Tues-Sun; 12.00-14.00hrs, Evening meals; Tues-Sat; 17.30-21.00hrs, Sun; 19.00-20.00hrs.

4   Crown & Anchor

   Tel: (01430) 449757

1 mile east along the B1230, in Newport, on the right. (Mansfield) Lunch; Tues-Sat; 12.00-14.00hrs, Sun; 12.00-15.00hrs, Evening meals; Tues-Sat; 17.00-20.30hrs.

5   Kings Arms

Tel: (01430) 440289

1 mile east along the B1230, in Newport, on the right. (Pubmaster) Lunch; 12.00-14.00hrs daily, Evening meals; Mon, Tues & Thu-Sat; 18.00-19.30hrs.

MOTORWAY ENDS

(TOTAL LENGTH OF MOTORWAY 106.0 MILES)

Large Hotel Small Hotel Motel Guest House/ Bed & Breakfast Disabled Facilities Childrens Facilities

OFF THE MOTORWAY
M69

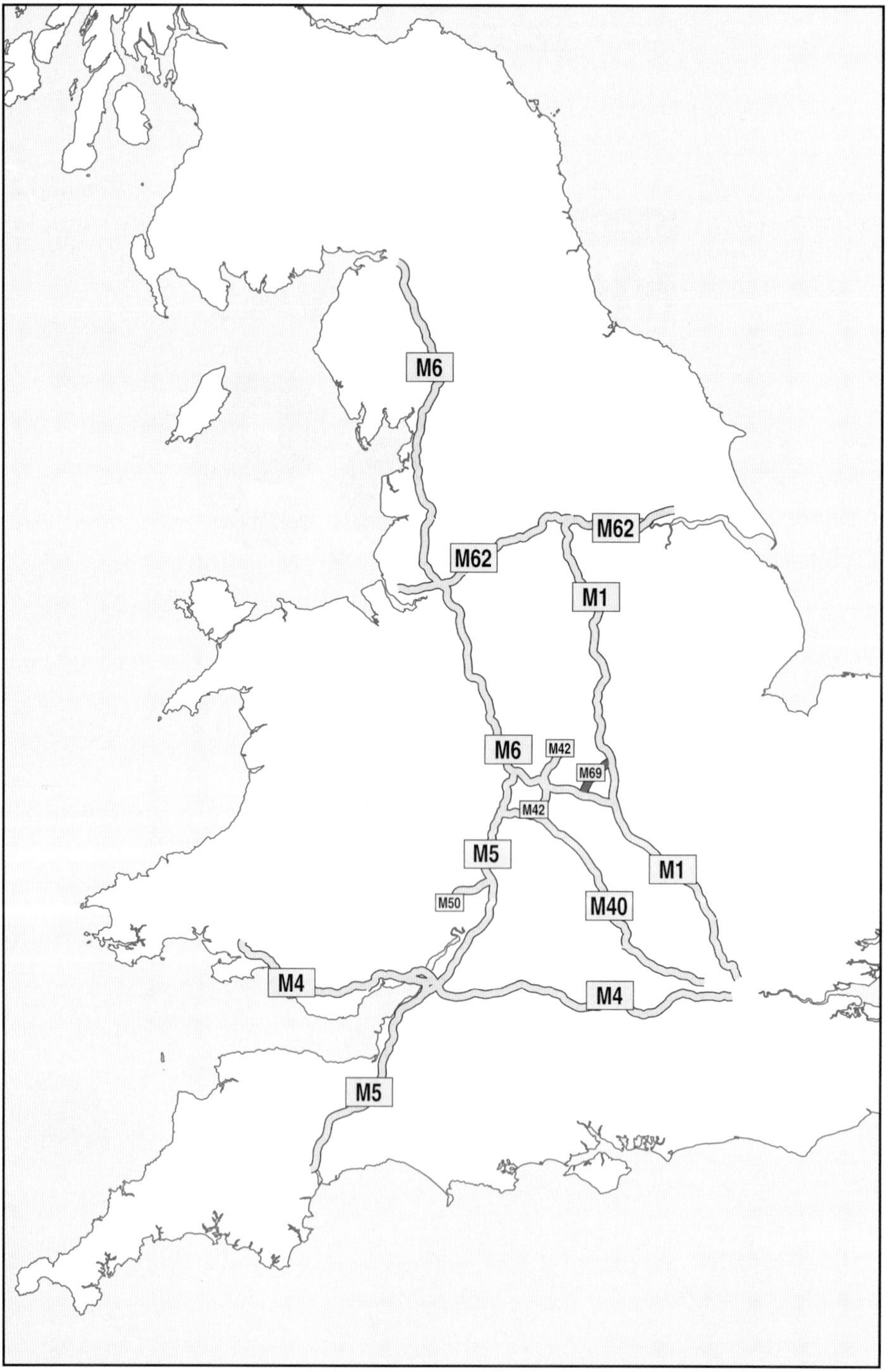
M6
M62
M62
M1
M6
M42
M69
M42
M5
M1
M50
M40
M4
M4
M5

The M69 Motorway

A short motorway linking Coventry and Leicester and providing a useful connection between the M1 and M6. Although there is little of note along the route, the junction with the M1 at the north end still remains as a monument to the world events that shaped strategic planning in the 1970s.

The earthworks were constructed in 1976 to lead to a flyover and clover leaf junction on the east side of the M1 (the site now partially occupied by Fosse Park) but, with a fuel crisis escalating daily, they were abandoned in the same year as it was felt that traffic levels would never warrant the investment!

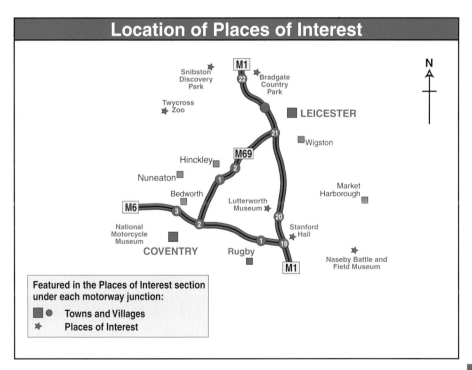

Location of Places of Interest

Snibston Discovery Park
M1
Bradgate Country Park
22
Twycross Zoo
LEICESTER
21
Wigston
Hinckley
M69
2
Nuneaton
1
Market Harborough
Bedworth
Lutterworth Museum
M6
3
20
National Motorcycle Museum
2
Stanford Hall
COVENTRY
Rugby
1
19
M1
Naseby Battle and Field Museum
N

Featured in the Places of Interest section under each motorway junction:

■ ● Towns and Villages
✶ Places of Interest

M69

M69 JUNCTION 1

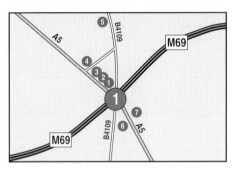

NEAREST NORTHBOUND A&E HOSPITAL

Leicester Royal Infirmary, Infirmary Square, Leicester LE1 5WW Tel: (0116) 254 1414

Proceed to Junction 21 (M1) and take the exit to Leicester. Follow the A5460 into the city and turn right along Upperton Road. Turn left at the end and the hospital is on the left. (Distance Approx 13.5 miles)

NEAREST SOUTHBOUND A&E HOSPITAL

George Eliot Hospital, College Street, Nuneaton CV10 7DJ Tel: (01203) 351351

Take the A5 west towards Atherstone and after about 2.9 miles turn left along the A47 to Nuneaton town centre. Follow the A444 south towards Coventry and turn right into College Street. The hospital is on the left. (Distance Approx 6.1 miles)

FACILITIES

1 The Gables

Tel: (01455) 632492
0.1 miles north along the A5, on the right

2 Star Three Pots Service Station (Texaco)

Tel: (01455) 620940
0.2 miles north along the A5, on the right. Access, Visa, Overdrive, All Star, Switch, Dial Card, Mastercard, Amex, Diners Club, Delta, Routex, BP Supercharge, Keyfuels, Texaco Cards.

3 The Three Pots Harvester

Tel: (01455) 615408
0.3 miles north along the A5, on the right. (Bass) Open all day. Bar Snacks served 12.00-21.00hrs daily, Restaurant Open; Sun-Fri 12.00-21.00hrs, Sat; 12.00-23.00hrs.

4 The Hinckley Knight

Tel: (01455) 610773
0.4 miles north along the A5, on the right. (Greenalls) Open all day. Meals served; 12.00-21.30hrs daily

5 Sketchley Grange Hotel

Tel: (01455) 251133
0.9 miles north along the B4109, on the left. The Terrace Bistro; Open; Mon-Sat; 11.00-23.00hrs, Sun; 11.00-22.30hrs. The Willow Restaurant; Open; Mon-Fri; 12.00-14.00hrs & 19.00-21.30hrs

6 Barnacles Restaurant

Tel: (01455) 633220
0.1 miles south along the A5, on the right. Open; Mon; 19.00-22.00hrs, Tues-Sat; 12.00-14.00hrs & 19.00-22.00hrs.

7 Hanover International Hotel & Restaurant

Tel: (01455) 631122
0.2 miles south along the A5, on the left. Brasserie Open; Breakfast; 07.00-10.00hrs daily, Lunch; Mon-Sat; 12.00-14.00hrs, Sun; 12.30-15.00hrs, Dinner; 19.00-22.00hrs daily. There is also an all day menu from 12.00-19.00hrs daily.

M69 JUNCTION 2

THIS IS A RESTRICTED ACCESS JUNCTION

There is no exit for northbound vehicles
There is no access for southbound vehicles

| Large Hotel | Small Hotel | Motel | Guest House/ Bed & Breakfast | Disabled Facilities | Childrens Facilities |

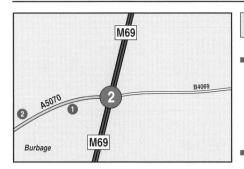

M1 — JUNCTION 21

MOTORWAY ENDS

(TOTAL LENGTH OF MOTORWAY 15.7 MILES)

NEAREST NORTHBOUND A&E HOSPITAL

Leicester Royal Infirmary, Infirmary Square, Leicester LE1 5WW Tel: (0116) 254 1414

Proceed to Junction 21 (M1) and take the exit to Leicester. Follow the A5460 into the city and turn right along Upperton Road. Turn left at the end and the hospital is on the left. (Distance Approx 10.6 miles)

NEAREST SOUTHBOUND A&E HOSPITAL

George Eliot Hospital, College Street, Nuneaton CV10 7DJ Tel: (01203) 351351

Proceed to Junction 1, take the A5 west towards Atherstone and after about 2.9 miles turn left along the A47 to Nuneaton town centre. Follow the A444 south towards Coventry and turn right into College Street. The hospital is on the left. (Distance Approx 8.7 miles)

FACILITIES

1   Woodside Garage

P **WC**   Tel: (01455) 632253
0.6 miles west along the A5070, on the left. Access, Visa, Overdrive, All Star, Switch, Dial Card, Mastercard, Amex, Diners Club, Delta, UK Fuelcard. Open; Mon-Fri; 06.00-20.00hrs, Sat; 06.30-19.00hrs, Sun; 08.00-19.00hrs. Attended Service.

2   Wynnes Motor Services (Jet)

P **WC**   Tel: (01455) 610213
1 mile west along the A5070, on the right. Access, Visa, Overdrive, All Star, Switch, Dial Card, Mastercard, Amex, Diners Club, Delta, Jet Cards. Open; Mon-Sat; 07.00-20.00hrs, Sun; 08.30-16.30hrs. Attended Service.

| Pets Welcome | £ Cash Dispenser | 24 Hour Facilities | Toilets | Alcoholic Drinks | Food and Drink | Fuel |

M69

VIEW FROM THE MOTORWAY

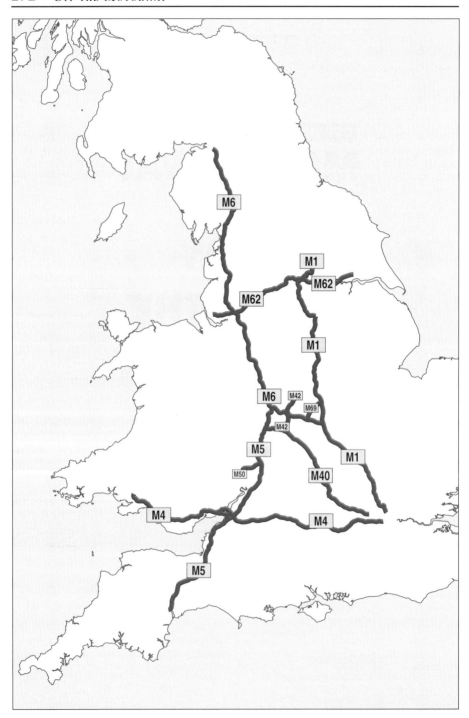

View from the Motorway

In recent years, whilst travelling up and down our motorway system, numerous things visible from the carriageways have exercised our curiosity, such as; what is the story behind the water tower adjacent to the M40, why was there a bridge that appeared to go absolutely nowhere between Junctions 1 and 2 on the M1, and, more curiously still, what caused the huge dent in a bridge between Junctions 4 and 5 northbound on the M5? This was especially puzzling as this was not the first bridge from Junction 5 but the second and logic would assume that an oversized load or vehicle would have not progressed beyond the first. In setting out to discover what did happen here, it occurred to us that others may have been curious too, and hence "View from the Motorway", a brief resume of points of interest in view from the carriageways, was formulated.

What follows is a summary of geographical and architectural landmarks, civil engineering constructions and an identification of the railway lines and rivers and canal systems that criss-cross the motorways with, where applicable, a few background notes. Adjacent marker post numbers have been identified but, with the exception of bridges which are more or less adjacent to these, they are purely a guide to place the reader in the correct section of the motorway in which to observe the item in question and, obviously, perspectives vary depending upon which direction the vehicle is travelling. Similarly, some views may be obstructed in the summer when the trees screening the carriageways have a full complement of leaves. In the main, the marker posts tally with the view when travelling from Junction 1.

We trust that this section may go some way to answering some of the questions posed in the past as to just "what is that?" and, perhaps, it may help to while away the time on an otherwise uninteresting motorway journey.

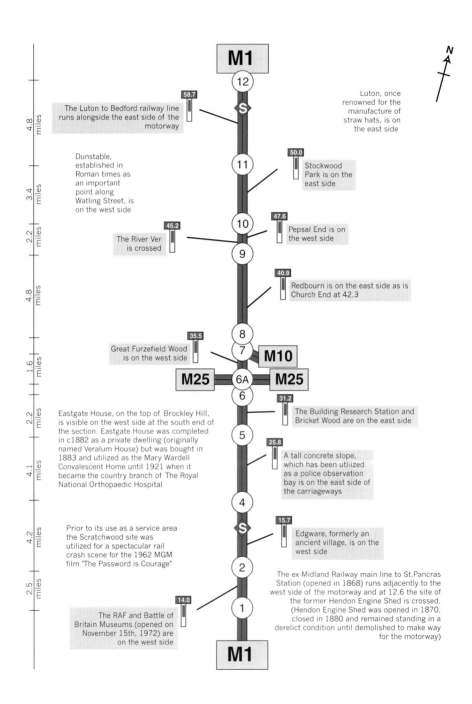

M1

12

58.7

The Luton to Bedford railway line runs alongside the east side of the motorway

Luton, once renowned for the manufacture of straw hats, is on the east side

11

50.0

Stockwood Park is on the east side

Dunstable, established in Roman times as an important point along Watling Street, is on the west side

10

47.6

Pepsal End is on the west side

45.2

The River Ver is crossed

9

40.9

Redbourn is on the east side as is Church End at 42.3

8

35.5

Great Furzefield Wood is on the west side

7 **M10**

M25 6A **M25**

6

31.2

The Building Research Station and Bricket Wood are on the east side

Eastgate House, on the top of Brockley Hill, is visible on the west side at the south end of the section. Eastgate House was completed in c1882 as a private dwelling (originally named Veralum House) but was bought in 1883 and utilized as the Mary Wardell Convalescent Home until 1921 when it became the country branch of The Royal National Orthopaedic Hospital

5

25.8

A tall concrete slope, which has been utliized as a police observation bay is on the east side of the carriageways

4

15.7

Edgware, formerly an ancient village, is on the west side

Prior to its use as a service area the Scratchwood site was utilized for a spectacular rail crash scene for the 1962 MGM film "The Password is Courage"

2

The ex-Midland Railway main line to St.Pancras Station (opened in 1868) runs adjacently to the west side of the motorway and at 12.6 the site of the former Hendon Engine Shed is crossed. (Hendon Engine Shed was opened in 1870, closed in 1880 and remained standing in a derelict condition until demolished to make way for the motorway)

14.0

The RAF and Battle of Britain Museums (opened on November 15th, 1972) are on the west side

1

M1

miles 4.8

miles 3.4

miles 2.2

miles 4.8

miles 1.6

miles 2.2

miles 4.1

miles 4.2

miles 2.5

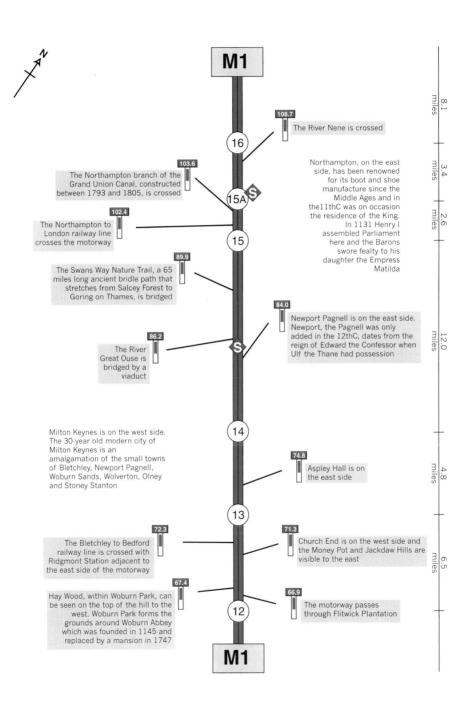

N

M1

108.7
The River Nene is crossed

16

103.6
The Northampton branch of the Grand Union Canal, constructed between 1793 and 1805, is crossed

Northampton, on the east side, has been renowned for its boot and shoe manufacture since the Middle Ages and in the11thC was on occasion the residence of the King. In 1131 Henry I assembled Parliament here and the Barons swore fealty to his daughter the Empress Matilda

15A S

102.4
The Northampton to London railway line crosses the motorway

15

89.9
The Swans Way Nature Trail, a 65 miles long ancient bridle path that stretches from Salcey Forest to Goring on Thames, is bridged

84.0
Newport Pagnell is on the east side. Newport, the Pagnell was only added in the 12thC, dates from the reign of Edward the Confessor when Ulf the Thane had possession

86.2
The River Great Ouse is bridged by a viaduct

S

Milton Keynes is on the west side. The 30-year old modern city of Milton Keynes is an amalgamation of the small towns of Bletchley, Newport Pagnell, Woburn Sands, Wolverton, Olney and Stoney Stanton

14

74.8
Aspley Hall is on the east side

13

72.3
The Bletchley to Bedford railway line is crossed with Ridgmont Station adjacent to the east side of the motorway

71.3
Church End is on the west side and the Money Pot and Jackdaw Hills are visible to the east

67.4
Hay Wood, within Woburn Park, can be seen on the top of the hill to the west. Woburn Park forms the grounds around Woburn Abbey which was founded in 1145 and replaced by a mansion in 1747

66.9
The motorway passes through Flitwick Plantation

12

M1

8.1 miles

3.4 miles

2.6 miles

12.0 miles

4.8 miles

6.5 miles

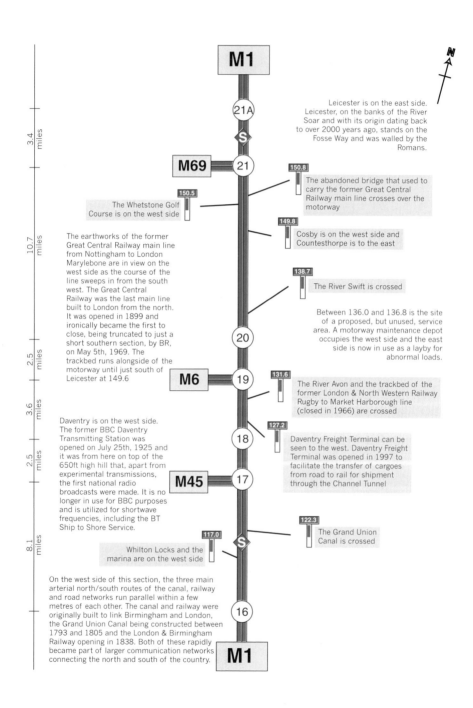

M1

21A

S

M69 — 21

Leicester is on the east side. Leicester, on the banks of the River Soar and with its origin dating back to over 2000 years ago, stands on the Fosse Way and was walled by the Romans.

3.4 miles

150.8

The abandoned bridge that used to carry the former Great Central Railway main line crosses over the motorway

150.5

The Whetstone Golf Course is on the west side

149.8

Cosby is on the west side and Countesthorpe is to the east

10.7 miles

The earthworks of the former Great Central Railway main line from Nottingham to London Marylebone are in view on the west side as the course of the line sweeps in from the south west. The Great Central Railway was the last main line built to London from the north. It was opened in 1899 and ironically became the first to close, being truncated to just a short southern section, by BR, on May 5th, 1969. The trackbed runs alongside of the motorway until just south of Leicester at 149.6

138.7

The River Swift is crossed

Between 136.0 and 136.8 is the site of a proposed, but unused, service area. A motorway maintenance depot occupies the west side and the east side is now in use as a layby for abnormal loads.

20

2.5 miles

M6 — 19

131.6

The River Avon and the trackbed of the former London & North Western Railway Rugby to Market Harborough line (closed in 1966) are crossed

3.6 miles

127.2

Daventry is on the west side. The former BBC Daventry Transmitting Station was opened on July 25th, 1925 and it was from here on top of the 650ft high hill that, apart from experimental transmissions, the first national radio broadcasts were made. It is no longer in use for BBC purposes and is utilized for shortwave frequencies, including the BT Ship to Shore Service.

18

Daventry Freight Terminal can be seen to the west. Daventry Freight Terminal was opened in 1997 to facilitate the transfer of cargoes from road to rail for shipment through the Channel Tunnel

2.5 miles

M45 — 17

122.3

The Grand Union Canal is crossed

117.0 S

Whilton Locks and the marina are on the west side

8.1 miles

On the west side of this section, the three main arterial north/south routes of the canal, railway and road networks run parallel within a few metres of each other. The canal and railway were originally built to link Birmingham and London, the Grand Union Canal being constructed between 1793 and 1805 and the London & Birmingham Railway opening in 1838. Both of these rapidly became part of larger communication networks connecting the north and south of the country.

16

M1

N

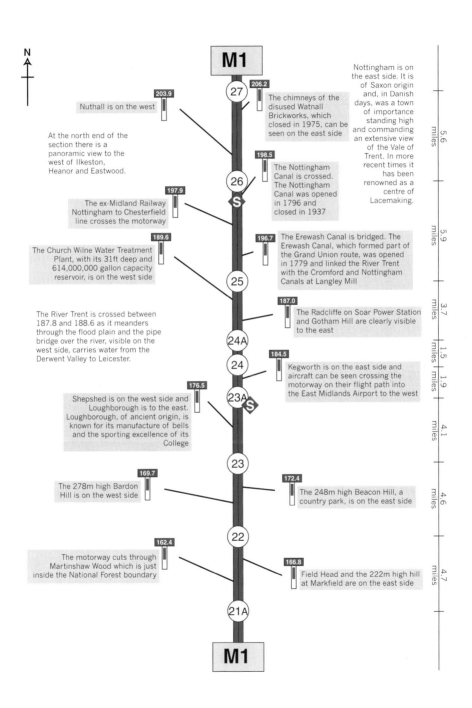

N

M1

27   206.2

Nuthall is on the west   203.9

Nottingham is on the east side. It is of Saxon origin and, in Danish days, was a town of importance standing high and commanding an extensive view of the Vale of Trent. In more recent times it has been renowned as a centre of Lacemaking.

The chimneys of the disused Watnall Brickworks, which closed in 1975, can be seen on the east side

At the north end of the section there is a panoramic view to the west of Ilkeston, Heanor and Eastwood.

198.5

26

S

The Nottingham Canal is crossed. The Nottingham Canal was opened in 1796 and closed in 1937

197.9

The ex-Midland Railway Nottingham to Chesterfield line crosses the motorway

196.7

The Erewash Canal is bridged. The Erewash Canal, which formed part of the Grand Union route, was opened in 1779 and linked the River Trent with the Cromford and Nottingham Canals at Langley Mill

189.6

The Church Wilne Water Treatment Plant, with its 31ft deep and 614,000,000 gallon capacity reservoir, is on the west side

25

187.0

The Radcliffe on Soar Power Station and Gotham Hill are clearly visible to the east

The River Trent is crossed between 187.8 and 188.6 as it meanders through the flood plain and the pipe bridge over the river, visible on the west side, carries water from the Derwent Valley to Leicester.

24A

24   184.5

Kegworth is on the east side and aircraft can be seen crossing the motorway on their flight path into the East Midlands Airport to the west

176.5

23A   **S**

Shepshed is on the west side and Loughborough is to the east. Loughborough, of ancient origin, is known for its manufacture of bells and the sporting excellence of its College

23

169.7

The 278m high Bardon Hill is on the west side

172.4

The 248m high Beacon Hill, a country park, is on the east side

22

162.4

The motorway cuts through Martinshaw Wood which is just inside the National Forest boundary

166.8

Field Head and the 222m high hill at Markfield are on the east side

21A

M1

5.6 miles

5.9 miles

3.7 miles

1.5 miles

1.9 miles

4.1 miles

4.6 miles

4.7 miles

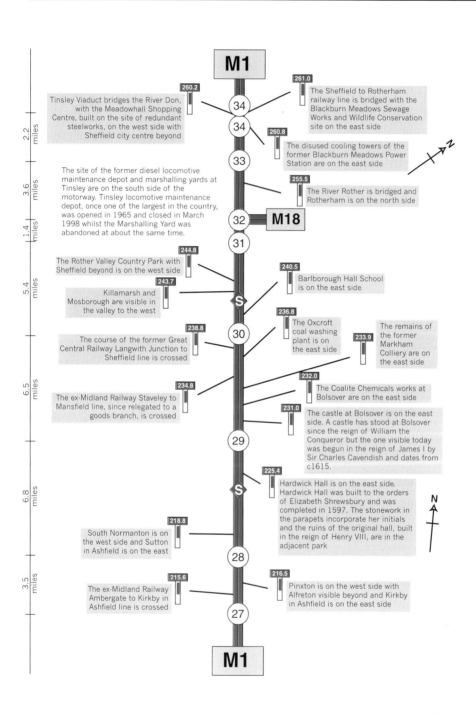

M1

261.0 The Sheffield to Rotherham railway line is bridged with the Blackburn Meadows Sewage Works and Wildlife Conservation site on the east side

260.2 Tinsley Viaduct bridges the River Don, with the Meadowhall Shopping Centre, built on the site of redundant steelworks, on the west side with Sheffield city centre beyond

34

34

260.8

260.8 The disused cooling towers of the former Blackburn Meadows Power Station are on the east side

33

The site of the former diesel locomotive maintenance depot and marshalling yards at Tinsley are on the south side of the motorway. Tinsley locomotive maintenance depot, once one of the largest in the country, was opened in 1965 and closed in March 1998 whilst the Marshalling Yard was abandoned at about the same time.

255.5 The River Rother is bridged and Rotherham is on the north side

32   **M18**

31

244.8

The Rother Valley Country Park with Sheffield beyond is on the west side

240.5 Barlborough Hall School is on the east side

243.7 Killamarsh and Mosborough are visible in the valley to the west

S

236.8 The Oxcroft coal washing plant is on the east side

233.9 The remains of the former Markham Colliery are on the east side

238.8 The course of the former Great Central Railway Langwith Junction to Sheffield line is crossed

30

232.0 The Coalite Chemicals works at Bolsover are on the east side

234.8 The ex-Midland Railway Staveley to Mansfield line, since relegated to a goods branch, is crossed

231.0 The castle at Bolsover is on the east side. A castle has stood at Bolsover since the reign of William the Conqueror but the one visible today was begun in the reign of James I by Sir Charles Cavendish and dates from c1615.

29

225.4 Hardwick Hall is on the east side. Hardwick Hall was built to the orders of Elizabeth Shrewsbury and was completed in 1597. The stonework in the parapets incorporate her initials and the ruins of the original hall, built in the reign of Henry VIII, are in the adjacent park

S

N

218.8 South Normanton is on the west side and Sutton in Ashfield is on the east

28

215.6 The ex-Midland Railway Ambergate to Kirkby in Ashfield line is crossed

216.5 Pinxton is on the west side with Alfreton visible beyond and Kirkby in Ashfield is on the east side

27

M1

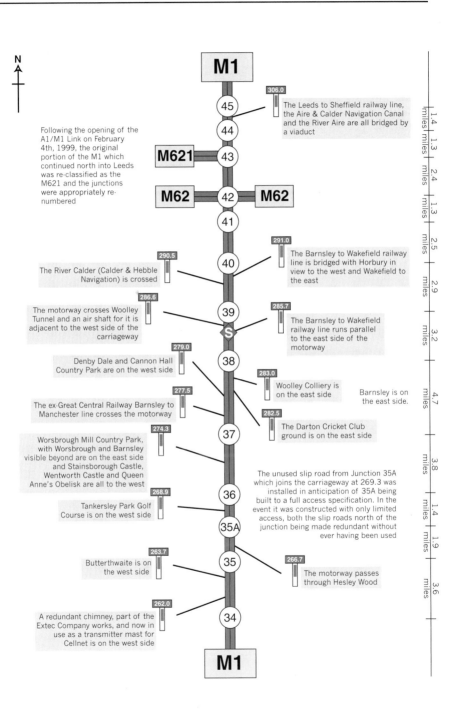

N

M1

45

306.0 The Leeds to Sheffield railway line, the Aire & Calder Navigation Canal and the River Aire are all bridged by a viaduct

44

Following the opening of the A1/M1 Link on February 4th, 1999, the original portion of the M1 which continued north into Leeds was re-classified as the M621 and the junctions were appropriately re-numbered

M621 43

M62 42 **M62**

41

40

291.0 The Barnsley to Wakefield railway line is bridged with Horbury in view to the west and Wakefield to the east

290.5 The River Calder (Calder & Hebble Navigation) is crossed

286.6 The motorway crosses Woolley Tunnel and an air shaft for it is adjacent to the west side of the carriageway

39

S

285.7 The Barnsley to Wakefield railway line runs parallel to the east side of the motorway

279.0 Denby Dale and Cannon Hall Country Park are on the west side

38

283.0 Woolley Colliery is on the east side

Barnsley is on the east side.

277.5 The ex-Great Central Railway Barnsley to Manchester line crosses the motorway

282.5 The Darton Cricket Club ground is on the east side

274.3 Worsbrough Mill Country Park, with Worsbrough and Barnsley visible beyond are on the east side and Stainsborough Castle, Wentworth Castle and Queen Anne's Obelisk are all to the west

37

268.9 Tankersley Park Golf Course is on the west side

36

The unused slip road from Junction 35A which joins the carriageway at 269.3 was installed in anticipation of 35A being built to a full access specification. In the event it was constructed with only limited access, both the slip roads north of the junction being made redundant without ever having been used

35A

263.7 Butterthwaite is on the west side

35

266.7 The motorway passes through Hesley Wood

262.0 A redundant chimney, part of the Extec Company works, and now in use as a transmitter mast for Cellnet is on the west side

34

M1

1.4 miles | 1.3 miles | 2.4 miles | 1.3 miles | 2.5 miles | 2.9 miles | 3.2 miles | 4.7 miles | 3.8 miles | 1.4 miles | 1.9 miles | 3.6 miles

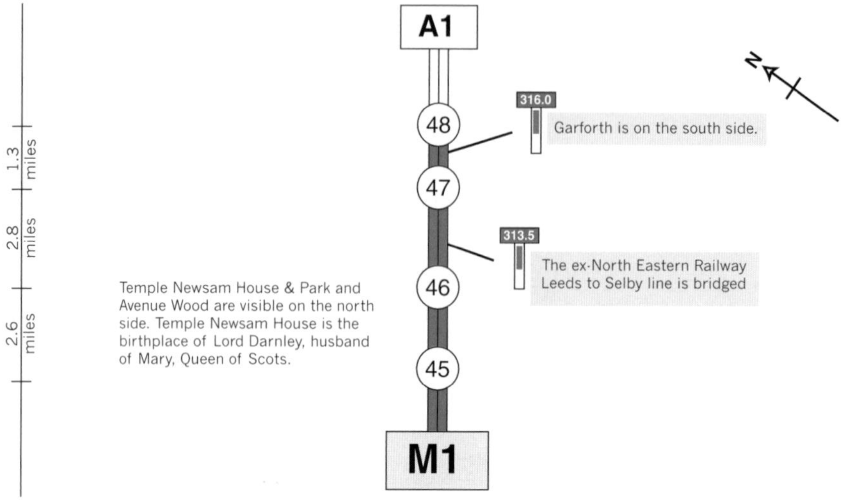

A1

48

316.0

Garforth is on the south side.

47

313.5

The ex-North Eastern Railway
Leeds to Selby line is bridged

Temple Newsam House & Park and
Avenue Wood are visible on the north
side. Temple Newsam House is the
birthplace of Lord Darnley, husband
of Mary, Queen of Scots.

46

45

M1

1.3 miles

2.8 miles

2.6 miles

N

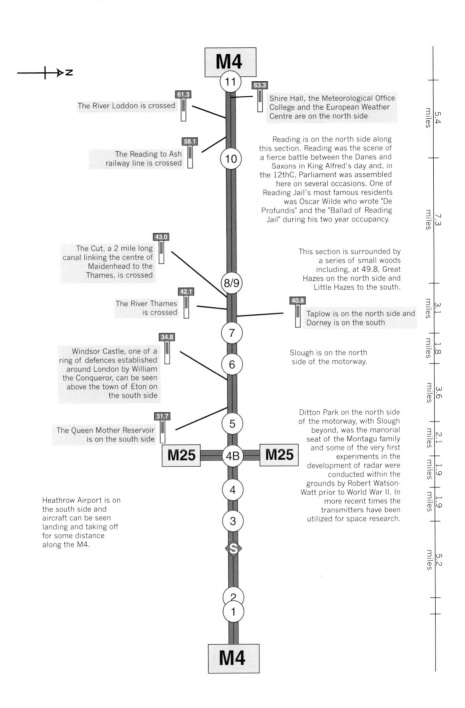

N

M4

11

61.3
The River Loddon is crossed

63.3
Shire Hall, the Meteorological Office College and the European Weather Centre are on the north side

58.1
The Reading to Ash railway line is crossed

10

Reading is on the north side along this section. Reading was the scene of a fierce battle between the Danes and Saxons in King Alfred's day and, in the 12thC, Parliament was assembled here on several occasions. One of Reading Jail's most famous residents was Oscar Wilde who wrote "De Profundis" and the "Ballad of Reading Jail" during his two year occupancy.

43.0
The Cut, a 2 mile long canal linking the centre of Maidenhead to the Thames, is crossed

8/9

This section is surrounded by a series of small woods including, at 49.8, Great Hazes on the north side and Little Hazes to the south.

42.1
The River Thames is crossed

40.8
Taplow is on the north side and Dorney is on the south

7

34.8
Windsor Castle, one of a ring of defences established around London by William the Conqueror, can be seen above the town of Eton on the south side

6

Slough is on the north side of the motorway.

5

31.7
The Queen Mother Reservoir is on the south side

M25 4B **M25**

Ditton Park on the north side of the motorway, with Slough beyond, was the manorial seat of the Montagu family and some of the very first experiments in the development of radar were conducted within the grounds by Robert Watson-Watt prior to World War II. In more recent times the transmitters have been utilized for space research.

4

Heathrow Airport is on the south side and aircraft can be seen landing and taking off for some distance along the M4.

3

S

2
1

M4

5.4 miles

7.3 miles

3.1 miles

1.8 miles

3.6 miles

2.1 miles

1.9 miles

1.9 miles

5.2 miles

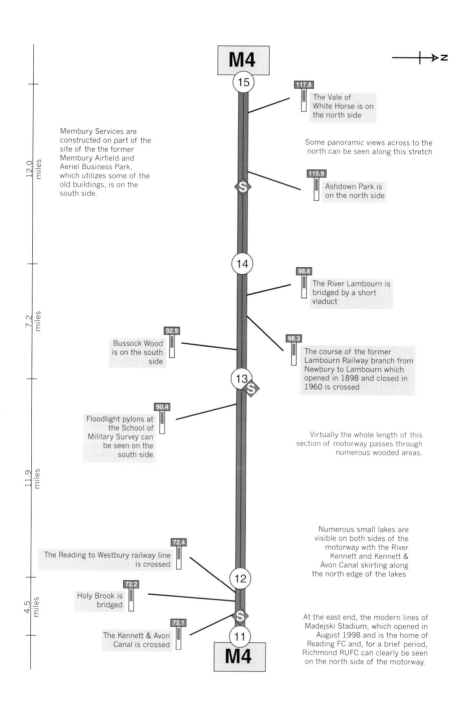

M4

15

117.8
The Vale of
White Horse is on
the north side

Membury Services are
constructed on part of the
site of the the former
Membury Airfield and
Aeriel Business Park,
which utilizes some of the
old buildings, is on the
south side.

Some panoramic views across to the
north can be seen along this stretch

S

115.9
Ashdown Park is
on the north side

14

98.6
The River Lambourn is
bridged by a short
viaduct

92.9
Bussock Wood
is on the south
side

98.3
The course of the former
Lambourn Railway branch from
Newbury to Lambourn which
opened in 1898 and closed in
1960 is crossed

13 S

90.4
Floodlight pylons at
the School of
Military Survey can
be seen on the
south side

Virtually the whole length of this
section of motorway passes through
numerous wooded areas.

Numerous small lakes are
visible on both sides of the
motorway with the River
Kennett and Kennett &
Avon Canal skirting along
the north edge of the lakes

72.4
The Reading to Westbury railway line
is crossed

12

72.2
Holy Brook is
bridged

S

72.1
The Kennett & Avon
Canal is crossed

11

M4

At the east end, the modern lines of
Madejski Stadium, which opened in
August 1998 and is the home of
Reading FC and, for a brief period,
Richmond RUFC can clearly be seen
on the north side of the motorway.

12.0 miles

7.2 miles

11.9 miles

4.5 miles

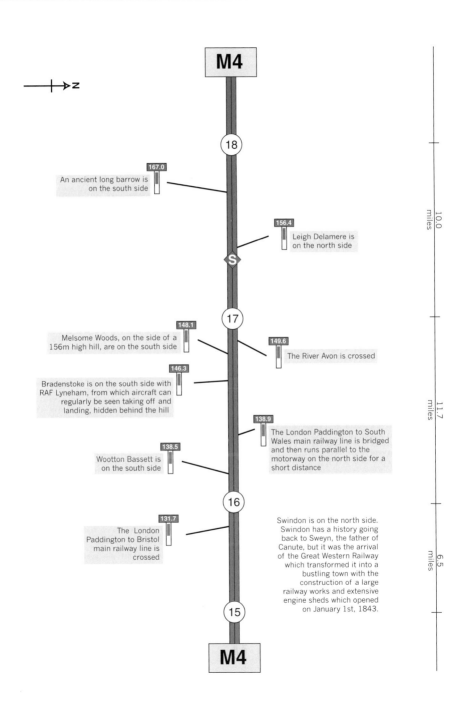

M4

18

167.0 An ancient long barrow is on the south side

156.4 Leigh Delamere is on the north side

S

17

148.1 Melsome Woods, on the side of a 156m high hill, are on the south side

149.6 The River Avon is crossed

146.3 Bradenstoke is on the south side with RAF Lyneham, from which aircraft can regularly be seen taking off and landing, hidden behind the hill

138.9 The London Paddington to South Wales main railway line is bridged and then runs parallel to the motorway on the north side for a short distance

138.5 Wootton Bassett is on the south side

16

131.7 The London Paddington to Bristol main railway line is crossed

Swindon is on the north side. Swindon has a history going back to Sweyn, the father of Canute, but it was the arrival of the Great Western Railway which transformed it into a bustling town with the construction of a large railway works and extensive engine sheds which opened on January 1st, 1843.

15

M4

10.0 miles

11.7 miles

6.5 miles

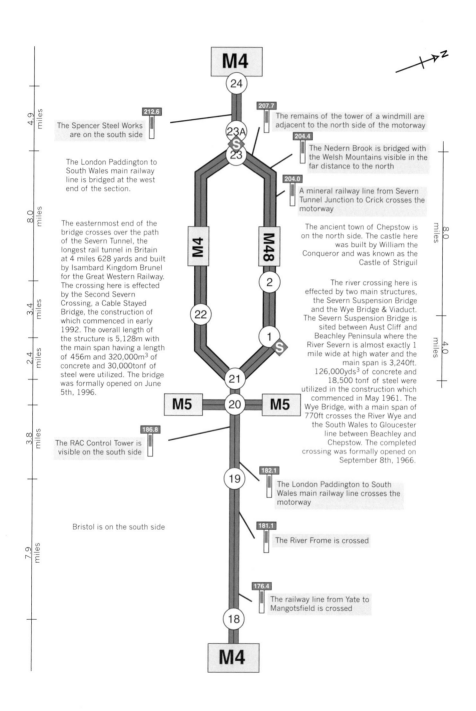

M4
24

212.6
The Spencer Steel Works
are on the south side

The London Paddington to
South Wales main railway
line is bridged at the west
end of the section.

The easternmost end of the
bridge crosses over the path
of the Severn Tunnel, the
longest rail tunnel in Britain
at 4 miles 628 yards and built
by Isambard Kingdom Brunel
for the Great Western Railway.
The crossing here is effected
by the Second Severn
Crossing, a Cable Stayed
Bridge, the construction of
which commenced in early
1992. The overall length of
the structure is 5,128m with
the main span having a length
of 456m and 320,000m³ of
concrete and 30,000tonf of
steel was utilized. The bridge
was formally opened on June
5th, 1996.

M4
22

M5
20

186.8
The RAC Control Tower is
visible on the south side

Bristol is on the south side

207.7
The remains of the tower of a windmill are
adjacent to the north side of the motorway

23A
S
23

204.4
The Nedern Brook is bridged with
the Welsh Mountains visible in the
far distance to the north

204.0
A mineral railway line from Severn
Tunnel Junction to Crick crosses the
motorway

The ancient town of Chepstow is
on the north side. The castle here
was built by William the
Conqueror and was known as the
Castle of Striguil

M48
2

The river crossing here is
effected by two main structures,
the Severn Suspension Bridge
and the Wye Bridge & Viaduct.
The Severn Suspension Bridge is
sited between Aust Cliff and
Beachley Peninsula where the
River Severn is almost exactly 1
mile wide at high water and the
main span is 3,240ft.
126,000yds³ of concrete and
18,500 tonf of steel were
utilized in the construction which
commenced in May 1961. The
Wye Bridge, with a main span of
770ft crosses the River Wye and
the South Wales to Gloucester
line between Beachley and
Chepstow. The completed
crossing was formally opened on
September 8th, 1966.

1
S

21

M5

19

182.1
The London Paddington to South
Wales main railway line crosses the
motorway

181.1
The River Frome is crossed

176.4
The railway line from Yate to
Mangotsfield is crossed

18

M4

4.9 miles

8.0 miles

3.4 miles

2.4 miles

3.8 miles

7.9 miles

8.0 miles

4.0 miles

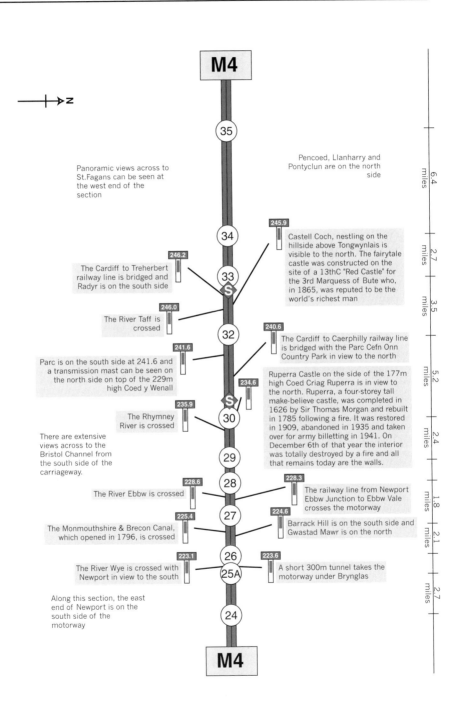

M4

N

35

Panoramic views across to St.Fagans can be seen at the west end of the section

Pencoed, Llanharry and Pontyclun are on the north side

34

245.9 Castell Coch, nestling on the hillside above Tongwynlais is visible to the north. The fairytale castle was constructed on the site of a 13thC "Red Castle" for the 3rd Marquess of Bute who, in 1865, was reputed to be the world's richest man

246.2 The Cardiff to Treherbert railway line is bridged and Radyr is on the south side

33 S

246.0 The River Taff is crossed

32

240.6 The Cardiff to Caerphilly railway line is bridged with the Parc Cefn Onn Country Park in view to the north

241.6 Parc is on the south side at 241.6 and a transmission mast can be seen on the north side on top of the 229m high Coed y Wenall

234.6 Ruperra Castle on the side of the 177m high Coed Criag Ruperra is in view to the north. Ruperra, a four-storey tall make-believe castle, was completed in 1626 by Sir Thomas Morgan and rebuilt in 1785 following a fire. It was restored in 1909, abandoned in 1935 and taken over for army billetting in 1941. On December 6th of that year the interior was totally destroyed by a fire and all that remains today are the walls.

235.9 The Rhymney River is crossed

S **30**

There are extensive views across to the Bristol Channel from the south side of the carriageway.

29

228.6 The River Ebbw is crossed

28

228.3 The railway line from Newport Ebbw Junction to Ebbw Vale crosses the motorway

225.4 The Monmouthshire & Brecon Canal, which opened in 1796, is crossed

27

224.6 Barrack Hill is on the south side and Gwastad Mawr is on the north

26

223.1 The River Wye is crossed with Newport in view to the south

25A

223.6 A short 300m tunnel takes the motorway under Brynglas

Along this section, the east end of Newport is on the south side of the motorway

24

M4

6.4 miles
2.7 miles
3.5 miles
5.2 miles
2.4 miles
1.8 miles
2.1 miles
2.7 miles

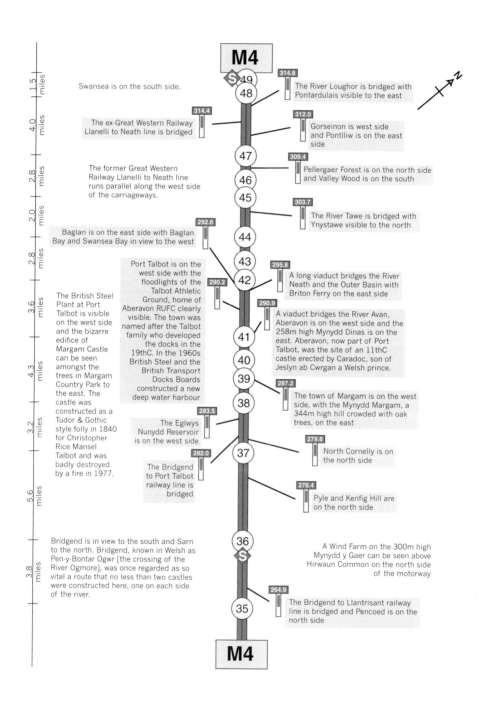

M4

314.8

Swansea is on the south side.

The River Loughor is bridged with Pontardulais visible to the east

314.4

The ex-Great Western Railway Llanelli to Neath line is bridged

312.0

Gorseinon is west side and Pontlliw is on the east side

The former Great Western Railway Llanelli to Neath line runs parallel along the west side of the carriageways.

309.4

Pellergaer Forest is on the north side and Valley Wood is on the south

303.7

The River Tawe is bridged with Ynystawe visible to the north

292.6

Baglan is on the east side with Baglan Bay and Swansea Bay in view to the west

295.8

Port Talbot is on the west side with the floodlights of the Talbot Athletic Ground, home of Aberavon RUFC clearly visible. The town was named after the Talbot family who developed the docks in the 19thC. In the 1960s British Steel and the British Transport Docks Boards constructed a new deep water harbour

290.2

A long viaduct bridges the River Neath and the Outer Basin with Briton Ferry on the east side

290.9

The British Steel Plant at Port Talbot is visible on the west side and the bizarre edifice of Margam Castle can be seen amongst the trees in Margam Country Park to the east. The castle was constructed as a Tudor & Gothic style folly in 1840 for Christopher Rice Mansel Talbot and was badly destroyed by a fire in 1977.

A viaduct bridges the River Avan, Aberavon is on the west side and the 258m high Mynydd Dinas is on the east. Aberavon, now part of Port Talbot, was the site of an 11thC castle erected by Caradoc, son of Jeslyn ab Cwrgan a Welsh prince.

287.2

The town of Margam is on the west side, with the Mynydd Margam, a 344m high hill crowded with oak trees, on the east

283.5

The Eglwys Nunydd Reservoir is on the west side

279.6

North Cornelly is on the north side

282.0

The Bridgend to Port Talbot railway line is bridged

276.4

Pyle and Kenfig Hill are on the north side

Bridgend is in view to the south and Sarn to the north. Bridgend, known in Welsh as Pen-y-Bontar Ogwr [the crossing of the River Ogmore], was once regarded as so vital a route that no less than two castles were constructed here, one on each side of the river.

A Wind Farm on the 300m high Mynydd y Gaer can be seen above Hirwaun Common on the north side of the motorway

264.9

The Bridgend to Llantrisant railway line is bridged and Pencoed is on the north side

M4

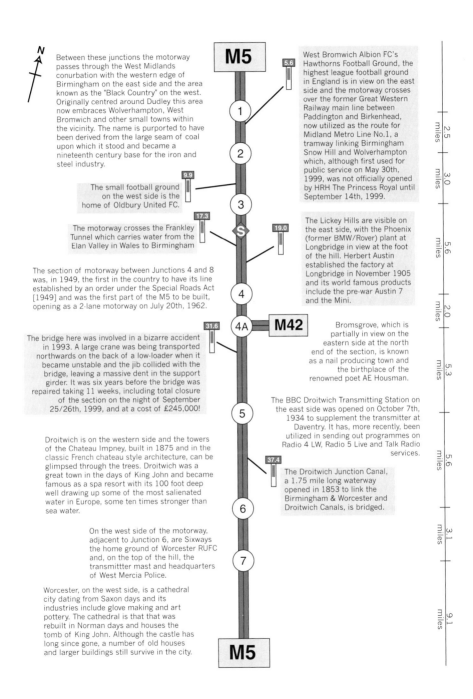

N

Between these junctions the motorway passes through the West Midlands conurbation with the western edge of Birmingham on the east side and the area known as the "Black Country" on the west. Originally centred around Dudley this area now embraces Wolverhampton, West Bromwich and other small towns within the vicinity. The name is purported to have been derived from the large seam of coal upon which it stood and became a nineteenth century base for the iron and steel industry.

M5

5.6

West Bromwich Albion FC's Hawthorns Football Ground, the highest league football ground in England is in view on the east side and the motorway crosses over the former Great Western Railway main line between Paddington and Birkenhead, now utilized as the route for Midland Metro Line No.1, a tramway linking Birmingham Snow Hill and Wolverhampton which, although first used for public service on May 30th, 1999, was not officially opened by HRH The Princess Royal until September 14th, 1999.

1

2

9.9

The small football ground on the west side is the home of Oldbury United FC.

3

17.3

The motorway crosses the Frankley Tunnel which carries water from the Elan Valley in Wales to Birmingham

S

19.0

The Lickey Hills are visible on the east side, with the Phoenix (former BMW/Rover) plant at Longbridge in view at the foot of the hill. Herbert Austin established the factory at Longbridge in November 1905 and its world famous products include the pre-war Austin 7 and the Mini.

The section of motorway between Junctions 4 and 8 was, in 1949, the first in the country to have its line established by an order under the Special Roads Act [1949] and was the first part of the M5 to be built, opening as a 2-lane motorway on July 20th, 1962.

4

31.6

4A   **M42**

Bromsgrove, which is partially in view on the eastern side at the north end of the section, is known as a nail producing town and the birthplace of the renowned poet AE Housman.

The bridge here was involved in a bizarre accident in 1993. A large crane was being transported northwards on the back of a low-loader when it became unstable and the jib collided with the bridge, leaving a massive dent in the support girder. It was six years before the bridge was repaired taking 11 weeks, including total closure of the section on the night of September 25/26th, 1999, and at a cost of £245,000!

5

The BBC Droitwich Transmitting Station on the east side was opened on October 7th, 1934 to supplement the transmitter at Daventry. It has, more recently, been utilized in sending out programmes on Radio 4 LW, Radio 5 Live and Talk Radio services.

Droitwich is on the western side and the towers of the Chateau Impney, built in 1875 and in the classic French chateau style architecture, can be glimpsed through the trees. Droitwich was a great town in the days of King John and became famous as a spa resort with its 100 foot deep well drawing up some of the most salienated water in Europe, some ten times stronger than sea water.

37.4

The Droitwich Junction Canal, a 1.75 mile long waterway opened in 1853 to link the Birmingham & Worcester and Droitwich Canals, is bridged.

6

On the west side of the motorway, adjacent to Junction 6, are Sixways the home ground of Worcester RUFC and, on the top of the hill, the transmitter mast and headquarters of West Mercia Police.

7

Worcester, on the west side, is a cathedral city dating from Saxon days and its industries include glove making and art pottery. The cathedral is that that was rebuilt in Norman days and houses the tomb of King John. Although the castle has long since gone, a number of old houses and larger buildings still survive in the city.

M5

2.5 miles

3.0 miles

5.6 miles

2.0 miles

5.3 miles

5.6 miles

3.1 miles

9.1 miles

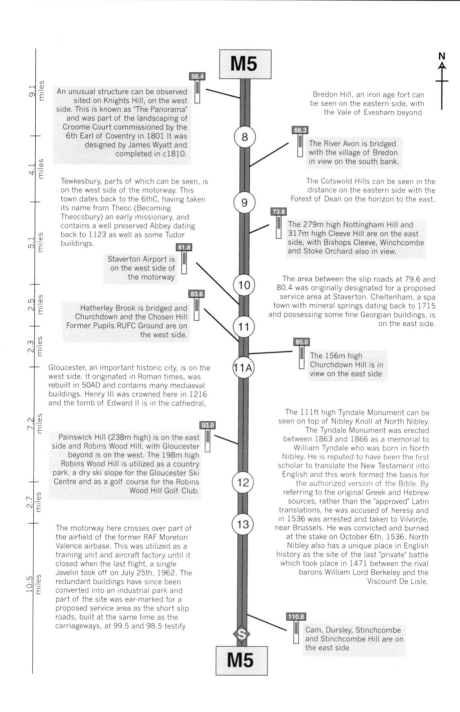

M5

56.4

N

An unusual structure can be observed sited on Knights Hill, on the west side. This is known as "The Panorama" and was part of the landscaping of Croome Court commissioned by the 6th Earl of Coventry in 1801 It was designed by James Wyatt and completed in c1810.

Bredon Hill, an iron age fort can be seen on the eastern side, with the Vale of Evesham beyond

66.3

8

The River Avon is bridged with the village of Bredon in view on the south bank.

Tewkesbury, parts of which can be seen, is on the west side of the motorway. This town dates back to the 6thC, having taken its name from Theoc (Becoming Theocsbury) an early missionary, and contains a well preserved Abbey dating back to 1123 as well as some Tudor buildings.

The Cotswold Hills can be seen in the distance on the eastern side with the Forest of Dean on the horizon to the east.

9

73.8

The 279m high Nottingham Hill and 317m high Cleeve Hill are on the east side, with Bishops Cleeve, Winchcombe and Stoke Orchard also in view.

81.8

Staverton Airport is on the west side of the motorway

The area between the slip roads at 79.6 and 80.4 was originally designated for a proposed service area at Staverton. Cheltenham, a spa town with mineral springs dating back to 1715 and possessing some fine Georgian buildings, is on the east side.

83.6

10

Hatherley Brook is bridged and Churchdown and the Chosen Hill Former Pupils RUFC Ground are on the west side.

11

85.0

The 156m high Churchdown Hill is in view on the east side

Gloucester, an important historic city, is on the west side. It originated in Roman times, was rebuilt in 50AD and contains many mediaeval buildings. Henry III was crowned here in 1216 and the tomb of Edward II is in the cathedral.

11A

The 111ft high Tyndale Monument can be seen on top of Nibley Knoll at North Nibley. The Tyndale Monument was erected between 1863 and 1866 as a memorial to William Tyndale who was born in North Nibley. He is reputed to have been the first scholar to translate the New Testament into English and this work formed the basis for the authorized version of the Bible. By referring to the original Greek and Hebrew sources, rather than the "approved" Latin translations, he was accused of heresy and in 1536 was arrested and taken to Vilvorde, near Brussels. He was convicted and burned at the stake on October 6th, 1536. North Nibley also has a unique place in English history as the site of the last "private" battle which took place in 1471 between the rival barons William Lord Berkeley and the Viscount De Lisle.

93.0

Painswick Hill (238m high) is on the east side and Robins Wood Hill, with Gloucester beyond is on the west. The 198m high Robins Wood Hill is utilized as a country park, a dry ski slope for the Gloucester Ski Centre and as a golf course for the Robins Wood Hill Golf Club.

12

13

The motorway here crosses over part of the airfield of the former RAF Moreton Valence airbase. This was utilized as a training unit and aircraft factory until it closed when the last flight, a single Javelin took off on July 25th, 1962. The redundant buildings have since been converted into an industrial park and part of the site was ear-marked for a proposed service area as the short slip roads, built at the same time as the carriageways, at 99.5 and 98.5 testify

110.8

Cam, Dursley, Stinchcombe and Stinchcombe Hill are on the east side

M5

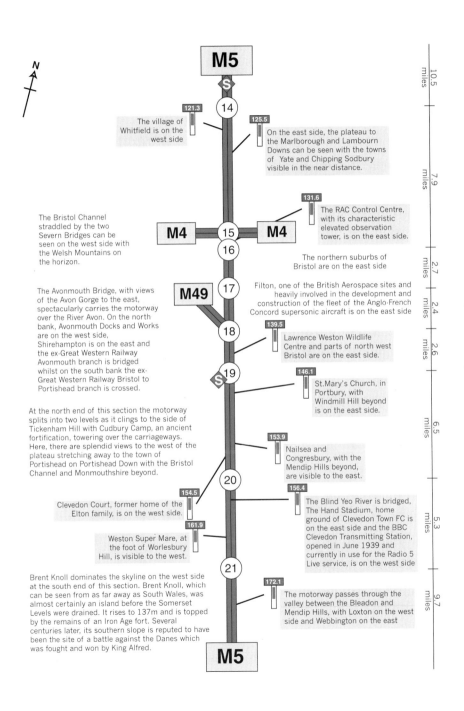

N

M5
S

121.3   14

The village of Whitfield is on the west side

125.5
On the east side, the plateau to the Marlborough and Lambourn Downs can be seen with the towns of Yate and Chipping Sodbury visible in the near distance.

The Bristol Channel straddled by the two Severn Bridges can be seen on the west side with the Welsh Mountains on the horizon.

131.6
The RAC Control Centre, with its characteristic elevated observation tower, is on the east side.

M4   15   **M4**

16

The northern suburbs of Bristol are on the east side

The Avonmouth Bridge, with views of the Avon Gorge to the east, spectacularly carries the motorway over the River Avon. On the north bank, Avonmouth Docks and Works are on the west side, Shirehampton is on the east and the ex-Great Western Railway Avonmouth branch is bridged whilst on the south bank the ex-Great Western Railway Bristol to Portishead branch is crossed.

M49   17

Filton, one of the British Aerospace sites and heavily involved in the development and construction of the fleet of the Anglo-French Concord supersonic aircraft is on the east side

18

139.5
Lawrence Weston Wildlife Centre and parts of north west Bristol are on the east side.

S   19

146.1
St.Mary's Church, in Portbury, with Windmill Hill beyond is on the east side.

At the north end of this section the motorway splits into two levels as it clings to the side of Tickenham Hill with Cudbury Camp, an ancient fortification, towering over the carriageways. Here, there are splendid views to the west of the plateau stretching away to the town of Portishead on Portishead Down with the Bristol Channel and Monmouthshire beyond.

153.9
Nailsea and Congresbury, with the Mendip Hills beyond, are visible to the east.

20

154.5
Clevedon Court, former home of the Elton family, is on the west side.

156.4
The Blind Yeo River is bridged, The Hand Stadium, home ground of Clevedon Town FC is on the east side and the BBC Clevedon Transmitting Station, opened in June 1939 and currently in use for the Radio 5 Live service, is on the west side

161.9
Weston Super Mare, at the foot of Worlesbury Hill, is visible to the west.

21

Brent Knoll dominates the skyline on the west side at the south end of this section. Brent Knoll, which can be seen from as far away as South Wales, was almost certainly an island before the Somerset Levels were drained. It rises to 137m and is topped by the remains of an Iron Age fort. Several centuries later, its southern slope is reputed to have been the site of a battle against the Danes which was fought and won by King Alfred.

172.1
The motorway passes through the valley between the Bleadon and Mendip Hills, with Loxton on the west side and Webbington on the east

M5

10.5 miles

7.9 miles

2.7 miles

2.4 miles

2.6 miles

6.5 miles

5.3 miles

9.7 miles

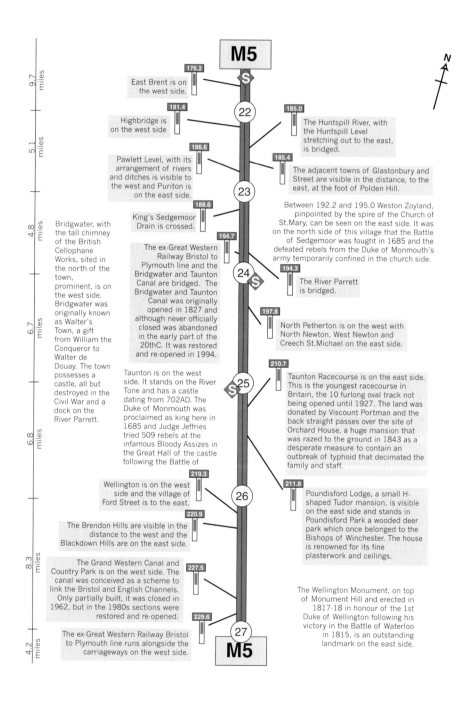

M5

N

176.2 East Brent is on the west side.

22

181.4 Highbridge is on the west side

185.0 The Huntspill River, with the Huntspill Level stretching out to the east, is bridged.

186.6 Pawlett Level, with its arrangement of rivers and ditches is visible to the west and Puriton is on the east side.

185.4 The adjacent towns of Glastonbury and Street are visible in the distance, to the east, at the foot of Polden Hill.

23

188.6 King's Sedgemoor Drain is crossed.

Between 192.2 and 195.0 Weston Zoyland, pinpointed by the spire of the Church of St.Mary, can be seen on the east side. It was on the north side of this village that the Battle of Sedgemoor was fought in 1685 and the defeated rebels from the Duke of Monmouth's army temporarily confined in the church side.

Bridgwater, with the tall chimney of the British Cellophane Works, sited in the north of the town, prominent, is on the west side. Bridgwater was originally known as Walter's Town, a gift from William the Conqueror to Walter de Douay. The town possesses a castle, all but destroyed in the Civil War and a dock on the River Parrett.

194.7 The ex-Great Western Railway Bristol to Plymouth line and the Bridgwater and Taunton Canal are bridged. The Bridgwater and Taunton Canal was originally opened in 1827 and although never officially closed was abandoned in the early part of the 20thC. It was restored and re-opened in 1994.

24

194.3 The River Parrett is bridged.

197.8 North Petherton is on the west with North Newton, West Newton and Creech St.Michael on the east side.

Taunton is on the west side. It stands on the River Tone and has a castle dating from 702AD. The Duke of Monmouth was proclaimed as king here in 1685 and Judge Jeffries tried 509 rebels at the infamous Bloody Assizes in the Great Hall of the castle following the Battle of

210.7 Taunton Racecourse is on the east side. This is the youngest racecourse in Britain, the 10 furlong oval track not being opened until 1927. The land was donated by Viscount Portman and the back straight passes over the site of Orchard House, a huge mansion that was razed to the ground in 1843 as a desperate measure to contain an outbreak of typhoid that decimated the family and staff.

25

219.3 Wellington is on the west side and the village of Ford Street is to the east.

26

211.8 Poundisford Lodge, a small H-shaped Tudor mansion, is visible on the east side and stands in Poundisford Park a wooded deer park which once belonged to the Bishops of Winchester. The house is renowned for its fine plasterwork and ceilings.

220.9 The Brendon Hills are visible in the distance to the west and the Blackdown Hills are on the east side.

The Grand Western Canal and Country Park is on the west side. The canal was conceived as a scheme to link the Bristol and English Channels. Only partially built, it was closed in 1962, but in the 1980s sections were restored and re-opened.

227.5

The Wellington Monument, on top of Monument Hill and erected in 1817-18 in honour of the 1st Duke of Wellington following his victory in the Battle of Waterloo in 1815, is an outstanding landmark on the east side.

229.6 The ex-Great Western Railway Bristol to Plymouth line runs alongside the carriageways on the west side.

27

M5

9.7 miles
5.1 miles
4.8 miles
6.7 miles
6.8 miles
8.3 miles
4.2 miles

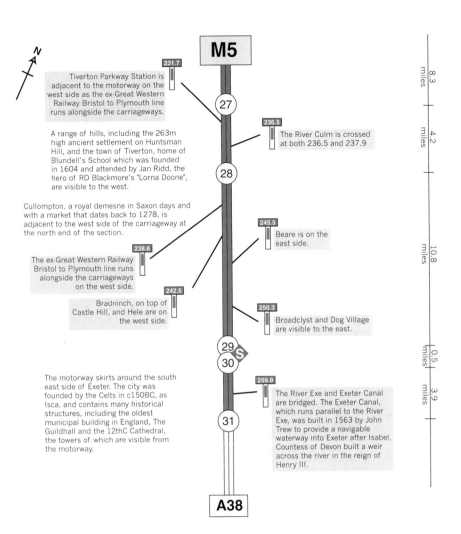

M5

27

28

29
30

31

A38

N

231.7
Tiverton Parkway Station is adjacent to the motorway on the west side as the ex-Great Western Railway Bristol to Plymouth line runs alongside the carriageways.

A range of hills, including the 263m high ancient settlement on Huntsman Hill, and the town of Tiverton, home of Blundell's School which was founded in 1604 and attended by Jan Ridd, the hero of RD Blackmore's "Lorna Doone", are visible to the west.

Cullompton, a royal demesne in Saxon days and with a market that dates back to 1278, is adjacent to the west side of the carriageway at the north end of the section.

238.6
The ex-Great Western Railway Bristol to Plymouth line runs alongside the carriageways on the west side.

242.5
Bradninch, on top of Castle Hill, and Hele are on the west side.

The motorway skirts around the south east side of Exeter. The city was founded by the Celts in c150BC, as Isca, and contains many historical structures, including the oldest municipal building in England, The Guildhall and the 12thC Cathedral, the towers of which are visible from the motorway.

236.5
The River Culm is crossed at both 236.5 and 237.9

245.5
Beare is on the east side.

250.3
Broadclyst and Dog Village are visible to the east.

259.9
The River Exe and Exeter Canal are bridged. The Exeter Canal, which runs parallel to the River Exe, was built in 1563 by John Trew to provide a navigable waterway into Exeter after Isabel, Countess of Devon built a weir across the river in the reign of Henry III.

8.3 miles

4.2 miles

10.8 miles

0.5 miles

3.9 miles

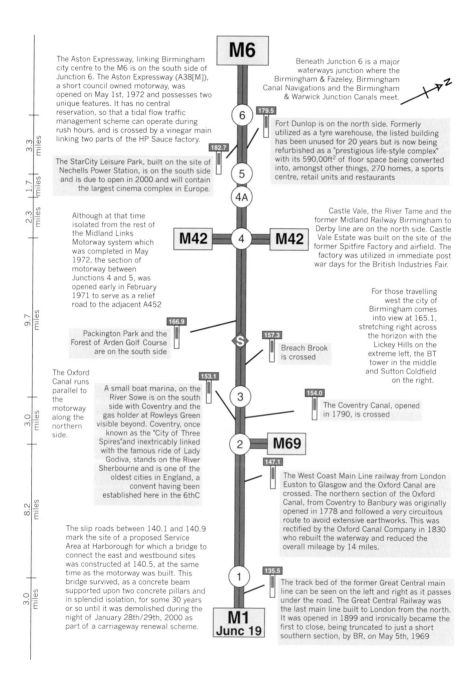

M6

The Aston Expressway, linking Birmingham city centre to the M6 is on the south side of Junction 6. The Aston Expressway (A38[M]), a short council owned motorway, was opened on May 1st, 1972 and possesses two unique features. It has no central reservation, so that a tidal flow traffic management scheme can operate during rush hours, and is crossed by a vinegar main linking two parts of the HP Sauce factory.

Beneath Junction 6 is a major waterways junction where the Birmingham & Fazeley, Birmingham Canal Navigations and the Birmingham & Warwick Junction Canals meet.

6 — 179.5

Fort Dunlop is on the north side. Formerly utilized as a tyre warehouse, the listed building has been unused for 20 years but is now being refurbished as a "prestigious life-style complex" with its 590,00ft² of floor space being converted into, amongst other things, 270 homes, a sports centre, retail units and restaurants

182.7

The StarCity Leisure Park, built on the site of Nechells Power Station, is on the south side and is due to open in 2000 and will contain the largest cinema complex in Europe.

5

4A

Although at that time isolated from the rest of the Midland Links Motorway system which was completed in May 1972, the section of motorway between Junctions 4 and 5, was opened early in February 1971 to serve as a relief road to the adjacent A452

M42　4　**M42**

Castle Vale, the River Tame and the former Midland Railway Birmingham to Derby line are on the north side. Castle Vale Estate was built on the site of the former Spitfire Factory and airfield. The factory was utilized in immediate post war days for the British Industries Fair.

For those travelling west the city of Birmingham comes into view at 165.1, stretching right across the horizon with the Lickey Hills on the extreme left, the BT tower in the middle and Sutton Coldfield on the right.

166.9

Packington Park and the Forest of Arden Golf Course are on the south side

S — 157.3

Breach Brook is crossed

The Oxford Canal runs parallel to the motorway along the northern side.

153.1

A small boat marina, on the River Sowe is on the south side with Coventry and the gas holder at Rowleys Green visible beyond. Coventry, once known as the "City of Three Spires"and inextricably linked with the famous ride of Lady Godiva, stands on the River Sherbourne and is one of the oldest cities in England, a convent having been established here in the 6thC

3 — 154.0

The Coventry Canal, opened in 1790, is crossed

2　**M69**

147.1

The West Coast Main Line railway from London Euston to Glasgow and the Oxford Canal are crossed. The northern section of the Oxford Canal, from Coventry to Banbury was originally opened in 1778 and followed a very circuitous route to avoid extensive earthworks. This was rectified by the Oxford Canal Company in 1830 who rebuilt the waterway and reduced the overall mileage by 14 miles.

The slip roads between 140.1 and 140.9 mark the site of a proposed Service Area at Harborough for which a bridge to connect the east and westbound sites was constructed at 140.5, at the same time as the motorway was built. This bridge survived, as a concrete beam supported upon two concrete pillars and in splendid isolation, for some 30 years or so until it was demolished during the night of January 28th/29th, 2000 as part of a carriageway renewal scheme.

1 — 135.5

The track bed of the former Great Central main line can be seen on the left and right as it passes under the road. The Great Central Railway was the last main line built to London from the north. It was opened in 1899 and ironically became the first to close, being truncated to just a short southern section, by BR, on May 5th, 1969

M1
Junc 19

3.3 miles

1.7 miles

2.3 miles

9.7 miles

3.0 miles

8.2 miles

3.0 miles

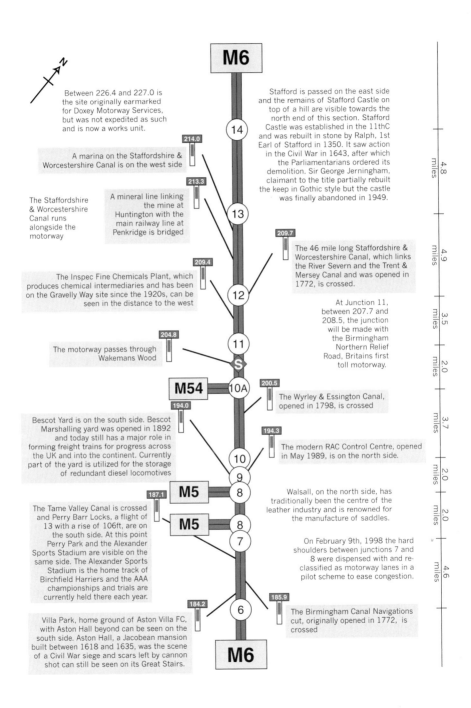

M6

14

13

12

11

S

M54 **10A**

M5 **8**

M5 **8**

9

10

7

6

M6

214.0

213.3

209.4

204.8

194.0

187.1

184.2

209.7

200.5

194.3

185.9

Between 226.4 and 227.0 is the site originally earmarked for Doxey Motorway Services, but was not expedited as such and is now a works unit.

A marina on the Staffordshire & Worcestershire Canal is on the west side

The Staffordshire & Worcestershire Canal runs alongside the motorway

A mineral line linking the mine at Huntington with the main railway line at Penkridge is bridged

The Inspec Fine Chemicals Plant, which produces chemical intermediaries and has been on the Gravelly Way site since the 1920s, can be seen in the distance to the west

The motorway passes through Wakemans Wood

Bescot Yard is on the south side. Bescot Marshalling yard was opened in 1892 and today still has a major role in forming freight trains for progress across the UK and into the continent. Currently part of the yard is utilized for the storage of redundant diesel locomotives

The Tame Valley Canal is crossed and Perry Barr Locks, a flight of 13 with a rise of 106ft, are on the south side. At this point Perry Park and the Alexander Sports Stadium are visible on the same side. The Alexander Sports Stadium is the home track of Birchfield Harriers and the AAA championships and trials are currently held there each year.

Villa Park, home ground of Aston Villa FC, with Aston Hall beyond can be seen on the south side. Aston Hall, a Jacobean mansion built between 1618 and 1635, was the scene of a Civil War siege and scars left by cannon shot can still be seen on its Great Stairs.

Stafford is passed on the east side and the remains of Stafford Castle on top of a hill are visible towards the north end of this section. Stafford Castle was established in the 11thC and was rebuilt in stone by Ralph, 1st Earl of Stafford in 1350. It saw action in the Civil War in 1643, after which the Parliamentarians ordered its demolition. Sir George Jerningham, claimant to the title partially rebuilt the keep in Gothic style but the castle was finally abandoned in 1949.

The 46 mile long Staffordshire & Worcestershire Canal, which links the River Severn and the Trent & Mersey Canal and was opened in 1772, is crossed.

At Junction 11, between 207.7 and 208.5, the junction will be made with the Birmingham Northern Relief Road, Britains first toll motorway.

The Wyrley & Essington Canal, opened in 1798, is crossed

The modern RAC Control Centre, opened in May 1989, is on the north side.

Walsall, on the north side, has traditionally been the centre of the leather industry and is renowned for the manufacture of saddles.

On February 9th, 1998 the hard shoulders between junctions 7 and 8 were dispensed with and re-classified as motorway lanes in a pilot scheme to ease congestion.

The Birmingham Canal Navigations cut, originally opened in 1772, is crossed

4.8 miles

4.9 miles

3.5 miles

2.0 miles

3.7 miles

2.0 miles

2.0 miles

4.6 miles

N

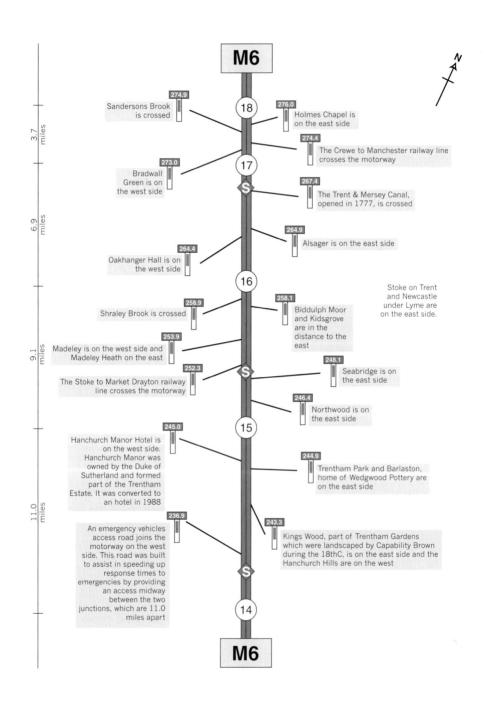

M6

N

18

274.9 Sandersons Brook is crossed

276.0 Holmes Chapel is on the east side

274.4 The Crewe to Manchester railway line crosses the motorway

17

273.0 Bradwall Green is on the west side

267.4 The Trent & Mersey Canal, opened in 1777, is crossed

264.9 Alsager is on the east side

264.4 Oakhanger Hall is on the west side

16

258.9 Shraley Brook is crossed

258.1 Biddulph Moor and Kidsgrove are in the distance to the east

Stoke on Trent and Newcastle under Lyme are on the east side.

253.9 Madeley is on the west side and Madeley Heath on the east

252.3 The Stoke to Market Drayton railway line crosses the motorway

248.1 Seabridge is on the east side

246.4 Northwood is on the east side

15

245.0 Hanchurch Manor Hotel is on the west side. Hanchurch Manor was owned by the Duke of Sutherland and formed part of the Trentham Estate. It was converted to an hotel in 1988

244.9 Trentham Park and Barlaston, home of Wedgwood Pottery are on the east side

236.9 An emergency vehicles access road joins the motorway on the west side. This road was built to assist in speeding up response times to emergencies by providing an access midway between the two junctions, which are 11.0 miles apart

243.3 Kings Wood, part of Trentham Gardens which were landscaped by Capability Brown during the 18thC, is on the east side and the Hanchurch Hills are on the west

14

M6

3.7 miles

6.9 miles

9.1 miles

11.0 miles

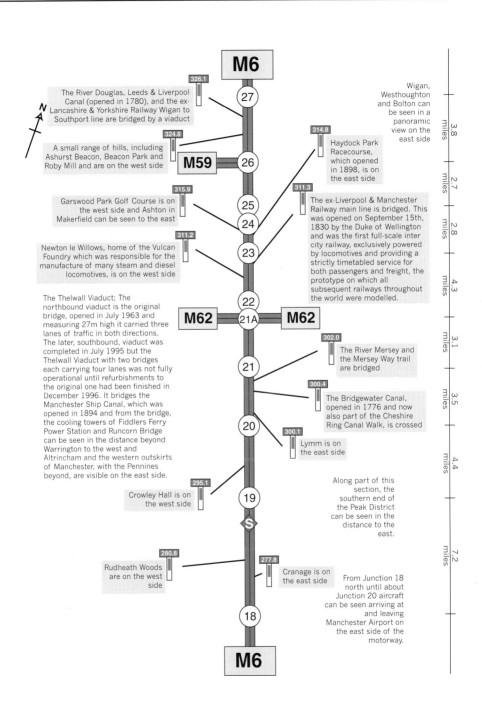

M6

27

326.1 The River Douglas, Leeds & Liverpool Canal (opened in 1780), and the ex-Lancashire & Yorkshire Railway Wigan to Southport line are bridged by a viaduct

Wigan, Westhoughton and Bolton can be seen in a panoramic view on the east side

3.8 miles

314.8 Haydock Park Racecourse, which opened in 1898, is on the east side

324.8 A small range of hills, including Ashurst Beacon, Beacon Park and Roby Mill and are on the west side

M59   26

2.7 miles

315.9 Garswood Park Golf Course is on the west side and Ashton in Makerfield can be seen to the east

25

24

311.3 The ex-Liverpool & Manchester Railway main line is bridged. This was opened on September 15th, 1830 by the Duke of Wellington and was the first full-scale inter city railway, exclusively powered by locomotives and providing a strictly timetabled service for both passengers and freight, the prototype on which all subsequent railways throughout the world were modelled.

2.8 miles

311.2 Newton le Willows, home of the Vulcan Foundry which was responsible for the manufacture of many steam and diesel locomotives, is on the west side

23

22

4.3 miles

The Thelwall Viaduct; The northbound viaduct is the original bridge, opened in July 1963 and measuring 27m high it carried three lanes of traffic in both directions. The later, southbound, viaduct was completed in July 1995 but the Thelwall Viaduct with two bridges each carrying four lanes was not fully operational until refurbishments to the original one had been finished in December 1996. It bridges the Manchester Ship Canal, which was opened in 1894 and from the bridge, the cooling towers of Fiddlers Ferry Power Station and Runcorn Bridge can be seen in the distance beyond Warrington to the west and Altrincham and the western outskirts of Manchester, with the Pennines beyond, are visible on the east side.

M62   21A   **M62**

21

3.1 miles

302.0 The River Mersey and the Mersey Way trail are bridged

300.4 The Bridgewater Canal, opened in 1776 and now also part of the Cheshire Ring Canal Walk, is crossed

20

3.5 miles

300.1 Lymm is on the east side

Along part of this section, the southern end of the Peak District can be seen in the distance to the east.

4.4 miles

295.1 Crowley Hall is on the west side

19

S

280.8 Rudheath Woods are on the west side

277.8 Cranage is on the east side

From Junction 18 north until about Junction 20 aircraft can be seen arriving at and leaving Manchester Airport on the east side of the motorway.

7.2 miles

18

M6

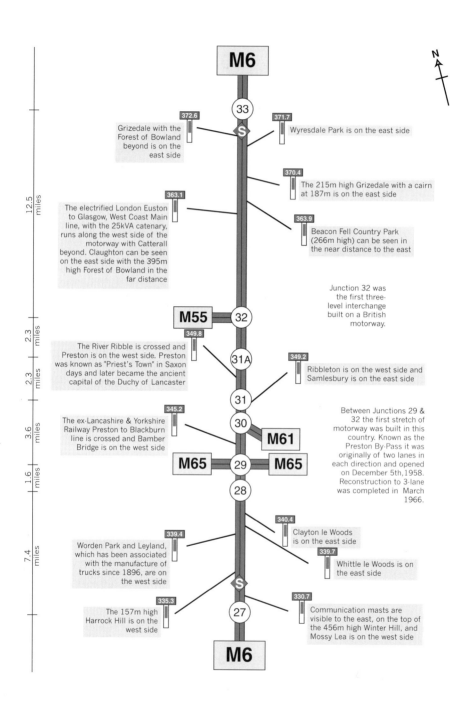

M6

33

372.6 Grizedale with the Forest of Bowland beyond is on the east side

371.7 Wyresdale Park is on the east side

370.4 The 215m high Grizedale with a cairn at 187m is on the east side

363.1 The electrified London Euston to Glasgow, West Coast Main line, with the 25kVA catenary, runs along the west side of the motorway with Catterall beyond. Claughton can be seen on the east side with the 395m high Forest of Bowland in the far distance

363.9 Beacon Fell Country Park (266m high) can be seen in the near distance to the east

Junction 32 was the first three-level interchange built on a British motorway.

M55 — 32

349.8 The River Ribble is crossed and Preston is on the west side. Preston was known as "Priest's Town" in Saxon days and later became the ancient capital of the Duchy of Lancaster

31A

349.2 Ribbleton is on the west side and Samlesbury is on the east side

31

345.2 The ex-Lancashire & Yorkshire Railway Preston to Blackburn line is crossed and Bamber Bridge is on the west side

30

M61

M65 — 29 — **M65**

Between Junctions 29 & 32 the first stretch of motorway was built in this country. Known as the Preston By-Pass it was originally of two lanes in each direction and opened on December 5th, 1958. Reconstruction to 3-lane was completed in March 1966.

28

340.4 Clayton le Woods is on the east side

339.4 Worden Park and Leyland, which has been associated with the manufacture of trucks since 1896, are on the west side

339.7 Whittle le Woods is on the east side

335.3 The 157m high Harrock Hill is on the west side

27

330.7 Communication masts are visible to the east, on the top of the 456m high Winter Hill, and Mossy Lea is on the west side

M6

12.5 miles

2.3 miles

2.3 miles

3.6 miles

1.6 miles

7.4 miles

N

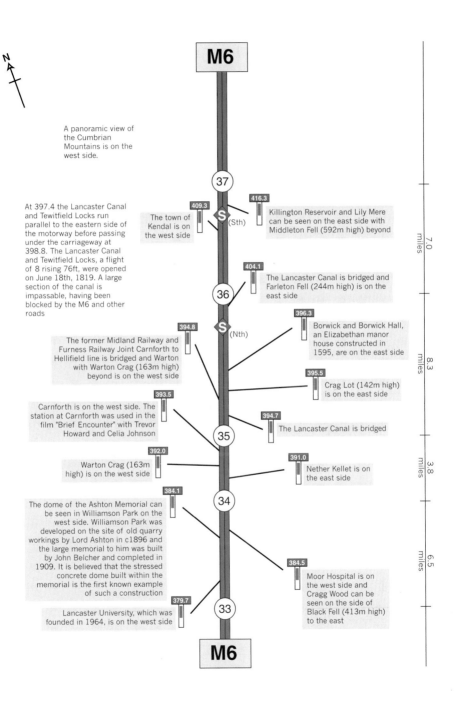

A panoramic view of the Cumbrian Mountains is on the west side.

At 397.4 the Lancaster Canal and Tewitfield Locks run parallel to the eastern side of the motorway before passing under the carriageway at 398.8. The Lancaster Canal and Tewitfield Locks, a flight of 8 rising 76ft, were opened on June 18th, 1819. A large section of the canal is impassable, having been blocked by the M6 and other roads

M6

37

409.3 The town of Kendal is on the west side

(Sth)

416.3 Killington Reservoir and Lily Mere can be seen on the east side with Middleton Fell (592m high) beyond

404.1 The Lancaster Canal is bridged and Farleton Fell (244m high) is on the east side

36

394.8 The former Midland Railway and Furness Railway Joint Carnforth to Hellifield line is bridged and Warton with Warton Crag (163m high) beyond is on the west side

(Nth)

396.3 Borwick and Borwick Hall, an Elizabethan manor house constructed in 1595, are on the east side

395.5 Crag Lot (142m high) is on the east side

393.5 Carnforth is on the west side. The station at Carnforth was used in the film "Brief Encounter" with Trevor Howard and Celia Johnson

394.7 The Lancaster Canal is bridged

35

392.0 Warton Crag (163m high) is on the west side

391.0 Nether Kellet is on the east side

384.1 The dome of the Ashton Memorial can be seen in Williamson Park on the west side. Williamson Park was developed on the site of old quarry workings by Lord Ashton in c1896 and the large memorial to him was built by John Belcher and completed in 1909. It is believed that the stressed concrete dome built within the memorial is the first known example of such a construction

34

384.5 Moor Hospital is on the west side and Cragg Wood can be seen on the side of Black Fell (413m high) to the east

379.7 Lancaster University, which was founded in 1964, is on the west side

33

M6

7.0 miles

8.3 miles

3.8 miles

6.5 miles

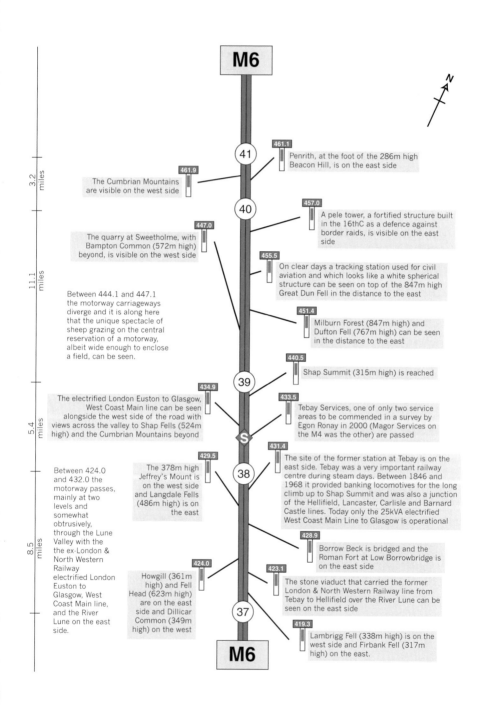

M6

3.2 miles

41

461.1 Penrith, at the foot of the 286m high Beacon Hill, is on the east side

461.9 The Cumbrian Mountains are visible on the west side

40

457.0 A pele tower, a fortified structure built in the 16thC as a defence against border raids, is visible on the east side

447.0 The quarry at Sweetholme, with Bampton Common (572m high) beyond, is visible on the west side

455.5 On clear days a tracking station used for civil aviation and which looks like a white spherical structure can be seen on top of the 847m high Great Dun Fell in the distance to the east

11.1 miles

Between 444.1 and 447.1 the motorway carriageways diverge and it is along here that the unique spectacle of sheep grazing on the central reservation of a motorway, albeit wide enough to enclose a field, can be seen.

451.4 Milburn Forest (847m high) and Dufton Fell (767m high) can be seen in the distance to the east

440.5 Shap Summit (315m high) is reached

39

434.9 The electrified London Euston to Glasgow, West Coast Main line can be seen alongside the west side of the road with views across the valley to Shap Fells (524m high) and the Cumbrian Mountains beyond

433.5 Tebay Services, one of only two service areas to be commended in a survey by Egon Ronay in 2000 (Magor Services on the M4 was the other) are passed

5.4 miles

S

431.4 The site of the former station at Tebay is on the east side. Tebay was a very important railway centre during steam days. Between 1846 and 1968 it provided banking locomotives for the long climb up to Shap Summit and was also a junction of the Hellifield, Lancaster, Carlisle and Barnard Castle lines. Today only the 25kVA electrified West Coast Main Line to Glasgow is operational

38

429.5 The 378m high Jeffrey's Mount is on the west side and Langdale Fells (486m high) is on the east

Between 424.0 and 432.0 the motorway passes, mainly at two levels and somewhat obtrusively, through the Lune Valley with the the ex-London & North Western Railway electrified London Euston to Glasgow, West Coast Main line, and the River Lune on the east side.

8.5 miles

428.9 Borrow Beck is bridged and the Roman Fort at Low Borrowbridge is on the east side

424.0 Howgill (361m high) and Fell Head (623m high) are on the east side and Dillicar Common (349m high) on the west

423.1 The stone viaduct that carried the former London & North Western Railway line from Tebay to Hellifield over the River Lune can be seen on the east side

37

419.3 Lambrigg Fell (338m high) is on the west side and Firbank Fell (317m high) on the east.

M6

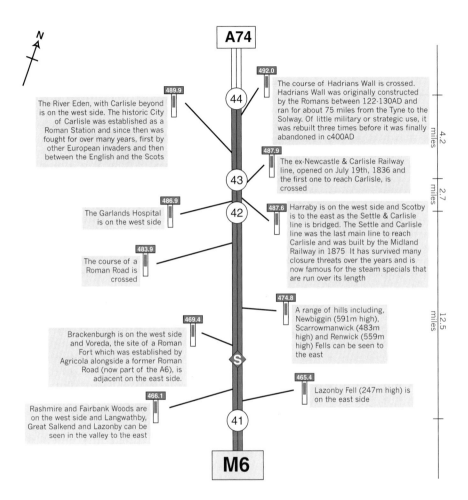

N

A74

492.0

The course of Hadrians Wall is crossed. Hadrians Wall was originally constructed by the Romans between 122-130AD and ran for about 75 miles from the Tyne to the Solway. Of little military or strategic use, it was rebuilt three times before it was finally abandoned in c400AD

489.9

44

The River Eden, with Carlisle beyond is on the west side. The historic City of Carlisle was established as a Roman Station and since then was fought for over many years, first by other European invaders and then between the English and the Scots

487.9

The ex-Newcastle & Carlisle Railway line, opened on July 19th, 1836 and the first one to reach Carlisle, is crossed

43

486.9

The Garlands Hospital is on the west side

42

487.6 Harraby is on the west side and Scotby is to the east as the Settle & Carlisle line is bridged. The Settle and Carlisle line was the last main line to reach Carlisle and was built by the Midland Railway in 1875 It has survived many closure threats over the years and is now famous for the steam specials that are run over its length

483.9

The course of a Roman Road is crossed

474.8

A range of hills including, Newbiggin (591m high), Scarrowmanwick (483m high) and Renwick (559m high) Fells can be seen to the east

469.4

Brackenburgh is on the west side and Voreda, the site of a Roman Fort which was established by Agricola alongside a former Roman Road (now part of the A6), is adjacent on the east side.

S

465.4

Lazonby Fell (247m high) is on the east side

466.1

Rashmire and Fairbank Woods are on the west side and Langwathby, Great Salkend and Lazonby can be seen in the valley to the east

41

M6

4.2 miles

2.7 miles

12.5 miles

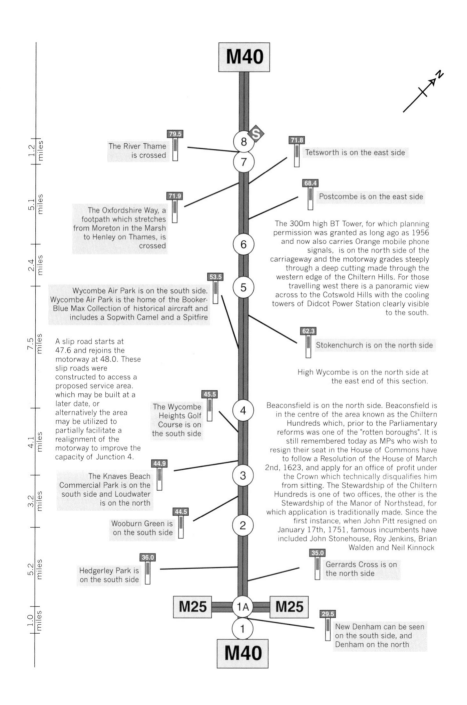

N

M40

79.5 **8** S
The River Thame is crossed

7

71.8
Tetsworth is on the east side

68.4
Postcombe is on the east side

71.9
The Oxfordshire Way, a footpath which stretches from Moreton in the Marsh to Henley on Thames, is crossed

6
The 300m high BT Tower, for which planning permission was granted as long ago as 1956 and now also carries Orange mobile phone signals, is on the north side of the carriageway and the motorway grades steeply through a deep cutting made through the western edge of the Chiltern Hills. For those travelling west there is a panoramic view across to the Cotswold Hills with the cooling towers of Didcot Power Station clearly visible to the south.

53.5 **5**
Wycombe Air Park is on the south side. Wycombe Air Park is the home of the Booker Blue Max Collection of historical aircraft and includes a Sopwith Camel and a Spitfire

62.3
Stokenchurch is on the north side

A slip road starts at 47.6 and rejoins the motorway at 48.0. These slip roads were constructed to access a proposed service area. which may be built at a later date, or alternatively the area may be utilized to partially facilitate a realignment of the motorway to improve the capacity of Junction 4.

High Wycombe is on the north side at the east end of this section.

45.5 **4**
The Wycombe Heights Golf Course is on the south side

Beaconsfield is on the north side. Beaconsfield is in the centre of the area known as the Chiltern Hundreds which, prior to the Parliamentary reforms was one of the "rotten boroughs". It is still remembered today as MPs who wish to resign their seat in the House of Commons have to follow a Resolution of the House of March 2nd, 1623, and apply for an office of profit under the Crown which technically disqualifies him from sitting. The Stewardship of the Chiltern Hundreds is one of two offices, the other is the Stewardship of the Manor of Northstead, for which application is traditionally made. Since the first instance, when John Pitt resigned on January 17th, 1751, famous incumbents have included John Stonehouse, Roy Jenkins, Brian Walden and Neil Kinnock

44.9 **3**
The Knaves Beach Commercial Park is on the south side and Loudwater is on the north

44.5 **2**
Wooburn Green is on the south side

36.0
Hedgerley Park is on the south side

35.0
Gerrards Cross is on the north side

M25 **1A** **M25**

29.5
New Denham can be seen on the south side, and Denham on the north

1

M40

1.2 miles
5.1 miles
2.4 miles
7.5 miles
4.1 miles
3.2 miles
5.2 miles
1.0 miles

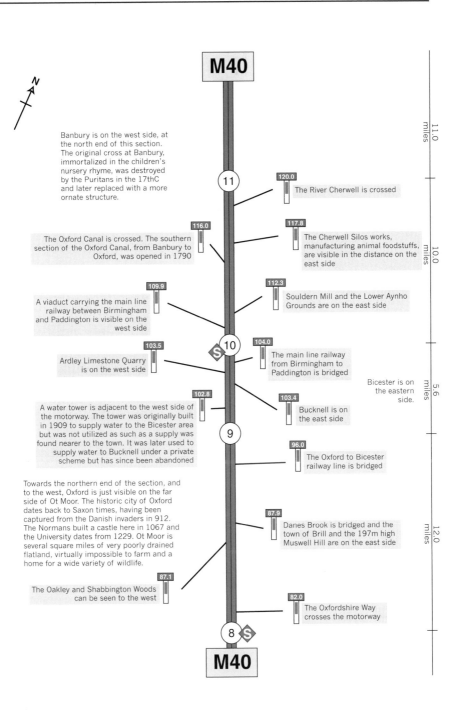

M40

Banbury is on the west side, at the north end of this section. The original cross at Banbury, immortalized in the children's nursery rhyme, was destroyed by the Puritans in the 17thC and later replaced with a more ornate structure.

11

120.0 The River Cherwell is crossed

116.0 The Oxford Canal is crossed. The southern section of the Oxford Canal, from Banbury to Oxford, was opened in 1790

117.8 The Cherwell Silos works, manufacturing animal foodstuffs, are visible in the distance on the east side

109.9 A viaduct carrying the main line railway between Birmingham and Paddington is visible on the west side

112.3 Souldern Mill and the Lower Aynho Grounds are on the east side

10

103.5 Ardley Limestone Quarry is on the west side

104.0 The main line railway from Birmingham to Paddington is bridged

Bicester is on the eastern side.

102.8 A water tower is adjacent to the west side of the motorway. The tower was originally built in 1909 to supply water to the Bicester area but was not utilized as such as a supply was found nearer to the town. It was later used to supply water to Bucknell under a private scheme but has since been abandoned

103.4 Bucknell is on the east side

9

96.0 The Oxford to Bicester railway line is bridged

Towards the northern end of the section, and to the west, Oxford is just visible on the far side of Ot Moor. The historic city of Oxford dates back to Saxon times, having been captured from the Danish invaders in 912. The Normans built a castle here in 1067 and the University dates from 1229. Ot Moor is several square miles of very poorly drained flatland, virtually impossible to farm and a home for a wide variety of wildlife.

87.9 Danes Brook is bridged and the town of Brill and the 197m high Muswell Hill are on the east side

87.1 The Oakley and Shabbington Woods can be seen to the west

82.0 The Oxfordshire Way crosses the motorway

8

M40

11.0 miles

10.0 miles

5.6 miles

12.0 miles

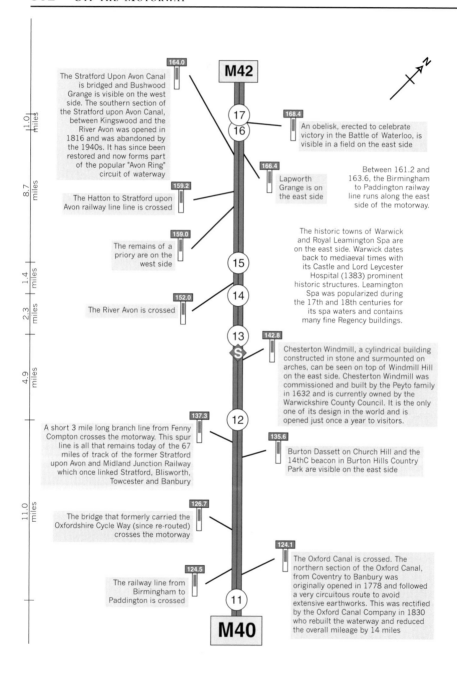

M42

164.0

The Stratford Upon Avon Canal is bridged and Bushwood Grange is visible on the west side. The southern section of the Stratford upon Avon Canal, between Kingswood and the River Avon was opened in 1816 and was abandoned by the 1940s. It has since been restored and now forms part of the popular "Avon Ring" circuit of waterway

17
16

168.4

An obelisk, erected to celebrate victory in the Battle of Waterloo, is visible in a field on the east side

166.4

Lapworth Grange is on the east side

Between 161.2 and 163.6, the Birmingham to Paddington railway line runs along the east side of the motorway.

159.2

The Hatton to Stratford upon Avon railway line line is crossed

159.0

The remains of a priory are on the west side

15

The historic towns of Warwick and Royal Leamington Spa are on the east side. Warwick dates back to mediaeval times with its Castle and Lord Leycester Hospital (1383) prominent historic structures. Leamington Spa was popularized during the 17th and 18th centuries for its spa waters and contains many fine Regency buildings.

14

152.0

The River Avon is crossed

13

142.8

S

Chesterton Windmill, a cylindrical building constructed in stone and surmounted on arches, can be seen on top of Windmill Hill on the east side. Chesterton Windmill was commissioned and built by the Peyto family in 1632 and is currently owned by the Warwickshire County Council. It is the only one of its design in the world and is opened just once a year to visitors.

12

137.3

A short 3 mile long branch line from Fenny Compton crosses the motorway. This spur line is all that remains today of the 67 miles of track of the former Stratford upon Avon and Midland Junction Railway which once linked Stratford, Blisworth, Towcester and Banbury

135.6

Burton Dassett on Church Hill and the 14thC beacon in Burton Hills Country Park are visible on the east side

126.7

The bridge that formerly carried the Oxfordshire Cycle Way (since re-routed) crosses the motorway

124.1

The Oxford Canal is crossed. The northern section of the Oxford Canal, from Coventry to Banbury was originally opened in 1778 and followed a very circuitous route to avoid extensive earthworks. This was rectified by the Oxford Canal Company in 1830 who rebuilt the waterway and reduced the overall mileage by 14 miles

124.5

The railway line from Birmingham to Paddington is crossed

11

M40

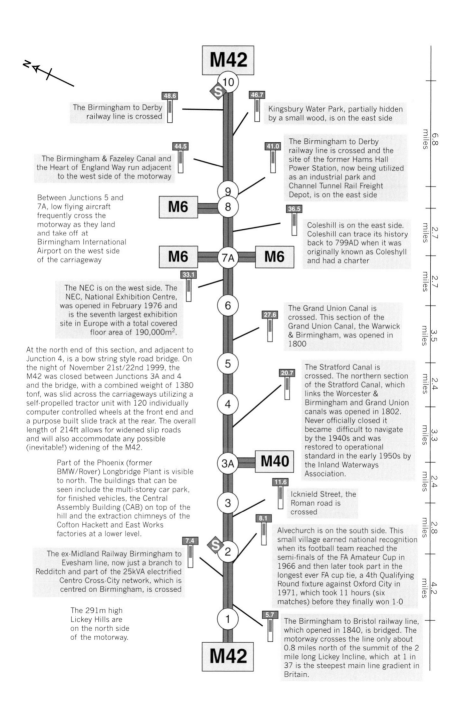

M42

⑩ S

48.6 The Birmingham to Derby railway line is crossed

46.7 Kingsbury Water Park, partially hidden by a small wood, is on the east side

44.5 The Birmingham & Fazeley Canal and the Heart of England Way run adjacent to the west side of the motorway

41.0 The Birmingham to Derby railway line is crossed and the site of the former Hams Hall Power Station, now being utilized as an industrial park and Channel Tunnel Rail Freight Depot, is on the east side

Between Junctions 5 and 7A, low flying aircraft frequently cross the motorway as they land and take off at Birmingham International Airport on the west side of the carriageway

M6 ⑨ ⑧

36.5 Coleshill is on the east side. Coleshill can trace its history back to 799AD when it was originally known as Coleshyll and had a charter

M6 ⑦A **M6**

33.1 The NEC is on the west side. The NEC, National Exhibition Centre, was opened in February 1976 and is the seventh largest exhibition site in Europe with a total covered floor area of 190,000m².

⑥

27.6 The Grand Union Canal is crossed. This section of the Grand Union Canal, the Warwick & Birmingham, was opened in 1800

At the north end of this section, and adjacent to Junction 4, is a bow string style road bridge. On the night of November 21st/22nd 1999, the M42 was closed between Junctions 3A and 4 and the bridge, with a combined weight of 1380 tonf, was slid across the carriageways utilizing a self-propelled tractor unit with 120 individually computer controlled wheels at the front end and a purpose built slide track at the rear. The overall length of 214ft allows for widened slip roads and will also accommodate any possible (inevitable!) widening of the M42.

⑤ ④

20.7 The Stratford Canal is crossed. The northern section of the Stratford Canal, which links the Worcester & Birmingham and Grand Union canals was opened in 1802. Never officially closed it became difficult to navigate by the 1940s and was restored to operational standard in the early 1950s by the Inland Waterways Association.

Part of the Phoenix (former BMW/Rover) Longbridge Plant is visible to north. The buildings that can be seen include the multi-storey car park, for finished vehicles, the Central Assembly Building (CAB) on top of the hill and the extraction chimneys of the Cofton Hackett and East Works factories at a lower level.

③A **M40**

11.6 Icknield Street, the Roman road is crossed

③

8.1 Alvechurch is on the south side. This small village earned national recognition when its football team reached the semi-finals of the FA Amateur Cup in 1966 and then later took part in the longest ever FA cup tie, a 4th Qualifying Round fixture against Oxford City in 1971, which took 11 hours (six matches) before they finally won 1-0

7.4 The ex-Midland Railway Birmingham to Evesham line, now just a branch to Redditch and part of the 25kVA electrified Centro Cross-City network, which is centred on Birmingham, is crossed

S ②

The 291m high Lickey Hills are on the north side of the motorway.

①

5.7 The Birmingham to Bristol railway line, which opened in 1840, is bridged. The motorway crosses the line only about 0.8 miles north of the summit of the 2 mile long Lickey Incline, which at 1 in 37 is the steepest main line gradient in Britain.

M42

6.8 miles
2.7 miles
2.7 miles
3.5 miles
2.4 miles
3.3 miles
2.4 miles
2.8 miles
4.2 miles

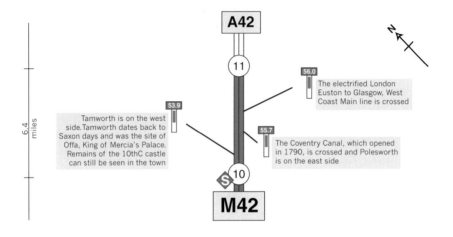

A42

11

56.0
The electrified London
Euston to Glasgow, West
Coast Main line is crossed

53.9
Tamworth is on the west
side. Tamworth dates back to
Saxon days and was the site of
Offa, King of Mercia's Palace.
Remains of the 10thC castle
can still be seen in the town

55.7
The Coventry Canal, which opened
in 1790, is crossed and Polesworth
is on the east side

S 10

M42

6.4 miles

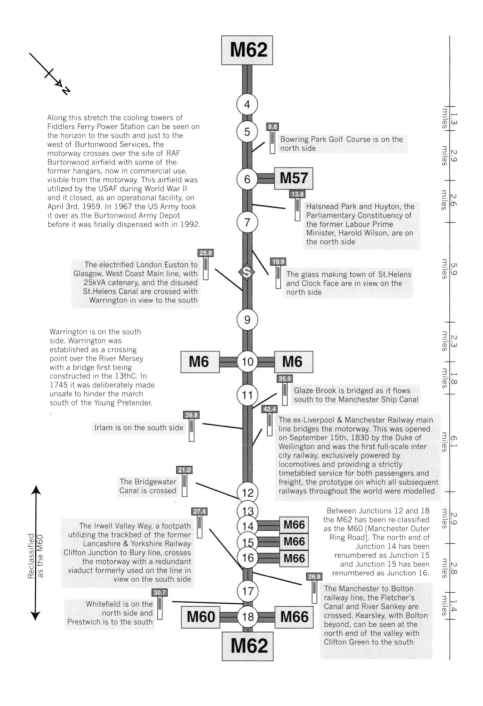

M62

4

Along this stretch the cooling towers of Fiddlers Ferry Power Station can be seen on the horizon to the south and just to the west of Burtonwood Services, the motorway crosses over the site of RAF Burtonwood airfield with some of the former hangars, now in commercial use, visible from the motorway. This airfield was utilized by the USAF during World War II and it closed, as an operational facility, on April 3rd, 1959. In 1967 the US Army took it over as the Burtonwood Army Depot before it was finally dispensed with in 1992.

5

8.8

Bowring Park Golf Course is on the north side

6   **M57**

13.8

Halsnead Park and Huyton, the Parliamentary Constituency of the former Labour Prime Minister, Harold Wilson, are on the north side

7

25.8

The electrified London Euston to Glasgow, West Coast Main line, with 25kVA catenary, and the disused St.Helens Canal are crossed with Warrington in view to the south

S   19.9

The glass making town of St.Helens and Clock Face are in view on the north side

9

Warrington is on the south side. Warrington was established as a crossing point over the River Mersey with a bridge first being constructed in the 13thC. In 1745 it was deliberately made unsafe to hinder the march south of the Young Pretender.

M6   10   **M6**

35.5

11

Glaze Brook is bridged as it flows south to the Manchester Ship Canal

39.8

Irlam is on the south side

42.4

The ex-Liverpool & Manchester Railway main line bridges the motorway. This was opened on September 15th, 1830 by the Duke of Wellington and was the first full-scale inter city railway, exclusively powered by locomotives and providing a strictly timetabled service for both passengers and freight, the prototype on which all subsequent railways throughout the world were modelled

21.0

The Bridgewater Canal is crossed

12

27.4

The Irwell Valley Way, a footpath utilizing the trackbed of the former Lancashire & Yorkshire Railway Clifton Junction to Bury line, crosses the motorway with a redundant viaduct formerly used on the line in view on the south side

13

14   **M66**

15   **M66**

16   **M66**

Between Junctions 12 and 18 the M62 has been re-classified as the M60 [Manchester Outer Ring Road]. The north end of Junction 14 has been renumbered as Junction 15 and Junction 15 has been renumbered as Junction 16.

26.8

17

The Manchester to Bolton railway line, the Fletcher's Canal and River Sankey are crossed. Kearsley, with Bolton beyond, can be seen at the north end of the valley with Clifton Green to the south

30.7

Whitefield is on the north side and Prestwich is to the south

M60   18   **M66**

M62

Reclassified as the M60

miles 1.3

miles 2.9

miles 2.6

miles 5.9

miles 2.3

miles 1.8

miles 6.1

miles 2.9

miles 2.8

miles 1.4

N

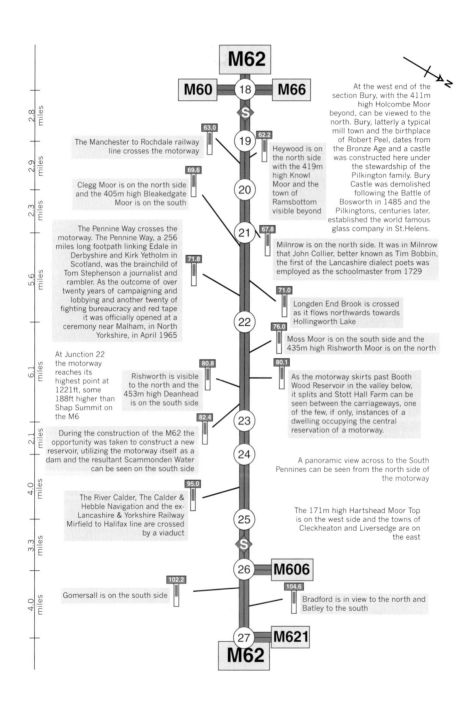

M62

M60 18 **M66**

S

63.0

The Manchester to Rochdale railway line crosses the motorway

19 62.2

Heywood is on the north side with the 419m high Knowl Moor and the town of Ramsbottom visible beyond

At the west end of the section Bury, with the 411m high Holcombe Moor beyond, can be viewed to the north. Bury, latterly a typical mill town and the birthplace of Robert Peel, dates from the Bronze Age and a castle was constructed here under the stewardship of the Pilkington family. Bury Castle was demolished following the Battle of Bosworth in 1485 and the Pilkingtons, centuries later, established the world famous glass company in St.Helens.

69.6

Clegg Moor is on the north side and the 405m high Bleakedgate Moor is on the south

20

21 67.8

The Pennine Way crosses the motorway. The Pennine Way, a 256 miles long footpath linking Edale in Derbyshire and Kirk Yetholm in Scotland, was the brainchild of Tom Stephenson a journalist and rambler. As the outcome of over twenty years of campaigning and lobbying and another twenty of fighting bureaucracy and red tape it was officially opened at a ceremony near Malham, in North Yorkshire, in April 1965

71.8

Milnrow is on the north side. It was in Milnrow that John Collier, better known as Tim Bobbin, the first of the Lancashire dialect poets was employed as the schoolmaster from 1729

71.0

Longden End Brook is crossed as it flows northwards towards Hollingworth Lake

22 76.0

Moss Moor is on the south side and the 435m high Rishworth Moor is on the north

At Junction 22 the motorway reaches its highest point at 1221ft, some 188ft higher than Shap Summit on the M6

80.8

Rishworth is visible to the north and the 453m high Deanhead is on the south side

80.1

As the motorway skirts past Booth Wood Reservoir in the valley below, it splits and Stott Hall Farm can be seen between the carriageways, one of the few, if only, instances of a dwelling occupying the central reservation of a motorway.

82.4 23

During the construction of the M62 the opportunity was taken to construct a new reservoir, utilizing the motorway itself as a dam and the resultant Scammonden Water can be seen on the south side

24

A panoramic view across to the South Pennines can be seen from the north side of the motorway

95.0

The River Calder, The Calder & Hebble Navigation and the ex-Lancashire & Yorkshire Railway Mirfield to Halifax line are crossed by a viaduct

25

The 171m high Hartshead Moor Top is on the west side and the towns of Cleckheaton and Liversedge are on the east

S

26 **M606**

102.2

Gomersall is on the south side

104.6

Bradford is in view to the north and Batley to the south

27 **M621**

M62

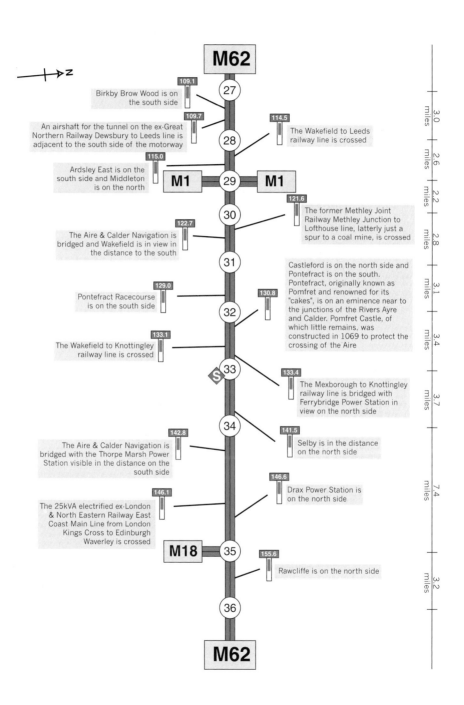

N

M62

109.1
27
Birkby Brow Wood is on the south side

109.7
28
An airshaft for the tunnel on the ex-Great Northern Railway Dewsbury to Leeds line is adjacent to the south side of the motorway

114.5
The Wakefield to Leeds railway line is crossed

115.0
Ardsley East is on the south side and Middleton is on the north

M1   29   **M1**

30
121.6
The former Methley Joint Railway Methley Junction to Lofthouse line, latterly just a spur to a coal mine, is crossed

122.7
The Aire & Calder Navigation is bridged and Wakefield is in view in the distance to the south

31

Castleford is on the north side and Pontefract is on the south. Pontefract, originally known as Pomfret and renowned for its "cakes", is on an eminence near to the junctions of the Rivers Ayre and Calder. Pomfret Castle, of which little remains, was constructed in 1069 to protect the crossing of the Aire

129.0
Pontefract Racecourse is on the south side

130.8
32

133.1
The Wakefield to Knottingley railway line is crossed

S 33
133.4
The Mexborough to Knottingley railway line is bridged with Ferrybridge Power Station in view on the north side

34

142.8
The Aire & Calder Navigation is bridged with the Thorpe Marsh Power Station visible in the distance on the south side

141.5
Selby is in the distance on the north side

146.6
Drax Power Station is on the north side

146.1
The 25kVA electrified ex-London & North Eastern Railway East Coast Main Line from London Kings Cross to Edinburgh Waverley is crossed

M18   35

155.6
Rawcliffe is on the north side

36

M62

3.0 miles
2.6 miles
2.2 miles
2.8 miles
3.1 miles
3.4 miles
3.7 miles
7.4 miles
3.2 miles

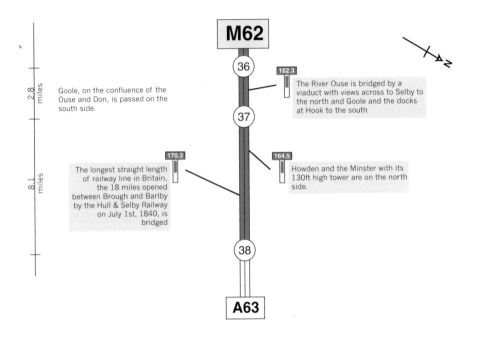

M62

36

162.3

The River Ouse is bridged by a viaduct with views across to Selby to the north and Goole and the docks at Hook to the south

37

Goole, on the confluence of the Ouse and Don, is passed on the south side.

2.8 miles

170.3

The longest straight length of railway line in Britain, the 18 miles opened between Brough and Barlby by the Hull & Selby Railway on July 1st, 1840, is bridged

8.1 miles

164.5

Howden and the Minster with its 130ft high tower are on the north side.

38

A63